HUMAN ADAPTATION

HUMAN ADAPTATION

ADAPTATION

Coping with Life Crises

Edited by
Rudolf H. Moos
Stanford University

D. C. HEATH AND COMPANY
Lexington, Massachusetts   Toronto

Clothbound edition published by
Lexington Books.

To
David Hamburg
whose iron fist was a velvet glove

Preface

This book describes how human beings cope with major life crises and transitions. It includes a broad overview of literature on coping and adaptation. Important dimensions of adaptation and of coping behavior are identified, and a general perspective on this area is provided.

The major portion of the book is organized from the perspective of various developmental life transitions. These life transitions, which all people must necessarily face, raise common relevant coping issues. The last several sections cover "accidental crises" and other types of life hazards that usually involve extreme stress. The material highlights the fact that human beings can successfully adapt to, and cope with, major life stresses, rather than the fact that severe symptoms and/or breakdowns sometimes occur. The importance of support from family, friends, and other sources of help in the community is emphasized. The selections discuss the specific coping tasks that must be handled in these developmental and unusual stress situations. Many of them include examples illustrating the material.

Human Adaptation: Coping with Life Crises should meet the needs of a diverse audience. The selections are useful as supplementary material for courses in the psychology of adjustment, personality theory and assessment, and abnormal psychology, and for other courses that emphasize a coping and adaptation perspective. They are also useful for selected introductory psychology and sociology courses that are oriented toward personality and/or social adjustment. In addition, many courses in developmental psychology have recently expanded their coverage to include not only the childhood and adolescent stages of development but also those of middle and late adulthood. This book will be useful in such courses, since it gives a developmental life span coverage to issues of coping and adaptation. There is also interest in these issues in the health sciences, particularly among medical students, nurses, social workers, and other paramedical personnel. This book should be of interest in selected courses taught in epidemiology, community health, family medicine, and psychiatry and behavioral sciences departments in schools of medicine, nursing, and public health.

The book has twelve parts. Part I sets the stage by reviewing five major areas of interest that have merged into a renewed focus on human growth and competence. In addition, important terms such as *adaptation, defense, mastery,* and *coping* are defined and contrasted. Part II presents articles dealing with developmental life transitions in the childhood years. Material relating to entry into kindergarten and to adaptation to separation experiences and to divorce is included. Part III deals with developmental life transitions with emphasis on bereavement in childhood. The articles discuss how children mourn, how they react to the death of a sibling, and how parents handle their children's adaptation to loss.

Part IV also deals with developmental life transitions, with special focus on the college years. The articles describe how high school students prepare for entry into college and how they cope with the new experiences of the first year of college. Material on the coping styles of black students on white campuses is included. Part V provides information on developmental life transitions with an emphasis on intimacy, marriage, and parenthood. These selections describe problems involved in living together, in adjusting to marriage, and in the birth of the first child. The subject of Part VI is developmental life transitions, particularly relocation and migration. The selections cover adaptation to residential change, to relocation and migration, and to exile.

Part VII discusses developmental life transitions with an emphasis on retirement and aging. The articles present information about the ways in which individuals cope with these stressful life transitions. Part VIII treats developmental life transitions that center on death and bereavement. The selections covers ways in which widows and widowers cope with the death of their spouses. A cross-cultural perspective is offered by an article discussing mourning in Japan.

Part IX provides information on those who cope with unusual stress, as exemplified by the prisoner of war. Examples are taken from the imprisonment of American men in the *Pueblo* incident in North Korea and from the reactions of American prisoners of war in Laos and Viet Nam. In Part X the material deals with coping with unusual stress, with emphasis on the Nazi concentration camps of World War II. The selections discuss both immediate and long-term effect of concentration camp experiences.

Coping with disasters is the subject of Part XI. The articles discuss psychological reactions to the atomic attack at Hiroshima and Nagasaki and individual reactions to a tornado and a flood. Finally, Part XII presents material on some exceptional stresses of modern life such as ways in which people cope in a skyjacked plane and human adjustment to the environment of a nuclear submarine. As an example of modern urban problems, the patterns of adaptation among victims of rape are discussed.

I have dedicated this book to David Hamburg, who for over a decade provided an environment in which I could cope and develop competence. His early descriptive papers of the specific ways in which people handle life crises are a model of clarity and relevance. He kindled my curiosity in this field and involved me in projects that enhanced my interest and efficacy in it. My many productive discussions with George Coelho have increased my understanding of this field, and thus have enriched this book. Erich Lindemann helped me clarify many of my formative concepts. My discussions with a small research group composed of John Adams, Paxton Cady, Alan Sidle, and Richard Stillman were perhaps more productive than any of us thought at the time. Richard Lazarus and David Mechanic have influenced me both by their writings and by our discussions. Erik Erikson and Robert White have influenced me mainly through their writings, which are unusually clear, cogent, and inspirational.

On a more practical level, Vivien Tsu searched for and helped select some of the articles. She coped with an overwhelming mass of material and shared responsibility with me in writing the commentaries for each section. Louise Doherty and Susanne Flynn efficiently performed the typing and secretarial tasks involved. The work was supported in part by Grant MH16026 from the National Institute of Mental Health, Grant AA00498 from the National Institute of Alcohol Abuse and Alcoholism, and by Veterans Administration Research project MRIS 5817-01 to the Social Ecology Laboratory, Department of Psychiatry and Behavioral Sciences, Stanford University and the Veterans Administration Hospital, Palo Alto.

Although Bernice coped with my working on this book, she never adapted to it. As usual, Karen and Kevin were delightful hindrances.

November 1975
Palo Alto

Contents

IX Coping with Unusual Stress: The Prisoner of War   305

X Coping with Unusual Stress: Concentration Camps   333

XI Coping with Unusual Stress: Disasters   373

XII Unusual Stresses of Modern Life   405

HUMAN ADAPTATION

Theory and Perspective

The two introductory selections present a broad conceptual overview of the major areas that have influenced the study of human competence and coping. The first selection, by Rudolf Moos and Vivien Tsu, briefly reviews the five major influences that have shaped current interest in this field: (1) Charles Darwin and evolutionary theory focused on the adaptation of animals and humans to their environment. Evolutionary theory emphasizes communal adaptation (that is, the adaptation of an entire group or society); however, successful communal adaptation presupposes successful individual adaptation. (2) Personality theorists have recently emphasized human fulfillment and growth. People have an exploratory drive, a need for varied sensory stimulation, a desire to engage in information processing, and so on. Human fulfillment theorists who emphasize these drives conceptualize the core tendency of man as self-actualization; that is, the inherent attempt of each person is to actualize or develop his or her capacities in ways that serve to maintain and enhance life.

(3) Personality theorists also emphasize a developmental life cycle focus. The notion is that important events occur during each individual's life, and that these events can be plotted along a "human life line" beginning with conception and ending with death. Erik Erikson's eight developmental stages of the life cycle provide a useful descriptive framework for understanding the general issues and transitions that occur during the life cycle. (4) The empirical underpinnings for the study of coping and adaptation come from naturalistic studies of coping behavior under extreme stress. Perhaps the most important observation is that most people manage to adequately handle even extreme life stress. Interest in coping behavior grew out of a natural curiosity to understand why this is so. Another influence involved studies demonstrating that clusters of life events that required changes in ongoing life adjustment (that is, coping) were associated with illness and disease. (5) Crisis theory asserts that life transitions or crises present

both an opportunity for personality growth and a danger of increased vulnerability to mental disorder. Crises are conceptualized as relatively short periods of disequilibrium in which people are more susceptible to influence by others than they are during periods of stable functioning. Thus life crises present unusual opportunities to positively influence coping ability and mental health. These five areas represent the major conceptual and empirical foundations underlying the study of human competence and coping. (See Coelho and others[1] for a more complete bibliography.)

In the second article, Robert White systematically describes various "strategies of adaptation." He illustrates the differences between various terms: *defense, mastery, coping,* and *adaptation.* He defines coping as referring to adaptation under relatively difficult conditions. White points out that living systems have a trend toward increased autonomy; they will attempt to maintain an adaptation permitting them to grow and to enhance their control over their environment. He lists three major tasks that must be met if successful adaptation is to be achieved: adequate information about the environment must be obtained, satisfactory internal conditions both for action and for processing information must be maintained, and freedom of movement or autonomy must be secured. It is essential in assessing the success of adaptive behavior to consider the long-term effects; as White points out, we should not "hastily conclude that the quickest strategies are necessarily the best."

REFERENCES

1. COELHO, G. V.; HAMBURG, D. A.; MOOS, R.; AND RANDOLPH, P., EDS. *Coping and adaptation: A behavioral sciences bibliography.* U.S. Department of Health, Education, and Welfare, Public Health Service Publication No. 2087, 1970.

Article 1   Human Competence and Coping

An Overview

RUDOLF H. MOOS

VIVIEN DAVIS TSU

The Western World, but particularly this nation, is now in the throes of a social upheaval of unparalleled proportions. . . . Standards of behavior, of individual freedom, of interpersonal relations, of the degree of acceptance of authority, of the experiences of the senses and of the very patterns of living are now undergoing the most profound alterations. Since the rate of change of events, of attitudes and of standards is so much more rapid than has ever occurred previously, the impact of these upheavals on society as a whole and on any specific individual is more profound than that experienced during earlier historical movements, which evolved over decades or centuries. Toffler has used the term "future shock" to describe the shattering stress and disorientation induced in individuals subjected to too much change in too short a time. . . . It is clear that if we are to survive the storm individually and collectively, our adaptive abilities will be strained to the utmost. . . . (pp. 1031, 1035)[1]

Alvin Toffler[22] has coined the term "future shock" to describe the stress and disorientation which result when people are subjected to too much rapid change. He points out that nothing is permanent, that the pace of life has accelerated, and that the increase of transience, novelty, and diversity literally makes us "strangers in our own society." The rapid rate of change means that we must often cope with situations to which our previous experience simply does not apply.

One index of change is represented by the greater impermanence of material objects. Paper wedding gowns which may be discarded, or converted into disposable kitchen curtains, symbolize our throw-away culture. Temporary classrooms, movable walls, modular apartments, and portable playgrounds symbolize transient architecture and the impermanent rootless society.

Rapid change is occurring in every facet of our life, and it strains our adaptive capacities to the breaking point. More people are moving than ever before; over 36 million Americans may move in a single year. More and more people change jobs or are transferred by their companies from one plant to another, thus necessitating a geographical relocation. Frequent moves mean that the duration of human relationships becomes shorter and less permanent. These changes all demand new adaptive approaches. Job turnover is increasing and the idea of "serial careers" and "job trajectories" in which individuals may change locations or professions from time to time is gaining popularity. But changing one's physical environment, changing interpersonal relationships, changing jobs, and so on, all involve some stress and thus necessitate new patterns of coping and adaptation.

It may be in alternative life styles and in changing patterns of family living that novelty and diversity have increased the most. Toffler lists a bewildering array of new choices. Men and women may be able to order babies with specified characteristics from assembly lines. Professional parents may take on full-time child-rearing functions for young married couples who want children, but do not want parenthood to tie them down. Corporate families, in which several adults who live and raise children in common would legally incorporate for economic and tax advantages, have been proposed. A series of temporary or trial marriages, in which the partners change throughout life as individual life styles vary, is a future possibility. These and other changes in our contemporary society present a multitude of new choices which threaten to overwhelm our coping capacities.

Lipowski[15] has linked the problem of identity crises to certain features of affluent societies that provide an overabundance of attractive stimuli and alternative life choices. He indicates that a society such as ours, which offers too many attractive options, promotes the development of conflict between incompatible approach tendencies. He suggests that a large number and diversity of possible options generates conflict. This gives rise to stimulus overload and to identity confusion. The sampling of different options, without any firm choice or sustained commitment, may lead to a sense of emptiness and futility, that is, a failure to cope.

Toffler feels that there are limits to the amount of change with which the human organism can cope. Since change and impermanence will continue in our open pluralistic society, he believes that we must learn more about the methods by which people cope with decision-making and change. New coping techniques or "strategies of survival" need to be developed for our society to continue to function adequately. He states:

> I gradually came to be appalled by how little is actually known about adaptivity either by those who call for and create vast changes in our society or by those who supposedly prepare us to cope with those changes. Earnest intellectuals talk bravely about educating for change or preparing people for the future. But we know virtually nothing about how to do it. In the most rapidly changing environment to which man has ever been exposed we remain pitifully ignorant of how the human animal copes.

Why is it that some people seek change and thrive on challenge, whereas others fear novelty and want consistency? Why is it that some people adequately handle serious life crises, whereas others break down under minor stress? This book is dedicated to these questions. It examines in detail how individuals cope with various life crises and transitions. Why is it that one child reacts to the challenge of kindergarten or first grade in an adaptive way, whereas another child is overcome by fear? Why is it that some students develop new ways of personal growth and competence in their freshmen year at college, whereas others are overwhelmed by feelings of inferiority, doubt, and anxiety? Why is it that one couple can develop an intimate supportive relationship that "sets the stage" for their mutual growth, whereas another couple bickers and fights in a fundamentally destructive manner? Why is it that some people can retire and live out their lives in reasonably productive ways, whereas others are overcome by the stress of boredom and apathy?

The study of coping and adaptation, particularly in relation to life crises and transitions, has recently become a growing area of interest (see Lazarus[13] for a more detailed review of the field). Five major influences have shaped current interest in this field.

EVOLUTION AND ADAPTATION

Charles Darwin focused on the adaptation of animals (and humans) to their environment in his evolutionary theory. The two essential elements in Darwinian theory are variation in the reproduction and inheritance of living organisms, and natural selection or the survival of the fittest. One factor is internal and one is external. The inner factor, variation, is conceptualized as positive and creative. It produces the variations needed for man's progress. The external factor, natural selection, is conceptualized as negative and destructive. It eliminates the harmful, the less fit, the less useful variations. It leaves the beneficial, the more fit, or the more useful variations to develop and multiply. Living organisms exist in the "web of life" in which they "struggle for existence" in relation to the environment. This process fits and adapts the species to the particular character of its environment.

Darwin's ideas resulted in the development of ecology, which is the study of the relation of organisms or groups of organisms to their environment. Human ecology is distinguished from plant and animal ecology specifically by the unique characteristics of man and the human community. Unlike plants and animals, human beings can construct or alter their own environment. They also have an elaborate technology and culture; and they are regulated by conscious controls— by rules, norms, laws, and formal organizations. As human ecology developed, it began to be seen as a general perspective useful for the scientific study of social life.[21]

Evolutionary thought and human ecology have been mainly concerned with the study of communal adaptation. Human beings cannot adapt to their environment alone. They are highly dependent on their fellow human beings and must make collective efforts at adaptation. Adaptation to the conditions of the physical environment is facilitated by adaptation to (and cooperation with) other living beings. It is this communal adaptation that is the distinctive subject matter of ecology. Human ecology, which studies aggregate populations of human beings rather than individuals, is distinctive in its hypothesis that the human community is an essential adaptive mechanism in man's relation to the environment.[7]

Thus, from an evolutionary perspective, it is the reproductive success of a population that constitutes adaptation. However, communal adaptation presupposes, and must be studied in conjunction with, individual adaptation.

> Although the emphasis in evolutionary theory is usually placed on populations of organisms, it is important to note that reproductive success of a population involves all phases of the life cycle of the *individual* organism and that such success is most commonly favored by individual adaptation to his particular environment . . . as complex organisms have evolved, behavior has become an exceedingly important factor in meeting adaptive tasks which contribute to species survival.[6]

HUMAN FULFILLMENT AND GROWTH

What is it that keeps people going? What are the basic motivations that impel man? Sigmund Freud felt that man's motivation was to reduce tension, and that the major drives arousing tension were sex and aggression. All energy and motivation were thought to be instinctual; all behavior was attributed to sexual and aggressive instincts. The id was the repository of the instincts, and the super ego of civilized morality and restraint. The ego developed out of the resolution of the conflict between id and super ego.

Various theorists developed objections to these ideas. What about ego functions such as perception, memory, language, thought, and attention? Can these functions be explained simply as arising out of the conflict between instinctual impulses and the restraints of the super ego? The ego psychologists did not think so. They conceptualized a "conflict free ego sphere" with its own "autonomous" energy. The idea was that ego functions had their own sources of energy, independent of instinctual energy. Furthermore, exercising ego functions such as perceiving and thinking was thought to be pleasurable and rewarding in itself. People obtain pleasure and reinforcement when their actions enable them to control and alter their environments. These ideas formed the basis of an entirely new set of theories about the development of human personality.

Human beings wish to reduce excessive tensions; however, they also dislike boredom and need at least some stimulation, novelty, and excitement. Anyone who has watched children play knows that it is "a joy to be a cause." Human beings have drives toward mastery and control of their environments. Robert

White[24] proposes that these different drives be categorized under the label of competence. Other terms that have been used to describe various aspects of competence include an instinct of curiosity, an exploratory drive, a positive need for varied sensory stimulation (even a "stimulus hunger"), a need to engage in information processing, and so on. "It is in connection with striving to attain competence that the activity inherent in living organisms is given its clearest representation—the power of initiative and exertion that we experience as a sense of being an agent in the living of our lives. This experience may be called a *feeling of efficacy*." [24]

Recent theorists, among whom Carl Rogers and Abraham Maslow are probably the best known, have organized their concepts of personality around human growth or fulfillment models. Rogers hypothesizes that the core tendency of man is self-actualization. Each person attempts to actualize or develop all of his or her capacities in ways that serve to maintain and enhance life.[19]

Maslow postulates two underlying patterns of motivation: growth motivation and deficiency or deprivation motivation. Deficiency motivation aims to decrease tension arising from unfulfilled tissue needs such as hunger and thirst. Like the tension reduction model discussed earlier, this reflects a motive to survive. Growth motivation, on the other hand, reflects the tendency toward self-actualization and involves the urge to enrich our experience and to expand our horizons. Maslow talks about creative living, peak experiences, and unselfish love as examples of this.

As might be expected, growth and fulfillment theories have led to studies of normal, healthy people and their patterns of adaptation and coping. Conflict and tension reduction models of personality were generally associated with the study of abnormal functioning and psychiatric symptomatology. Once self-actualization and growth motivation were emphasized, the systematic study of healthy, mature, self-actualized people began. Maslow[16] informally studied a number of outstanding contemporary and historical figures. For example, self-actualized individuals, according to Maslow, included Thomas Jefferson, Abraham Lincoln, Albert Einstein, Jane Addams, and Eleanor Roosevelt. (Franklin Roosevelt and Sigmund Freud were thought to have fallen somewhat short of ideal self-actualization.)

What were some of the characteristics by which these and other self-actualized people coped with their environments? Maslow identified them as follows: an efficient perception of reality and comfortable relations with it, a problem-centered focus, and a broad view of life in the widest frame of reference. Maslow describes these self-actualized people as spontaneous in their behavior, as having a strong social interest and a genuine desire to help others, as usually having a small but intimate circle of friends, and as being somewhat independent of their culture and environment.

Other investigators have addressed themselves to more normal mentally healthy people. Roy Grinker[5] studied a group of young male college students whom he labeled homoclites (people following the common rule, that is, healthy, ordinary people). Grinker found that these students did reasonably well academically and were able to play well and with satisfaction. They were adequately adjusted to reality, had a firm sense of identity, felt good, and had hope for the future. They had warm human relationships with parents, teachers, and friends (both male and female). Their aspirations were to be successful, to make some contribution, and to be liked. Grinker identifies a cluster of conditions associated with this mentally healthy group, such as sound health, average intelligence, satisfactory relationships with both parents, parental agreement and cooperation in child raising, definite and known limitations or boundaries placed on behavior, sound early religious training, early work history, action orientation, and so on.

There have been many recent studies of normal adolescent development, of the development of maturity in college students, of outstanding individuals such as the first astronauts, and of other mentally healthy individuals such as Peace Corps volunteers. These studies have focused on how these individuals coped with their life situation and handled specific relatively stressful experiences (for example, being a Peace Corps volunteer in a foreign country). Interest in studying normal healthy individuals and their adaptation and coping patterns has also given rise to a multitude of new psychological assessment techniques. For example, techniques have been developed to measure variables such as self-esteem, self-actualization, ego identity, exploratory behavior, novelty needs, stimulus-seeking motivation, feelings of efficacy and competence, and so on.[17]

A DEVELOPMENTAL LIFE CYCLE FOCUS

Early psychoanalytic theorists suggested that life events in infancy essentially determined the structure of an individual's adult personality. Anxieties, conflicts, and hostilities that arose during a person's later life were thought to be a direct result of initial stages in the individual's development. Interest in the development of the ego and ego functions, and in the study of normal healthy individuals, indicated that events early in infancy did not necessarily determine an individual's personality, nor his reactions to crises and stress. Thus the flexibility and variability of individual development over the life cycle were stressed, as were significant life events and relevant environmental characteristics.

A useful conception is that of a "human life line," which begins with conception and ends with death. Important events occurring during each individual's life can be plotted along this life line. Kimmel describes this clearly:

> One useful analogy is to think of the life line as a representation of a journey with a number of interesting places and crucial junctions along the way. Some years ago when travel was slower roads were commonly marked with *milestones* or mileposts to mark off each mile traveled. These milestones were important because the traveler was progressing toward a goal through time and his progress was measured by the number of miles traveled in a unit of time (such as a day). Not surprisingly, humans tend to mark off their progress through their life cycle in a very similar way. We are consciously aware

of our progress during the past year, and we have a sense of moving too slowly or very quickly along the path toward our goal based on a measure of time . . . we . . . in fact often celebrate these milestones (such as graduation, marriage or retirement). (pp. 11, 12)[11]

There are several theories about the broad course of human development and aging (see Chapter 1 of Kimmel), but the most comprehensive and relevant for our purposes is that developed by Erik Erikson.[4] Erikson describes eight stages of man. Each stage in the life cycle represents a new challenge, a new transition, a new "crisis" that must be negotiated successfully for the individual to be in a position to handle the next stage well. Broad coping skills learned in infancy and adolescence affect the success of coping in adulthood and old age.

Erikson's first four stages occur in the years between birth and puberty.

1. *Basic trust versus basic mistrust.* Erikson theorizes that the "quality of the maternal relationship" early in infancy determines the extent to which an individual develops basic trust. Because the infant is helpless, other people, primarily the mother, must nurture, protect, and reassure it with a sense of consistency and continuity. If the basic needs are fulfilled then, the infant, and later the child, will presumably perceive the world in a generally trustful manner. If the infant is handled inconsistently, or is partially or severely deprived, he or she will probably be mistrustful.
2. *Autonomy versus shame and doubt.* In this stage, which occurs approximately in the infant's second year of life, muscular maturation and motor locomotion begin. The child tests out his or her own powers and begins to act with some autonomy and independence. If the child is encouraged in this, the seeds of later independence are sown. Harsh arbitrary punishment of a child's development of autonomy and independence will, according to Erikson, engender later attitudes of shame and doubt.
3. *Initiative versus guilt.* This stage usually occurs between the ages of three and five, during which the child has a surplus of energy, obtains pleasure in "attack and conquest," and basically enjoys his or her new powers of locomotion and manipulation. Erikson postulates that the danger of this stage is the development of a sense of guilt over "the goals contemplated and the acts initiated in one's exuberant enjoyment of new locomotor and mental power." Again here the issue is the degree and quality of outside (parental) regulation and control. If regulation is mutual, consistent, and within the context of a trusting relationship, the child is likely to develop initiative and responsibility. If regulation (that is, punishment) is overly harsh and strict, the child will probably develop feelings of resignation, guilt, and unworthiness.
4. *Industry versus inferiority.* During this stage the maturing child learns to win recognition by producing things. Working at given skills and tasks, the child learns the value and pleasure of attention, diligence, skill, the

accomplishment of specific tasks, and so on. The successful resolution of this fourth stage results, according to Erikson, in the traits of methods and competence. Thus the child, on entering the stage of puberty, has passed through four critical life stages and has hopefully developed the basis for trust, autonomy, initiative, and industry.

Erikson postulates four additional stages characterizing the life cycle from puberty to death.

5. *Identity versus role confusion.* Erikson's fifth stage revolves around the task of settling on an identity. This includes making choices about entering college or university, about an occupational or vocational career, about one's basic values and beliefs, and usually about a marriage partner. Decisions essential to one's developing ego identity are usually made during this stage. If these decisions are not resolved in a reasonably positive way then, the result may be "identity confusion," a lack of clarity about one's role in life.

6. *Intimacy versus isolation.* Erikson believes that an individual is ready to merge his or her identity with that of another individual only after developing a sense of identity. One is then ready for intimacy, which Erikson defines as the capacity to commit oneself "to concrete affiliations and partnerships and to develop the ethical strength to abide by such commitments even though they may call for significant sacrifices and compromises." The notion is that one must have a sense of identity (of who one is) before one can merge one's identity with another (see Part IV).

7. *Generativity versus stagnation.* This stage includes not only the production of, and caring for, one's offspring, but also the development of productivity and creativity in occupational or vocational spheres. Thus this stage may extend from young adulthood until old age, and may play a central role in achieving a sense of fulfillment in life. The danger of this stage is a feeling of stagnation, whereas its success results in traits of productivity and caring.

8. *Ego integrity versus despair.* This final stage of life may begin as early as middle age and is accentuated by an increasing awareness of aging and the close of one's life. The issue of the meaning of one's life may be triggered by retirement, death of a spouse, or declining health.

Erikson states:

Only in him who in some way has taken care of things and people and has adapted himself to the triumphs and disappointments adherent to being the originator of others or the generator of products and ideas—only in him may gradually ripen the fruit of these seven stages. I know no better word for it than ego integrity . . . it is the acceptance of one's one and only life cycle as something that had to be and that by necessity permitted no substitutions. . . . In such final consolidation death loses its sting.

The lack or loss of this accrued ego integration is signified by fear of death: the one and only life cycle is not accepted as the ultimate of life. Despair expresses the feeling that the time is now short, too short for the attempt to start another life and to try out alternate roads to integrity. (pp. 268–69)[8]

Erikson's theory of the eight developmental stages of man provides a useful descriptive framework for understanding some of the general issues and transitions that occur during the life cycle. Erikson's important point is that transitions and crises occur throughout the life cycle, rather than only in infancy and pre-adolescence, and that the resolution of one stage leaves a legacy important for the resolution of subsequent stages. This notion forms an important rationale for "crisis theory," and for the development of interventions to help individuals progress more adequately through the life cycle.

COPING WITH EXTREME LIFE STRESS

Two areas have provided empirical data for the development of this field. The first includes detailed naturalistic and interview studies of coping behavior under extreme stress. Probably the most unusual of these studies are the observational accounts of coping and adaptive behavior in the Nazi concentration camps (see Part X). This was a situation of hardship and stress, comparable to the most harrowing conditions ever endured by man. Some people managed to survive these conditions, and to observe and write about them in objective and even creative terms. An obvious question arose from these accounts. How was it that these individuals were able to cope with a life-threatening crisis of this magnitude?

Accounts of camps like Buchenwald [12] and Treblinka[20] have shown that people try to achieve some control over their ultimate fate even under the most hellish conditions. An extensive underground organization of inmates in Buchenwald, for example, was successful in controlling work assignments and reassignments of inmates, and in hiding and protecting valuable members of the underground. The inmates in Treblinka managed to stage a camp revolution under what can only be regarded as impossible circumstances. Thus a new interest in coping and adaptation grew in part out of man's appalling inhumanity to man.

Other studies along these lines included detailed observations of how severely burned patients coped with their injuries, of how survivors coped with the sudden death of their spouses or close relatives (see Parts III and VIII), of how patients coped with major surgery, of how parents coped with experiencing the slow death of one of their children, and of how individuals and communities coped with extreme disasters (see Part XI). The thrust of these studies was to detail the adaptive aspects of individual coping to psychiatrists and psychologists who were primarily concerned with anxieties, symptom formation, neuroses, and individual pathology. The fact that people could cope with stresses of this magnitude was something to be reckoned with. Perhaps if one could identify the processes that these individuals used, one could teach other people to cope more adequately with their own life situations.

The second empirical trend developed out of an emphasis on life changes in relation to health and disease. A well-known psychiatrist, Adolph Meyer, developed the "life chart" as a tool in medical and psychiatric diagnosis. Meyer emphasized the importance of events such as migration, school entrance, graduations, job changes or failures, important births and deaths, and so on. Meyer believed that "regular" life events and transitions provided relevant clues to the development of symptoms and diseases. Others have also emphasized the important role of the environment in facilitating or inhibiting good physical and mental health.[18,9]

Holmes and his colleagues used the Meyerian life chart to study life events empirically observed to cluster at the time of disease onset. They developed the Social Readjustment Rating Scale (SRRS), which consisted of different life events scaled according to the amount of "readjustment" they were judged to require; for example, death of spouse (100 Life Change Units), divorce (73 Units), marital separation (65 Units), jail term (63 Units), marriage (50 Units), retirement (45 Units), pregnancy (40 Units), gain of a new family member (39 Units), change to a different line of work (36 Units), son or daughter leaving home (29 Units), begin or end school (26 Units), change in residence (20 Units), and so on. Social readjustment was defined to include the amount and duration of change in one's accustomed pattern of life resulting from various life events, regardless of the desirability of the events.

Holmes and his colleagues found that clusters of life events that required changes in ongoing life adjustment were associated with illness onset. Their studies have shown, for example, that life events cluster significantly in the two-year period preceding the onset of tuberculosis, heart disease, skin disease, and hernia.

In a conceptual link to the crisis and transition area, Holmes and his colleagues defined a life crisis "as any clustering of life change events whose individual values summed to 150 Life Change Units (LCU) or more in one year."[8] They found a direct relationship between the magnitude of the life crisis and the risk of health change. Specifically, 37 percent of mild life crises were associated with health changes, 51 percent of moderate life crises were associated with health changes, whereas 79 percent of major life crises were associated with health changes. Thus, the greater the magnitude of life change (or life crisis) the greater the probability of some associated illness or disease. Holmes and his colleagues link their work to coping and adaptation literature by postulating "that life change events, by evoking adaptive efforts by the human organism that are faulty in kind and duration, lower bodily resistance and enhance the probability of disease occurrence."

TRANSITIONS, CRISES, AND PREVENTIVE INTERVENTION

The above four trends influenced the development of "crisis theory," which is concerned with how individuals can cope with life transitions and crises. Crisis theory has become a conceptual framework for preventive psychiatry and for the community mental health movement. The fundamental ideas were developed by Erich Lindemann,[14] who vividly described the process of grief and mourning in the survivors of those killed in the Coconut Grove nightclub fire in Boston. Lindemann felt that the bereaved family members could be helped to cope with the loss of their loved ones through the assistance of community caretakers such as clergymen and friends (see Part VIII).

Caplan[2] points out that when one examines the history of some psychiatric patients it seems that they have dealt with major life crises in a maladjusted manner, emerging less healthy than before. However, changes during life crises may also be toward increased health and maturity, and in this sense a life transition may be a period of opportunity. A life crisis is a transitional period, a turning point, presenting an individual both with an opportunity for personality growth and with the danger of increased vulnerability to mental disorder.

Crisis theory asserts that people generally operate in consistent patterns, in equilibrium with their environment, solving problems with minimal delay by habitual mechanisms and reactions. When the usual problem-solving mechanisms do not work, tension arises and feelings of discomfort or strain occur. The individual experiences anxiety, fear, guilt or shame, a feeling of helplessness, some disorganization of functioning, and possibly other symptoms. Thus a crisis is essentially a disturbance of the equilibrium, an "upset in a steady state." Various factors influence the outcome of a crisis: previous experience with similar crises, the degree of seriousness of the crisis, the degree of support available in the environment, the degree to which family members facilitate or inhibit the healthy resolution of a crisis, the degree of help available from significant others in the community, and so on.

A crisis is a relatively short period of disequilibrium in which a person tries to work out new ways of handling a problem through sources of strength in himself and in his environment. His new level of functioning may be more or less healthy than his precrisis pattern. Every crisis presents both an opportunity for psychological growth and a danger of psychological deterioration. During the crisis a person is more susceptible to influence by others than during periods of stable functioning. Thus, a relatively minor intervention may drastically change the ultimate outcome. Therefore, crises present an unusual opportunity to positively influence coping ability and mental health. The successful mastery of life crises can constitute an important growth experience.[10] The opportunity and growth potential aspect of life transitions and crises support a focus on preventive intervention and on facilitating effective coping behavior.

Several factors affect the quality of a person's coping response. For example, the environment may offer practical and emotional resources, such as social agencies and support from family and friends. The environment also sets limits, in the form of cultural norms and expectations, within which the individual must operate. In addition, the nature of the stressor, such as its intensity or duration, has a major influence. Handling one's first year at college (see Part IV) involves a different set of responses and a different expectation of success than does coping with imprisonment in a prisoner of war camp (see Part IX). Finally, the person's own nature, her or his past experience and abilities, have a significant impact on coping behavior.

Caplan[3] has identified seven characteristics of effective coping behavior that seem to cut across different types of life transitions and crises. These characteristics, one or more of which are identified in many of the selections below, are as follows:

1. Active exploration of reality issues and search for information.
2. Free expression of both positive and negative feelings and a tolerance of frustration.
3. Active invoking of help from others.
4. Breaking problems down into manageable bits and working them through one at a time.
5. Awareness of fatigue and tendencies toward disorganization with pacing of efforts and maintenance of control in as many areas of functioning as possible.
6. Active mastery of feelings where possible and acceptance of inevitability where not. Flexibility and willingness to change.
7. Basic trust in oneself and others and basic optimism about outcome.

Specific coping strategies differ in appropriateness according to the particular combination of situation and personality variables involved, but Caplan's characteristics can serve as general criteria in judging the effectiveness of a chosen pattern of coping behavior.

In reality, coping refers to two distinct but related tasks. One must respond to the requirements of the external situation and also to one's own feelings about that situation. In a disaster, for example, people may get themselves out of immediate danger, see to the security of their families, find temporary housing, and so on. They must also deal with feelings of loss, depression, inadequacy or guilt, and other threats to their self-esteem (see Part XI).

These two tasks are not necessarily dealt with simultaneously, for the coping process is dynamic, with the demands of the situation and the strategies of the individual changing as time passes. The overall pattern tends to fall into two phases: an acute phase in which energy is directed at minimizing the impact of the stress, and a reorganization phase in which the new reality is faced and accepted. In the acute period feelings may be denied while attention is directed at practical matters. This allows people to ration out their limited physical and emotional energy, while giving them some time to adjust to the change in their lives. The reorganization phase involves the gradual return to normal functioning and to the achievement of a new equilibrium in which changed circumstances and new feelings are integrated into the individual's life and self-image.

The coping process that we are exploring on the individual level, with its opportunities for growth and its threat of harm, has an interesting parallel on the societal level. The British historian Arnold Toynbee[23] points out that although people have lived on earth for at least two million years, the earliest recorded civilizations date back only about six thousand years. In his analysis of the development of higher civilizations Toynbee suggests that the histories of individuals, of communities, and of entire civilizations fall into successive stages. In each stage some groups of people are confronted by a specific challenge that imposes an ordeal (crisis). Some people discover a response to the challenge that not only allows them to cope with the ordeal of the moment but also puts them in a favorable position for undergoing the next ordeal.

Toynbee points out that the conditions offered to man by the environments that have been the birthplaces of civilizations have been unusually difficult, not unusually easy. In his view, man needs to be exposed to challenge from either the physical or the human environment. If the challenge is too weak (that is, the environment is too easy), human potential will remain unfulfilled. If the challenge is excessive, human attempts to cope with it will result in failure and decline. When the challenge is optimal, human beings will be stimulated to new creative heights. Challenge and change may stimulate individuals to creative growth.

Our concern here is with the individual, rather than with the entire society. In the final analysis, each individual must cope with rapid technological and social change. Each individual must select from the available choices of "alternative life styles," and then cope with his or her selection; thus we focus here on individuals adapting to major life changes. Societal coping patterns, however, are partly an outgrowth of creative solutions developed by a few individuals. In this sense the study of how individuals cope with life crises can suggest alternative strategies that may ultimately benefit society.

REFERENCES

1. BRAUNWALD, E. Future shock in academic medicine. *New England Journal of Medicine,* 1972, 286, 1031, 1035.
2. CAPLAN, G. Emotional crises. In A. Deutsch, ed., *The encyclopedia of mental health,* Vol. 2. New York: Franklin Watts, 1963.
3. CAPLAN, G. *Principles of preventive psychiatry.* New York: Basic Books, 1964.
4. ERIKSON, E. H. *Childhood and Society* (2nd Ed.). New York: W. W. Norton, 1963.
5. GRINKER, R. R. "Mentally Healthy" young males (homoclites): A study. *Archives of General Psychiatry,* 1962, 6, 405–53.
6. HAMBURG, D. A.; COELHO, G. V.; AND ADAMS, J. E. Coping and adaptation: Steps toward a synthesis of biological and social perspectives. In G. V.

Coelho, D. A. Hamburg, and J. E. Adams, eds., *Coping and adaptation*. New York: Basic Books, 1974.

7. HAWLEY, A. *Human ecology*. New York: Ronald Press, 1950.

8. HOLMES, T. H., AND MASUDA, M. Life change and illness susceptibility. In B. S. Dohrenwend and B. P. Dohrenwend, *Stressful life events: Their nature and effects*. New York: John Wiley, 1974.

9. INSEL, P. M., AND MOOS, R. H., eds. *Health and the social environment*. Lexington, Mass.: D. C. Heath, 1974.

10. KALIS, B. L. Crisis theory: Its relevance for community psychology and directions for development. In D. Adelson and B. L. Kalis, eds., *Community psychology and mental health: Perspectives and challenges*. Scranton, Pa.: Chandler, 1970.

11. KIMMEL, D. C. *Adulthood and aging: An interdisciplinary, developmental view*. New York: John Wiley, 1974. Pp. 11, 12.

12. KOGON, E. *The theory and practice of hell*. New York: Berkeley Medallion Books, 1958.

13. LAZARUS, R. S. *Psychological stress and the coping process*. New York: McGraw-Hill, 1966.

14. LINDEMANN, E. Symptomatology and management of acute grief. *American Journal of Psychiatry*, 1944, *101*, 141–48.

15. LIPOWSKI, Z. J. The conflict of Buridan's Ass or some dilemmas of affluence: The theory of attractive stimulus overload. *American Journal of Psychiatry*, 1970, *127*, 273–79.

16. MASLOW, A. *Motivation and personality*. New York: Harper, 1954.

17. MOOS, R. H. Psychological techniques in the assessment of adaptive behavior. In G. V. Coelho, D. A. Hamburg, and J. E. Adams, eds., *Coping and adaptation*. New York: Basic Books, 1974.

18. MOOS, R. H., AND INSEL, P. M., eds. *Issues in social ecology: Human milieus*. Palo Alto, Calif.: National Press, 1974.

19. ROGERS, C. *On becoming a person*. Boston: Houghton Mifflin, 1961.

20. STEINER, J. F. *Treblinka* (H. Weaver, trans.). New York: Simon and Schuster, 1967.

21. THEODORSON, G., ed. *Studies in human ecology*. Evanston, Ill.: Row Peterson, 1961.

22. TOFFLER, A. *Future shock*. New York: Random House, 1970. P. 4.

23. TOYNBEE, A. *The study of history*, Vols. 1 and 2. *The geneses of civilizations*, Parts 1 and 2. New York: Oxford University Press, 1962.

24. WHITE, R. W. Motivation reconsidered: The concept of competence. *Psychological Review*, 1959, *66*, 297–333.

Article 2 Strategies of Adaptation

An Attempt at Systematic Description

ROBERT W. WHITE

At the outset of this inquiry we are confronted by four commonly used words with overlapping claims upon the territory to be discussed. The words are *adaptation, mastery, coping,* and *defense.* No attempt at systematic description is likely to prosper if these words are left in a state of free competition, jostling for the thinly scattered grains of truth that might nourish their meaning. Their peaceful coexistence requires, as in any well-regulated hen yard, the establishment of some sort of pecking order that everyone observes and fully understands. The first step in this direction is simple; clearly the boss hen is *adaptation.* This is the master concept, the superordinate category, under which the other three words must accept restricted meanings. Descriptions of mastery, of defense, or of coping alone cannot be systematic in any large sense, but they can become part of a system if they are ordered under the heading of *strategies of adaptation.*

PRELUDE ON TERMINOLOGY

The concept of *defense,* to take it first, is an obvious one, signifying response to danger or attack, but it comes to us with a somewhat swollen meaning because of the position it has been given in psychoanalytic theory. Freud's genius as an observer, so apparent in his unveiling of sexual and aggressive inclinations, never burned more brightly than in his perception of what came to be called the *mechanisms of defense*: repression, projection, undoing, and the other devices

ROBERT W. WHITE, PH.D., is Professor Emeritus, Clinical Psychology, Harvard University, Cambridge, Mass.

Source: Pages 47–61 from Chapter 4, "Strategies of Adaptation: An Attempt at Systematic Description," by Robert W. White in *Coping and Adaptation*, Edited by George V. Coelho, David A. Hamburg and John E. Adams, © 1974 by Basic Books, Inc., Publishers, New York.

whereby danger was parried and peace restored in the frightened psyche. Psycho-analytic therapists, following this lead, became expert at scenting anxiety in the free associations of their patients; expert, moreover, at unraveling the ramified operations whereby security was achieved. Presently these operations were seen to have worked over long periods of time, producing such complex results as character armor and the protective organization of personality. Unwary theorists even jumped to the generalization that development was a simple counterpoint between instinctual craving and defense. It became necessary after Freud's death for those who called themselves "ego psychologists" to restore explicitly the concept of adaptation and to confine defense to those instances of adaptation in which present danger and anxiety were of central importance.

The concept of *mastery*, perhaps an equally obvious concept, has never enjoyed the same vogue among psychologists. When used at all, it has generally been applied to behavior in which frustrations have been surmounted and adaptive efforts have come to a successful conclusion. The alternatives suggested by the word are not, as with defense, danger and safety, but something more like defeat and victory. This might imply a limiting definition, but in fact the concept of mastery has been used with no sense of limits. The English language, loved by poets for its flexibility, offers only pitfalls to the systematic thinker. There is nothing wrong with saying that danger and anxiety have to be mastered, which allows us to classify defense mechanisms as a form of mastery. It is equally correct to say that efforts at mastery serve as a defense against anxiety, which permits us to consider counteractive struggle a mechanism of defense. If mastery is to be used in any limited technical sense it should probably be confined to problems having a certain cognitive or manipulative complexity, but which at the same time are not heavily freighted with anxiety.

Where does the concept of *coping* stand? We can find out what we mean by it by noticing the kinds of situation chosen for studies of coping behavior. Sometimes these situations represent an acute dislocation of a person's life: serious crippling sickness, the death of close relatives, financial disaster, the necessity to live in a radically new environment. Sometimes the situation is less drastic, but it is still unusual in the subject's life: going to school for the first time, going to visit the child psychologist, or making the transition from high school to college. Nobody has chosen going to school for the sixty-third time as an occasion for coping. The freshman year at college, with all its new experience, clearly qualifies as coping, as does the sophomore year now that we are alert to the possibility of "sophomore slump" and dropping out, but nobody has yet detected any large-scale common problems that would justify choosing the junior year for an investigation of coping behavior. For "when the sea was calm," said Shakespeare, "all boats alike show'd mastership in floating"; only in a storm were they obliged to cope. It is clear that we tend to speak of coping when we have in mind a fairly drastic change or problem that defies familiar ways of behaving, requires the production of new behavior, and very likely gives rise to uncomfortable affects like anxiety, despair, guilt, shame, or grief, the relief of which forms part of the needed adaptation. Coping refers to adaptation under relatively difficult conditions.

This discussion of terms demonstrates the necessity of making *adaptation* the central concept. It may well be that in stressful situations things happen that have no counterpart in easier circumstances, but some of what happens is likely to come straight from the repertoire that is common to all adaptive behavior. There is a sense in which all behavior can be considered an attempt at adaptation. Even in the smoothest and easiest of times behavior will not be adequate in a purely mechanical or habitual way. Every day raises its little problems: what clothes to put on, how to plan a timesaving and step-saving series of errands, how to schedule the hours to get through the day's work, how to manage the cranky child, appease the short-tempered tradesman, and bring the long-winded acquaintance to the end of his communication. It is not advisable to tell a group of college students that they have no problems, nothing to cope with, during the happy and uneventful junior year. They will quickly tell you what it takes to get through that golden year, and as you listen to the frustrations, bewilderments, and sorrows as well as the triumphs and joys you will have a hard time conceptualizing it all as well-adapted reflexes or smoothly running habits. Life is tough, they will tell you, in case you have forgotten; life is a never-ending challenge. Every step of the way demands the solution of problems and every step must therefore be novel and creative, even the putting together of words and sentences to make you understand what it is like to cope with being a college junior.

Adaptation, then, is the only firm platform on which to build a systematic description. What is needed is an ordered account of *strategies of adaptation*, ranging from the simplest ways of dealing with minor problems and frustrations to the most complex fabric of adaptive and defensive devices that has ever been observed from the chair at the head of the psychoanalytic couch. If this can be done, the uses to be made of defense, mastery, and coping can be much more readily decided.

ADAPTATION AS COMPROMISE

There is another preliminary issue that is likely to get in the way if we do not deal with it at the start. Perhaps we can put a little blame on Freud for having started something that often crops up today as an unwitting tendency to think of adaptive behavior in a dichotomy of good and bad. Uncensorious as he was toward the neurotic behavior that circumstance and a repressive society had forced upon the patient, Freud was a stern and moral man who would not call a patient well until all neurotic anxieties were understood, all defense mechanisms abandoned, and all behavior brought under control of the clear-eyed ego that perceives everything exactly as it is. But this heroic prescription was meant to apply only to neurotic anxieties, legacies of childhood that did not correspond to present dangers. The ideal patient, issuing from his analysis cleansed of all anxiety, was really cleansed only of defenses against dangers that no longer existed. So we must attach less blame to the fastidious Freud than to the careless popularizers of mental health wisdom who have communicated the thought, utterly bizarre in one of the most frightening periods of human history, that the mentally healthy person is free from all anxiety and meets life with radiant confidence. Of course we all know better than that when we stop to think, but in the psychological and psychiatric literature there lies a concealed assumption that dangers must be faced because they are not really there, that any delay, avoidance, retreat, or cognitive distortion of reality is in the end a reprehensible piece of cowardice. We must march forward, ever forward, facing our problems, overcoming all obstacles, masters of our fate, fit citizens of the brave new world. Foolish as it is, this unwitting assumption sufficiently pervades our professional literature so that we really do have to stop and think. What are the transactions that actually take place between a person and his environment?

In actuality, of course, there are many situations that can be met only by compromise or even resignation. Events may occur that require us to give in, relinquish things we would have liked, perhaps change direction or restrict the range of our activities. We may have no recourse but to accept a permanent impoverishment of our lives and try to make the best of it. Furthermore, when dangers are real and information incomplete it is in no sense adaptive to march boldly forward. History provides many examples, none better than General Braddock in our own colonial days, who marched his column of British regulars through the forests of Pennsylvania straight into a French and Indian ambush. Described not inappropriately in military metaphors, adaptation often calls for delay, strategic retreat, regrouping of forces, abandoning of untenable positions, seeking fresh intelligence, and deploying new weapons. And just as recuperation from serious illness is not the work of a day, even though in the end it may be completely accomplished, so recovery from a personal loss or disaster requires a long period of internal readjustment that may not be well served at the start by forceful action or total clarity of perception. Sometimes adaptation to a severely frustrating reality is possible only if full recognition of the bitter truth is for a long time postponed.

The element of compromise in adaptive behavior can be well illustrated from that rich storehouse of information provided by Lois Murphy[9] in her study of young children in Topeka. She describes a number of three-year-olds brought for the first time from their homes to her study center, where the business of the day is to meet a psychologist and engage in some activities that constitute a test of intelligence. Her first two illustrations, boys named Brennie and Donald, present us at once with a striking contrast. Brennie appears to be confidence incarnate. He climbs happily into the car, alertly watches and comments upon the passing scene, charms everyone with his smile, walks into the testing room with perfect poise, accepts each proffered task with eager interest, makes conversation and asks for appropriate help from adults, and finally leaves the scene with a polite expression of thanks. Brennie might be judged a paragon of mental health, and any three-year-old so easy to deal with is certain to be a psychologist's delight. In contrast, the day of Donald's visit is a taxing one for the staff. The child comes accompanied by his mother, described as "warm and ample," and he utters not a word either during the ride, when entering the building, or for some time after he enters the psychologist's office. Invited to sit down, he stands resolutely beside his mother, his feet spread slightly apart. He will string beads only when his mother has done so first, and once embarked on this operation, he refuses to be diverted by the psychologist, who would like to get on with the test. Slowly he warms up enough to dispense with his mother's mediation and deal directly with the psychologist, but the testing still drags because Donald becomes involved in, for instance, building-block constructions of his own instead of those required for the test. The session ends with the assessment far from complete.

It is easy to imagine what Donald's session would look like in the records of a typical guidance agency. He has displayed two highly disquieting symptoms. He has a bad case of separation anxiety, clinging to his mother when he should be facing reality, and he also displays withdrawal and introversion by building with blocks according to his fantasy instead of responding properly to social stimulation. But before we hurry Donald into psychotherapy let us look at the situation from a child's point of view. As adults we know something that he does not know: we know that Mrs. Murphy and her staff are full of kindness and patience and that they will go to great lengths to keep discomfort and anxiety at a minimum. Donald can know only that he is being taken to a strange place for a purpose he cannot fathom. Many children by the age of three have been to the pediatrician's office, to the barber, and perhaps even to the dentist, and they may well have noticed a credibility gap between parental assurances and the discomforts actually experienced during these visits. Now they are being taken to play games with a nice lady—a likely story indeed! If such conditions existed for Donald, he exhibits commendable common sense in sticking close to his mother, the one familiar object, until he can figure out the nature of the racket. It is his good fortune that his principal observer, Mrs. Murphy, understands his position, perceives him not as anxiously dependent but as a "sturdy boy," and appreciates his strategy of adaptation. She says:

> Over the years we have seen Donald, this pattern has continued: cautious, deliberate, watchful entrance into a new situation, keeping his distance at first, quietly, firmly maintaining his right to move at his own pace, to make his own choices, to set his own terms, to cooperate when he got ready. These tendencies persisted long after he became able to separate from his mother. (p. 32).[9]

And what of the perfectly adjusted Brennie? In reviewing Mrs. Murphy's book, I likened Brennie to a genial cocker spaniel who welcomes friend and burglar with equal joy. He seems to trust everyone without discrimination. This is fine as long as he stays in a highly restricted circle consisting of family, nursery school teachers, and sympathetic psychological researchers, who support him lovingly and demand a minimum amount of compromise. But eventually Brennie is going to find out that life is not a rose garden. Before long he will be entering what Harry Stack Sullivan described as the "juvenile era," a time when crude competition and aggression among peers are only slowly brought under the control of ripening social understanding. He will find that there are adults who do not respect children and may even take advantage of them. In his teens some of his contemporaries will urge him not to trust anyone over thirty. It is easy to project his career line further into the still competitive adult world with the self-seeking, scandals, and rackets that fill the daily newspapers. By that time Brennie may have been badly burned for his innocent credulity and thus learned to be circumspect, but if we compare him with Donald at the age of three, we reach the painful conclusion that it is the cautious Donald who is better adapted to the average expectable human environment.

This is a long introduction to the main task of this paper, but it will not have been wasted if we now start that task with a clear realization of these points: (1) that the described phenomena of coping, mastery, and defense belong in the more general category of strategies of adaptation, as part of the whole tapestry of living; and (2) that adaptation does not mean either a total triumph over the environment or total surrender to it, but rather a striving toward acceptable compromise.

THE TREND TOWARD INCREASED AUTONOMY

The point of departure for a systematic description of strategies of adaptation should be the broadest possible statement. Let us put it this way: adaptation is something that is done by living systems in interaction with their environments. It is important to emphasize both the noun *systems* and the adjective *living*. Our whole enterprise can founder at the very start if the basic image is allowed to be mechanical rather than organismic. It is characteristic of a system that there is interaction among its various parts, so that changes in one part are likely to have considerable consequences in at least several other parts. A system, furthermore, tends to maintain itself as intact as possible and thus displays more or less extensive rebalancing processes when injured or deformed. This much is true of inanimate systems as well as animate ones, which makes it necessary to qualify the systems under discussion here as *living*. For it is characteristic of living systems that they do something more than maintain themselves. Cannon's historic studies of homeostasis have familiarized us with the remarkable mechanisms whereby animal and human living systems maintain internal steady states, such as body temperature and fluid content, and restore such states when circumstances have forced a temporary departure. But Cannon was well aware that maintaining homeostasis was not the whole story; he saw it as a necessary basis from which living systems could get on with their more important business. This further business consists of growth and reproduction. Living systems do not stay the same size. They grow dramatically larger: the puppy that you once held in your hands becomes the big dog that you can no longer hold in your lap. This increase eventually reaches its limit in any one system, but not until arrangements have been made to start a whole fresh lot of tiny living systems on their way toward maximum growth.

The fundamental property of growth in living systems was well described in 1941 by Andras Angyal. Looking for "the general pattern which the organismic total process follows," Angyal pictured the living system as partially open to the environment and as constantly taking material from the environment to become a functioning part of itself:

> It draws incessantly new material from the outside world, transforming alien objects into functional parts of its own. Thus the organism *expands* at the expense of its surroundings. The expansion may be a material one, as in the case of bodily growth, or a psychological one as in the case of the assimilation of experiences which result in mental growth, or a functional one as when one acquires skill, with a resulting increase of efficiency in dealing with the environment. (pp. 27–28)[1]

Thus the life process necessarily entails expansion, but Angyal carried the matter further. Living systems, he pointed out, exhibit *autonomy*. They are in part governed from inside, and are thus to a degree resistant to forces that would govern them from the outside. If this were not true, the whole concept of adaptation would be impossible. Angyal then describes the direction of the organismic process as one toward an *increase of autonomy*:

> Aggressiveness, combativeness, the urge for mastery, domination, or some equivalent urge or drive or trait is assumed probably by all students of personality. All these various concepts imply that the human being has a characteristic tendency toward self-determination, that is, a tendency to resist external influences and to subordinate the heteronomous forces of the physical and social environment to its own sphere of influence. (p. 49)[1]

It was an evil day, we may imagine, for the inanimate world when living systems first broke loose upon it. Conservative boulders doubtless shook their heads and predicted gloomily that if this subversive trend gained strength the day might come when living systems would overrun the earth. And this is indeed exactly what has happened. Most of the land surface is completely buried by living systems, and even the oceans are full of them. When we consider this outrageous imperialism it is small wonder that the expansion of peoples and of nations has been a besetting problem throughout human history. And even when we concentrate on strategies of adaptation, we must keep it in mind that human beings are rarely content with maintaining a personal homeostasis. Unless they are very old they are almost certain to be moving in the direction of increased autonomy. It can be a threat of disastrous proportions to discover in the midst of life that all avenues are blocked to further personal development.

Living creatures, in short, will constantly strive for an adaptive compromise that not only preserves them as they are, but also permits them to grow, to increase both their size and their autonomy. Consider an animal as it steps forth in the morning from where it has been sleeping and moves into its daytime environment. If all goes well, it will ingest a portion of that environment, maintain its visceral integrity by homeostatic processes and by eliminating waste material, add a tiny increment to its size, explore a little and thus process some fresh information about its environment, gain a bit in muscular strength and coordination, bask in the warm sunshine, and return at night to its den a little bigger, a little wiser, a little stronger, and a little more contented than it was in the morning. If the season is right, it may also have found an opportunity to set those processes in motion whereby a number of offspring will come into existence. A day like this can be described as one of maximum animal self-actualization. If all does not go well, the animal may return to the den hungry, cold, perhaps battered and bruised, yet still essentially intact as a living system, capable of recuperating during the night and setting forth again in the morning. Of course, it may have failed to keep itself intact or even alive, but we can be sure disaster occurred only because the animal's adaptive repertoire, employed with the utmost vigor, has not been equal to the circumstances. Animals try to go up; if they go down, they go down fighting.

SOME VARIABLES OF ADAPTIVE BEHAVIOR

The adaptive capacities of any species of animal are to some extent represented in bodily structure, the product of natural selection. Protective coloring, great weight and strength, or such features as the rabbit's powerful hind legs that enable it to make bewildering hairpin turns in the course of its flight are part of the inherited equipment that favor certain styles of adapting. When we speak of strategies of adaptation, however, we are referring more particularly to the realm of behavior, the realm that is directly controlled by the nervous system and that is in various degrees open to learning through experience. This realm is traditionally broken down into receptive processes, central storage and organizing processes, and motor processes that lead to further sensory input. In the case of animals, whose inner experience, whatever its nature, remains forever closed to us, strategies of adaptation have to be described in behavioral language. They have to be described in terms of what can go on in a behavioral system of receptors, central structures, and effectors, not overlooking, of course, the contributions of the autonomic nervous system and the input of information from inside the body. How can we best describe the possibilities of adaptive control and regulation in an animal's behavioral system?

We could start with a flourish of analytic logic by talking sequentially about regulation in the sensory, the central, and the motor spheres. But this is a dangerous piece of abstraction; in actuality the whole thing operates not as a sequence but as a system. What happens when we surprise a squirrel feeding on the ground? There is a whisk of tail and before we know it, the animal has darted up a tree and is sitting on a branch chattering angrily at us. You might judge from a carelessly written mental health tract that the squirrel's behavior was neurotic and deplorable, inasmuch as it retreated instead of facing reality. But the squirrel is facing reality all right; it has simply elected to face it from a position of strength rather than from one of weakness. When you are on the cluttered ground and a huge creature is approaching, fear and flight are adaptive. When the cognitive field has thus been changed so that you are above the huge creature and have at your disposal all the escape routes provided by the branches of a tree, it is adaptive to sit down, be angry, and try the power of scolding. The squirrel has regulated the cognitive field, but has done so in large part by motor activity, and this is surely typical of adaptive strategies in the animal world.

Because the living animal is a system, adaptive behavior entails managing several different things at once. The repertoire by which this management is carried out can be conceptualized at this point in terms of action. One possibility is simply orientation with a minimum of locomotion. When locomotion is employed, it can consist of approach, avoidance while still observing the object of interest, or flight, and a final option is the complete immobility of hiding. Those are the possibilities stated in the most general terms. In order to behave adaptively the animal must use this repertoire to produce what prove to be, even in simple instances, fairly complex results. It seems to me that there are at least three variables that are regularly involved in the process, three aspects of the total situation no one of which can be neglected without great risk. If the animal is to conduct a successful transaction with the environment, perhaps leading to enhancement and growth, but in any event not resulting in injury or destruction, it must (1) keep securing adequate information about the environment, (2) maintain satisfactory internal conditions both for action and for processing information, and (3) maintain its autonomy or freedom of movement, freedom to use its repertoire in a flexible fashion.

I shall enlarge upon these three variables in a moment, but let us first place them in concrete form in order to secure the point that they must all be managed as well as possible at the same time. When a cat hears a strange noise in a nearby thicket, locomotion stops, eyes and ears are pointed in the direction of sound, and the animal's whole being seems concentrated on obtaining cognitive clarity. But if this were the only consideration, the cat might now be expected to move straight into the thicket to see what is there; instead, it explores very slowly and with much circumspection, for it is combining the third variable with the first, maintaining a freedom of movement that would be lost in the thicket. If the noise turns out to have come from a strange cat intruding on the territory, and there ensues a battle of vocal and hair-raising threats leading to an exchange of blows, the second variable becomes decisive. We know from the work of the ecologists that animal battles rarely go on to the death. The animal that sustains injury, feels incompetent, or is slowed by fatigue, shifts its tactics from approach to flight, and wisely lives to fight another day. This result is more probable if the first and third variables have been sufficiently heeded so that the animal is not cornered and has kept escape routes open.

Information

Securing adequate information about the environment is an obvious necessity for adaptive behavior. Action can be carried on most successfully when the amount of information to be processed is neither too small nor too great. If the channels are underloaded, there will be no way to decide what to do, as we would express it in adult conscious experience. If the channels are overloaded, there will again be no way to decide what to do, this time because the number of possibilities creates confusion. Of course this is not just a quantitative matter; what really counts is the meaning of the information in terms of potential benefits and harms. With this modification, however, it is permissible to use a quantitative metaphor and say that there is a certain rate of information input that is conducive to unconfused, straightforward action, and that both higher rates and lower rates will tend, though for different reasons, to make action difficult. Adaptive behavior requires that the cognitive field have the right amount of information to serve as a guide to action. Depending on circumstances, then, adaptation may take the form either of seeking more information or of trying to cut down on the existing input. The cautiously exploring cat illustrates the former process, behaving as though it asked the question, "What is it?" But if the same cat is in the nursery and is exposed to the affection and curiosity of several children, it will try to get away from some of the overwhelming input and might be imagined to ask, "What is all this, anyway?"

Departure in either direction from the preferred level of information is illustrated in Murphy's descriptions of the Topeka three-year-olds. There is likely to be a shortage of information before the children arrive at the testing center, and this is not easily dispelled by adult explanations. Once they have arrived, however, the children are flooded by an input that, because of its newness, they cannot easily put in order. There is a new room, a psychologist, an observer, and a collection of more or less unfamiliar materials. We have already seen how the sturdy Donald dealt with this situation, standing close to his mother, surveying the scene with alert eyes, and consenting to take action only when he had structured the cognitive field sufficiently to isolate an activity he felt competent to undertake. Another of Murphy's procedures was to have groups of children come to her home for a party, a situation quite new, strange, and bewildering to them and possibly a little odd even to adult eyes, inasmuch as each child had an observer assigned to keep account of everything he did. Donald faced this situation with his characteristic determination to get the cognitive field straight. After a long silent survey, he discriminated a zone of likely competence in the toys in the garage and went there to examine them. Later, he picked out a safe entry into the social scene and ended the afternoon in fairly active participation.

In this he was more daring than another boy who found his first manageable zone to be building blocks in a corner of the garage and stayed with it the whole afternoon.

Internal Organization

Working on the cognitive field alone will not guarantee adaptive behavior if the internal organization of the system gets too far out of balance. This is crudely obvious if an animal is injured in a fight, weakened by loss of blood, or exhausted in a long struggle. It is clear also in the lowered alertness, curiosity, and effort of children who are feeling sick. Even in young children it is possible to detect another form of internal disorganization that can seriously hamper adaptive behavior: the disorganization produced by strong unpleasant affects such as anxiety, grief, or shame. Some of the Topeka children confronted their first session with the psychologist with a degree of emotion that made it difficult for them to make use of the available information. One little girl, for instance, became tearful and inert, as if drained of energy. When able to try the tasks at all, she could scarcely muster enough force to attend, handle objects, or speak above a whisper, and her most characteristic movement was to push the material gently away. The inhibition vanished magically when she started for home. A normally active boy showed the paralyzing effect of anxiety first by keeping close to his mother, avoiding contact with examiner and test materials, and then by tentative work on the tasks with quick giving up in the face of difficulties. He was able by these tactics to control the anxiety and work up to an active part in the testing. As his internal organization came back to its usual balance, he spoke more loudly, moved more vigorously, explored the materials more boldly, initiated conversation, and became increasingly master of the situation.

Autonomy

Even if the internal organization is in good balance and the cognitive field is being dealt with competently, adaptive behavior may come to grief if freedom of action is not to some extent maintained. Animals, we may suppose, often enough get trapped in situations from which they cannot escape, but to a remarkable extent they seem to avoid this mishap, as if they were constantly monitored by a small built-in superego reminding them to keep their escape routes open. Once when I kept hens, I was worried to see a large hawk circling high above the yard, but my neighbor reassured me that a hawk would go without its dinner rather than drop down into a narrow, high-fenced pen that might hamper its return to the realms of safe soaring; and, sure enough, no hens were taken. Preserving space in which to maneuver is always an important consideration in strategy.

Among the Topeka children Donald again comes to mind as one who kept initiative in his own hands by refusing to be drawn into situations until he had given them a thorough scrutiny. Tactics of delay and refusals to participate, frustrating as they may be to the psychologist and thus all too readily given a derogatory tag like "anxious avoidance" and "withdrawal," may actually be in the highest tradition of adaptive behavior, following the adage to "look before you leap." Especially adept at maintaining autonomy was a girl named Sheila, not quite three, who after looking at the test materials announced that she did not want to watch them and instead would play with the toys on the floor. There was no sign of anxiety, and very quickly she involved the examiner in her game with the toys. Momentarily intrigued by a performance test set up before her, Sheila began to play with it, but when gently pressed to follow the examiner's rules rather than her own, she returned to the floor, announcing, "I want to do *this*. We don't like the game we had." Murphy comments as follows:

> Here we see a child who in the face of continuing and skillfully applied adult pressures maintained her own autonomy. And it was not merely a matter of refusing and rejecting; it was a matter of doing this without allowing the pressures to depress her mood or to restrict her freedom of movement. Instead, during most of the time, the pressures served to stimulate her to her own best efforts in structuring the situation and obtaining enjoyment from it and from the relationship with the adult. (p. 82)[9]

In adult life, Sheila possibly will become one of those who regard psychological tests as an invasion of privacy, and we should hesitate to criticize her for this because it may be part of a courageous career in the cause of civil rights.

Adaptive behavior, in short, involves the simultaneous management of at least three variables: securing adequate information, maintaining satisfactory internal conditions, and keeping up some degree of autonomy. Whatever the specific nature of the problem may be, those other considerations can never be safely neglected. But if we think in these terms, it becomes clear that strategies of adaptation typically have a considerable development over time. The temporal dimension is of the utmost importance for our problem.

THE TIME DIMENSION IN ADAPTIVE BEHAVIOR

I doubt if any serious student of behavior has thought about the adaptive process without considering it to be extended over time. Yet it seems to me quite common in clinical assessments to look for samples of such behavior, for instance the client's initial reaction to the examiner or the way inkblots are dealt with on first meeting, and then jump to the generalization that these are the client's characteristic ways of meeting his problems. Undoubtedly this is one of the reasons for the well-known fact that psychological assessments based on tests picture everyone, even the healthiest, as a clinical case needing some kind of improvement. The client's characteristic ways of meeting a problem the first time may not be how he meets them the second and third times, still less the twentieth time. On their first visits to the study center Donald and Sheila would not have been recorded as secure children; in neither case could the psychologist come anywhere near to completing the examination. Fortunately, they were studied over a long period of time, and we know that both are strong sturdy specimens of humanity with sense enough to take their own time and deal with things in their own way. Strategy is not created on the instant. It develops over time and is progressively modified in the course of time.

If illustration of this principle were needed, the Topeka children could again furnish us with vivid examples. There is the little girl at the party, physically slowed by a slight orthopedic defect, who at the outset cannot manage the jumping board even with help, and at the end jumps joyfully entirely by herself. There are the two sisters, age five and three, who have to accommodate themselves to the awesome prospect of moving to another city. Their stategies are traced through the months of anticipation and preparation, through the move itself, and through the first few weeks of finding security and satisfactions in the new environment (Murphy,[9] pp. 69–75, 168–170, 178–185). But if we think in terms of the three variables just discussed, the importance of the time dimension becomes self-evident. The values of the variables are not likely to stay long unchanged. Perhaps clearer cognition will reveal danger, increase fear, and precipitate flight, but a good many of the situations encountered by children are simply new. There is, of course, always a little risk in newness, so what is required is a cautious approach allowing time to assess both the risk and the possibility of benefits. The input of information may lead to sharper discrimination of the field, discovery of areas of likely competence and enjoyment, quieting of disturbing affects in favor of pleasurable excitement, and a lowering of the premium on maintaining strict autonomy. All such rebalancing of the variables implies processes of learning extended over time. A profitable familiarity with the cognitive field can be gained simply by protracted inspection with no motor involvement beyond moving the eyes and head. Closer familiarity requires the making of behavioral tests, discovering one's competence to deal with promising portions of the environment. There are great individual differences among children in the speed and apparent ease with which they deal with the newness of the world around them. Considering the many and varied adaptations that have to be made, we should not hastily conclude that the quickest strategies are necessarily the best.

HUMAN COMPLICATIONS OF THE ADAPTIVE PROCESS

Up to this point we have described strategies of adaptation almost wholly in behavioral terms. Illustrations have been confined to the behavior of animals and quite young children. The purpose of this maneuver has been to lay down a descriptive framework—one might say, a sort of biological grid—upon which to place the vastly more extensive strategies available to human adults. The human brain makes possible a transcendence of the immediate present that we do not suppose to exist in even the most intelligent subhuman primates. This is partly a matter of language and communication. As Alfred Kroeber memorably expressed it:

> A bird's chirp, a lion's roar, a horse's scream, a man's moan express subjective conditions; they do not convey objective information. By objective information, we mean what is communicated in such statements as: "There are trees over the hill," "There is a single tree," "There are only bushes," "There were trees but are no longer," "If there are trees, he may be hiding in them," "Trees can be burned," and millions of others. All postinfantile, nondefective human beings can make and constantly do make such statements, though they may make them with quite different sounds according to what speech custom they happen to follow. But no subhuman animal makes *any* such statements. All the indications are that no subhuman animal even has any impulse to utter or convey such information. (p. 41)[5]

The result of this capacity to talk about and think about things that are not immediately present is in the end an immense extension of the human horizon. Asch describes this in the following words:

> Men live in a field that extends into a distant past and into a far future; the past and the future are to them present realities to which they must constantly orient themselves; they think in terms of days, seasons, and epochs, of good and bad times. . . . Because they can look forward and backward and perceive causal relations, because they can anticipate the consequences of their actions in the future and view their relation to the past, their immediate needs exist in a field of other needs, present and future. Because they consciously relate the past with the future, they are capable of representing their goals to themselves, to aspire to fulfill them, to test them in imagination, and to plan their steps with a purpose.
>
> An integral part of man's extended horizon is the kind of object he becomes to himself. In the same way that he apprehends differentiated objects and their properties he becomes aware of himself as an individual with a specific character and fate; he becomes *self*-conscious. . . . Because he is conscious of himself and capable of reflecting on his experiences, he also takes up an attitude to himself and takes measures to control his own actions and tendencies. The consequence of having a self is that he takes his stand in the world as a person. (pp. 120–122).[2]

It is this vastly expanded world of experience that human beings must devise their strategies of adaptation.

One's first thought may be that the ensuing complexities are certain to drown us. I believe, however, that we stand to gain by fitting human strategies as far as possible into the three behavioral variables deduced from animals and young children. Take first the second variable, the maintaining, and if possible the enhancing, of the system's internal organization. It is here that awareness of the remote, the past, and the future, and especially awareness of oneself as a person, most dramatically expand the meaning of the variable. Clearly there is much more to be maintained than bodily integrity and control over disruptive affects. One thing that must be enhanced if possible, and desperately maintained if necessary, is the level of self-esteem. In part this shows itself as a struggle to keep intact a satisfactory self-picture, in part as attempts to preserve a sense of competence, an inner assurance that one can do the things necessary for a satisfactory life. Wide are the ramifications of keeping up one's self-esteem. Almost any situation that is not completely familiar, even casual and superficial contacts with new people, even discussing the day's news, can touch off internal questions like, "What sort of impression am I making?" "How well am I dealing with this?" "What kind of a person am I showing myself to be?" When self-esteem is tender or when the situation is strongly challenging, such questions, even if only vaguely felt, can lead to anxiety, shame, or guilt with their threat of further disorganization. No adaptive strategy that is careless of the level of self-esteem is likely to be any good. We certainly regard it as rare, unusually mature, and uncommonly heroic when after an unfortunate happening that diminishes his importance or shows him to be wrong, a person quietly lowers his estimate of himself without making excuses or seeking to lodge the blame elsewhere.

Less dramatic but still important are the expanded meanings of the other two variables. Securing adequate information is no longer confined to the immediate cognitive field. Information about things absent assumes increasing significance, especially when it bears on courses of action that extend into the future. Resources for information are also much richer: other people can be asked, relevant reading matter can be sought, and in some cases it is possible to send friends, employees, or students out to increase the scope of one's informational net. The maintaining of autonomy similarly gains a future dimension and a wider meaning. Looking ahead into the future and frequently making plans of one kind or another, we soon learn to be at least somewhat careful about committing ourselves. We feel better if things can be left a little open, if there are options; if, for example, in taking a job we see room for varying its duties or believe that in any event it will be a good springboard toward other jobs. "If I take this job," so many have asked themselves, "with its demands for teaching or medical service, will I have time for my own research?" Preserving an acceptable level of freedom of movement continues to be an important consideration even when the present physical field of the exploring animal has expanded into the imagined future social field of the human adult.

REFERENCES

1. ANGYAL, A. *Foundations for a science of personality.* New York: The Commonwealth Fund, 1941.
2. ASCH, S. E. *Social psychology.* Englewood Cliffs, N. J.: Prentice-Hall, 1952.
3. FREUD, A. *The ego and the mechanisms of defence.* London: Hogarth Press, 1937.

4. HAMBURG, D. A., AND ADAMS, J. E. A perspective on coping behavior: Seeking and utilizing information in major transitions. *Archives of General Psychiatry*, 1967, *17*, 277–384.

5. KROEBER, A. *Anthropology*. Rev. ed. New York: Harcourt, Brace & World, 1948.

6. KROEBER, T. C. The coping functions of the ego mechanisms. In R. W. White, ed., *The study of lives*. New York: Atherton Press, 1963.

7. LEVENTHAL, H.; WATTS, J. C.; AND PAGANO, F. Effects of fear and instructions on how to cope with danger. *Journal of Personality and Social Psychology*, 1967, *6*, 313–321.

8. LINDEMANN, E. Symptomatology and management of acute grief. *American Journal of Psychiatry*, 1944, *101*, 141–148.

9. MURPHY, L. B. *The widening world of childhood: Paths toward mastery*. New York: Basic Books, 1962.

10. SILBER, E.; HAMBURG, D. A.; COELHO, G. V.; MURPHEY, E. B.; ROSENBERG, M., AND PEARLIN, L. D. Adaptive behavior in competent adolescents: Coping with the anticipation of college. *Archives of General Psychiatry*, 1961, *5*, 354–365.

Developmental Life Transitions: The Early Years

The early years of childhood may present several kinds of "crises." Some of these are common developmental transitions such as entry into school or short separations from mother or father. Less common are the children who must deal with the extended absence of a parent (usually the father), or whose parents separate or divorce. Crises such as physical abuse by parents (the battered child syndrome), the early death of a parent, or removal from the parents' home and placement in an institution or adoptive home are less frequent, but have more severe effects on a child.

We deal first with a normal developmental transition, the child's entry into kindergarten. Does school entry constitute a "crisis" or at least a severe stress in a child's life? Several studies suggest that it may, and that children are not equally successful in dealing with this challenge. Kellam and Schiff [3] found that 70 percent of first-grade children were rated by their teachers as maladapted (for example, timid, obstinate, thumb-sucking, or fidgety) and that almost 10 percent were observed to have significant symptoms of psychiatric disturbance. Klein and Ross, [4] focusing on the first two months of the kindergarten year, found evidence of temporary behavioral disturbances such as increased dependence on the mother, exaggerated uncooperativeness or hostility, and an increase in behaviors like crying, vomiting, bed-wetting, and thumb-sucking. On the other hand, the children were also becoming more independent of their families (such as playing further from home) and more grown-up in their behavior (taking on more responsibility, for example, with younger siblings).

This transition can be difficult for parents as well. They must deal not only with their child's responses to the situation (both disturbance and growth), but also with their own adjustment to the change. In the first article Karen Signell describes a group discussion program designed to help the parents of children entering kindergarten in understanding their child's efforts at coping with this "separation crisis" and their own feelings of apprehension and uneasiness. Parents must deal with their feelings of loss as their child becomes more independent of them. They often feel a sense of rivalry with the teacher, who is a new authority figure for the child. They may be nervous about their child's performance, which they see as evidence of their own success or failure as parents. These parental reactions create extra tensions for children who are already experiencing a new environment requiring innovations in their behavior. The premise of Signell's program is that by anticipating and discussing these normal reactions and concerns, parents *and* children will be helped to handle this and subsequent crises better.

Divorce constitutes a more permanent separation (usually from the father), and even young children, with their "exquisite sensitivity," are strongly affected by the upheaval and the need for major readjustments. In the second article John McDermott examines the effects of parental divorce on nursery school children. This study is particularly interesting because it is based on a normal sample of children whose parents were in the process of getting a divorce, rather than on the possibly biased samples of psychiatric patients used in other studies (as in Crumley and Blumenthal.[2]) Based on weekly observations made by nursery-school teachers, McDermott divided the children into four groups. The first group (three of sixteen children) remained unchanged at school, apparently getting enough help and understanding at home that they were able to work out their feelings about the divorce without serious upset. Another group of three (all girls) also handled the experience without marked behavior changes, but did so with a "pseudo-adult" manner (bossing and lecturing their peers), which hinted at more severe personality disorders not directly related to the divorce.

Ten of the sixteen children showed acute behavior changes (hitting and pushing other children, or giving up when faced with the slightest difficulty) but only two required psychiatric intervention. McDermott describes the largest group (seven boys and one girl) as the "sad, angry children." Their behavior changes were related to vague feelings of anger and grief, which they seemed unable to express through their usual games. They would smash toys, bully other children, or break into tears for no apparent reason. These children began to readjust about six to eight weeks after the onset of their disturbance. The final group of two children displayed severe signs of detachment (for example, wandered aimlessly) and were referred to a mental health clinic. When the teachers were alerted to the divorce by the parents, they were often able to help the children by giving sympathetic support and by encouraging them to express the anger and fear that the divorce aroused. Children often held mistaken feelings of personal responsibility for their parents' breakup. McDermott[5] argues that dealing with this anxiety and confusion early can prevent more permanent maladjustment, such as delinquency or psychiatric symptoms.

Beyond the realm of preventive intervention is the "battered child," who has been physically abused, or the child who has been emotionally or physically neglected to the extent that a major disturbance in development is evident. Rene

Spitz's[8] classic study of Founding Home vs. Nursery related maternal and social deprivation to increased infant mortality, susceptibility to disease, retardation in growth, and failure to achieve developmental milestones. John Bowlby and others[1] also studied the effects of disrupting the mother-child relationship by lengthy separation and argued that the emotional effects were serious and permanent. More recent work suggests that the damage need not be permanent. Rutter[6] reviews studies of maternal deprivation, concluding that the quality and strength of a relationship (not just of the maternal relationship but of others as well) will determine whether or not lengthy separation will break the bond of affection. The continuity of these bonds, he argues, and not the physical presence of key people, is a determinant of a child's emotional development.

In the third article Alfred Kadushin reports a study of ninety-one children who were adopted when five years or older. These children had been removed from their natural homes by the courts on the grounds of parental abuse and/or neglect, and most had lived in several institutions or foster homes before having been placed by adoption. In interviews with the adoptive parents several years after the adoption, Kadushin found that 78 percent of the children felt that their adoptions were successful. Kadushin concludes that even children who experienced the severe trauma of early permanent separation from their parents are frequently able to make a successful adaptation to their circumstances when placed in the supportive environment of an adoptive home. The study would have been even more interesting if Kadushin had interviewed the children themselves or others, such as teachers, with whom they came in contact, or if he had been able to do longer term follow-up, but there are other studies which corroborate his findings on reversibility.

For example, Skeels[7] studied two groups of mentally retarded, institutionalized children. One group was placed in an enriched institutional environment and eventually in adoptive homes, the other in a relatively nonstimulating orphanage. Following them through adolescence and into young adulthood, Skeels found, like Kadushin, that the children given an enriched setting were able to achieve well within the normal range of development (for example, they were self-supporting), while over half the control group remained institutionalized. Kadushin and others argue that the majority of people can cope with and reverse severe disturbances if they are placed in a supportive setting that reinforces appropriate behavior.

REFERENCES

1. Bowlby, J. Ainsworth, M.; Boston, M.; and Rosenbluth, D. The effects of mother-child separation: A follow-up study. *British Journal of Medical Psychology*, 1956, 29, 211–47.
2. Crumley, F. E., and Blumenthal, R. S. Children's reactions to temporary loss of the father. *American Journal of Psychiatry*, 1973, 130, 778–82.
3. Kellam, S. G., and Schiff, S. K. Adaptation and mental illness in the first-

grade classrooms of an urban community. *Psychiatric Research Reports*, 1967, *21*, 79–91.

4. KLEIN, D. C., AND ROSS, A. Kindergarten entry: A study of role transition. In H. J. Parad, ed., *Crisis intervention: Selected readings*. New York: Family Service Association of America, 1965.

5. McDERMOTT, J. F. Divorce and its psychiatric sequelae in children. *Archives of General Psychiatry*, 1970, 23, 421–27.

6. RUTTER, M. *Maternal deprivation: Reassessed*. Baltimore: Penguin, 1972.

7. SKEELS, H. Adult status of children with contrasting early life experiences: A follow-up study. *Monographs of the Society for Research in Child Development*, 1966, *31*, 1–65.

8. SPITZ, R. A. Hospitalism, I and II. *The Psychoanalytic Study of the Child*, 1945, 1946, Vols. 1 and 2.

Article 3 Kindergarten Entry

A Preventive Approach to Community Mental Health

KAREN A. SIGNELL

Kindergarten is often the first major separation between mother and child. At our community mental health center we were aware that successful handling of this crisis could prepare families for coping with subsequent separations, so we considered it a priority for preventive effort.

We approached the local school system and developed a series of small group discussions for parents called "Preparing Your Child for Kindergarten Entry." The project was based on Donald Klein and Ann Ross' [3] findings on kindergarten entry, with additions such as the use of indigenous nonprofessionals as coleaders.

One purpose was to educate "hard to reach" parents in suburbia. Many of these parents, who have middle class incomes but were raised in working class or lower middle class homes, have long felt alienated from schools. In their "suburban ghettos," parents are isolated from the support of extended family or even their own neighbors, and the schoolteacher becomes the main person to whom they can turn for help with their child. But they are afraid to ask questions of a teacher. Attendance at any meetings, such as the PTA, is low. To reach this population and insure responsiveness, the mental health center needed to find them at a time of crisis. Therefore, intervention was begun at the most critical time of the initial separation crisis—the first week of kindergarten.

KAREN A. SIGNELL, PH.D., is a clinical psychologist, North County Mental Health Center, San Mateo Mental Health Services Division, Daly City, Calif. The author acknowledges with appreciation the encouragement and contributions of Antoinette K. Palladino, M.P.H., and Patricia A. Scott, M.S.W., North County Mental Health Center, San Mateo Mental Health Services.

Source: This article first appeared in *Community Mental Health Journal,* Vol. 8 (1), 1972, Behavioral Publications, 72 Fifth Avenue, New York, N.Y. 10011. Current revision was made possible by a grant from the National Institute of Mental Health, U.S. Public Health Service (1 T21 MH1394) to Wright Institute, Berkeley, Calif.

NEGOTIATION WITH THE SCHOOL SYSTEM

The project was presented to school guidance personnel. They were immediately interested, but had reservations and questions. "Will you just spearhead the program and then pull out? Don't expect us to lead parents' groups!" It was explained that the clinic has direct responsibility for children's mental health in the community and such problems as separation anxiety. Therefore the decision was made for mental health staff to lead groups and the school to convene them by sending letters to parents. There was also concern that the presence of teachers and school counselors at the meetings might elicit parents' criticism of teachers or questions about curriculum, distracting attention from discussion about separation. And might not such brief contact between parents and teachers actually *increase* mutual fears and stereotyping? However, it was evident that parents would want to meet the teachers, and these meetings could provide them with an opportunity to establish a positive bond with an important care giver. Therefore, mental health staff met with teachers to discuss the natural competition that exists between school and parents and to exchange ideas on how to refocus school questions on the underlying separation anxiety.

NONPROFESSIONAL PARTICIPATION

The participation of parents was enlisted to expand what the mental health staff could do. Staff noted that often parents could more easily accept, or even hear, a message when it came from another parent rather than from a professional. For example, in our first small group meetings, parents often turned to the "mother of eight" in the group to hear her experiences.

A long-range goal of this parent participation was to train mothers as mental health resources for their community. Most parents in this tract-house community are young, without anyone to whom they can turn for help with such crises as hospitalization of a child or a parent, death in the family, or teenage deviance. This training of parents might be an initial step toward providing neighborhood help for crises and mental health education.

Therefore, parents who had been the most constructive contributors to small group discussion were invited back in subsequent years as coleaders for the groups of new kindergarten parents. Often the best leaders were those who had been very anxious during their own child's kindergarten entry. They served as an ignition for the group, vividly recounting their worries and how they had survived. These "experienced" mothers and fathers offered their homes for meetings. In the spring they walked up and down waiting lines of new parents at kindergarten registration, telling how the group discussions had helped them and informing them of the program in fall.

TRAINING OF MENTAL HEALTH STAFF

Training was essential. (See Signell.[7]) Seasoned clinicians required it as well as graduate students in psychology. They needed orientation to crisis theory in general and the crisis of kindergarten entry in particular. Although it was stressed that the task was only *crisis resolution*, the staff persistently misheard this as "screening"—early detection of childhood disorder, referral for treatment, or diagnosing premature development for kindergarten. *Crisis resolution* meant the very limited goal of helping parents cope successfully with this particular separation crisis, no matter what the usual level of functioning in the family or what the chronic symptomatology in the child.

Staff needed redefinition of their usual clinical role. We stressed that the contact with parents was *educational*. It was not therapy. The staff role-played how to answer questions, especially those questions most tantalizing for clinicians to pursue in depth. (For example, "Do they have separate bathrooms for boys and girls?") Educational as opposed to therapeutic responses were demonstrated. Educational as opposed to therapeutic expectations were also contrasted; that is, the staff would feel disappointed at the degree of denial and anger in the parents and at their unwillingness to talk, if the staff members had expected the kind of openness and responsiveness of a patient asking for help. Therefore, it was anticipated with the staff that many parents would deny concerns as part of the crisis process and might only work through the separation vicariously by talking about their child or hearing other parents talk about their anxiety.

The role of group discussion leader was close to that of a very active mental health consultant in the tradition of Caplan,[2] Berlin,[1] Signell and Scott[6]; that is, he had the functions of conveying general mental health principles and helping the group share feelings and find their own solutions. Parents would not be as eager as patients to talk. Leaders needed to be more active, asking questions such as:

"Are you going to walk your child all the way to school?"

"Does your girl or boy know that she or he shouldn't get in a car with strangers?"

"What about bad language at school?"

The staff required anticipatory guidance themselves. The role was a new one, requiring very subtle and complex clinical skills. The staff needed support. They shared their feelings and solutions to such fears as:

"What if no one shows up?"

"What if no one talks?"

"What if a father starts berating a teacher?"

"What if I don't know enough about child development?"

Fears of the staff diminished when the role was defined as being principally a *catalyst*, with the responsibility for participation and ideas shared by the entire group. A staff member's main contribution, then, would be in group process rather than content. Of course, he might make an occasional contribution to the group as a mental health resource, especially with information on ways of coping with separation. To further lessen staff members' anxiety before they began their first group, the goals were defined within the spirit of exploration of approaches and alternatives rather than an imperative for success in the first attempt.

Staff were introduced to standard clinical-educational interventions:

1. Initiate topics ("Some parents wonder if. . . .")
2. Listen (with involvement, not detachment)
3. Reflect feeling ("Parents generally worry when. . . ." "I get afraid when. . . .")
4. Facilitate sharing ("Is anyone else here concerned about . . . ?")
5. Confront the worst fear ("What would actually happen . . . ?")
6. Confront cherished hopes ("What would you really like for your boy or girl?")
7. Link present fear to other experience ("What did she do in the past, when you left her somewhere, such as her grandmother's house?" "How is bad language handled in the neighborhood?")
8. Explore consequences ("How will it affect your child . . . ?" "How is he likely to react?")
9. Educate about crises and general coping mechanisms, and how to comfort a child.
10. Referral to outside resources ("You can ask the teacher about special learning or counseling problems.")
11. Maintaining the crisis topic ("Now let's get back to separation: How do you think Tommy will do his first day?")

To insure that the staff would not use their automatic therapy skills, they role-played responses to typical parent questions, such as: "Why did they have screening in spring, and how do I know if he passed?" "What do you do when she keeps pestering you without letting up?" "Is there a bathroom in the classroom?" [4]

FIRST SESSION

Anticipatory Guidance

The first meeting took place in the evening a week before school began. The general format was a brief talk to the total group of about 24 parents, then discussion in small groups. The brief talk outlined the transitory symptoms or normal separation responses to be expected of children: crying, stomach upset, thumb sucking, nightmares, irritability, and clinging to the mother. It also outlined typical reactions of parents: general feelings of fearfulness, feeling criticized by the teacher, feeling left out, wanting to stay at the school, and finding themselves nervous in a very quiet house. They could also expect to have particular worries about specific children, such as: "I'm afraid my child will get lost; he wanders." "My child is too shy." "My child gets bullied." General advice was also given: "It is all right to allow a boy to cling for a few days; it won't make him a sissy. Crisis reactions are time limited and don't necessarily persist."

Small Group Discussions

The parents had an opportunity to work through their particular fears in small groups. Group size, we discovered through experience over the years, was crucial: for good discussions, the optimal size is four to six parents.

Loss. The parents' feelings of separation loss were often couched in terms of their child's fragility:

"Bonnie said to let her go by herself, but I think she'll get lost."

"My child has asthma attacks."

"Susie is the baby of the family; she always had her brothers to do things for her."

"My wife just can't let him go. . . . I think he is a bit shy too."

One father began by belligerently attacking schools about not teaching and not caring, then ended up by reminiscing about his oldest son's graduation from high school and going to Viet Nam, and his daughters leaving home to marry. Another parent mentioned death in the family reactivated by this crisis. Such traumas were dealt with briefly, and the discussion was refocused on the current crisis.

We avoided false reassurance about loss, which would have colluded with parents' own denial. Instead, they were encouraged to discuss feelings of loss and to share them with others. Often in their intervention, the staff took the lead in expressing strong feelings such as "I find separation one of the hardest things in life to go through." After exploring such feelings, the reassurance inevitably emerged from the group that one could survive such separations. Some interventions consisted of exploring how to help a child. The group leader would say that it is good to give extra comfort to a child in crisis. One mother said that she helped her youngest child cope by giving her "mother's kerchief" to take to school the first day.

Projected Fear. Parents brought up concerns about how the teacher would handle children in the classroom. These could be understood as fear of criticism about how they themselves handled their child at home. "If he gets in fights, can they spank him?" "I'm going to give a belt to his teacher; she'd better take it because that's what my child understands."

Several parents believed that their children had been put in the "dumb class," although there was no such class in existence.

Interventions followed the consultation procedure of talking in the third person; that is, the leader went along with the projection and then directly educated the parent about himself. For example, "The teacher probably will make some mistakes, just as parents do." "The teacher has her own style of doing things, just as you and your spouse have different styles."

Sex and Aggression. At this time of crisis, anxiety was so intense that parents expressed strong fears about sex and also about violence:

"Do they have to raise their hands to go to the bathroom?"
"I told my girl never to get in a car with a stranger."
"This boy in our neighborhood likes to dress up in girls' clothes. Can he do that at school?"
"I don't want my child having any sex education in school!"
"What if there's an accident at school? Can they call an ambulance?"
"I heard there's marijuana in the elementary school now."

A topic that ignited discussion was children's using bad language at school. Such a concern seemed trivial to mental health group leaders, but it stirred strong emotions in these parents. Therefore, leaders explored with the parents how bad language might affect their child. They used direct interventions to the center of the parents' concern: "What might the teacher think if your child used 'that F word'?" Eventually parents would joke about how all children pick up bad language at home or anywhere they hear it. Seeing that the teachers also understood this, parents worked through some of their unrealistic fears about this first authority outside the home.

As a general rule, when parents brought up strong fears, we asked how they usually handled such situations in the neighborhood or among inlaws, in this manner placing the problems within their experience in an everyday framework.

SECOND SESSION

General Discussion

The second session took place in the evening the day after school began. There were general discussions, an anticipatory guidance talk, and small group discussions. According to Klein and Ross' experience,[3] a shift in mood was expected by this time. Parents were more comfortable and talked more readily, but the tone was mild. There was a change from high tension to lower tension—relief and mild depression, a giggling tense tone, and some anger toward the school.

Discussion began with "What happened the first day?" There was some closure on the separation loss, parents' sharing their success and acknowledging the restoration of equilibrium. They laughed about common feelings, such as the mother who felt she should not want to follow her child to school, but did anyway. Traumatic events were brought up, such as the child who got on the wrong bus going home. A teacher recounted how one mother stood on the ledge and squashed her nose against the window pane until she was told to leave. The group felt a great release of tension in hearing of this more extreme, but understandable, behavior.

Anticipatory Guidance

Anticipatory guidance covered what parents could expect in the next few weeks, and how they might re-establish their role and general communication with their child. Parents were told what they could expect of their child: delayed separation reactions, independence, and special fears.

Delayed Separation Anxiety and Crisis Reactions. After accepting separation well the first week, a child might suddenly cry or refuse to take his coat off at school. This could be expected to occur especially if other crises occurred. Unexpected events, such as a teacher leaving, could reactivate early separation anxiety. There were explanations about crises in general—that a child does have strong reactions when relatives die, his grandmother goes to the hospital, or a friend moves away. There were explanations of the purpose of regression: for example, that crying or clinging during a crisis makes a child feel safer. A child sometimes feels younger when he is suddenly expected to act older and before he acquires the capability that goes along with the new role.

Independence. The child might say, "Leave me alone! I can do it myself." Independence might also take the form of being more reticent, having a private world, saying such things as "I don't know what goes on in school," or "They just sing at school." This was explained as a natural development of independence, not necessarily indicating less loyalty to the parent.

Fears. The child could be expected to express his own fears in the form of stubbornness or blaming. For example, he might say, "I don't like the children at school," which might express the underlying fear that other children will not like him.

The parents were also told what they might expect of themselves. They could expect to feel sad, relieved, and also ashamed that they felt relieved. They might feel annoyed at the child and the teacher for not telling them what happens in the child's life when they had always known every detail of his life. They could expect to feel annoyed at the child holding up the teacher as an authority: "But Miss Jones says. . . ."

Small Group Discussions

Estrangement. One of the main themes was feeling left out by the child and the school: "His younger brother keeps asking, 'Where's Jerry?'" "I asked him, 'What did you do in school today?' and he just said, 'Nothing.'" "If *my* child said 'I hate you,' I'd give him a good slap and tell him to pack his bags."

It emerged that one reason a child is reluctant to tell about school is confusion, and also a lack of cognitive and verbal skills. Then, too, a child can sense his mother's strong feelings about being left out, and will handle his own dual loyalties by silence.

Competence as Parents. The parents can become quite anxious about the child meeting competition in the outside world. An underlying concern here is their competence as mothers and fathers.

"I never learned much in school, but I don't want him to hate it."

"We tried to be good parents!"

"The teacher last year didn't even call me."

"I think they should teach them reading in kindergarten. How do you get in college if they don't teach such things!"

"Susie has begun pronouncing her B's like V's and it drives me nuts. It's as if she's trying purposely to bug me."

Sometimes these questions found answers in the group and at the level at which they were asked. Limited probing sometimes revealed concern that parents' own early experiences with teachers might be repeated for their child; then they could examine whether this was a realistic concern. Sometimes there was an underlying concern about whether a teacher could accept the child when they themselves could not accept certain things about a child. Such a concern might be made conscious and worked through. Sometimes there were no direct interventions about such things. Instead, the group leaders would see if others had similar concerns and, if so, facilitated the sharing of those feelings.

Specific concerns about children meeting competition took the form of parents' worry that their child was different. He might be stigmatized, labeled, or nicknamed. One mother said, "She mimics me and others. Will she do that at school?" Others said, "He walks on his tiptoes and he is very shy." "She is spoiled and can't do things." "He's a bully." It was often sufficient reassurance to have these concerns shared with others who had similar fears. Sometimes the group leader had them confront realistic fears, examine how they usually coped, and apply their coping to this new situation. The group leader might ask, "What happens in the neighborhood when your child is called a name because of his ethnic background? What do you do about it as mothers? Then what is a teacher likely to do about it?"

Mother-Teacher Rivalry. Many questions had an underlying theme about mother-teacher rivalry. We interpreted this as natural rivalry in turning the child over for the first time to another authority in the community. Involved in this was a legitimate concern that a child's individuality would be lost. Group leaders expressed the general feeling that parents care deeply that the school continue their investment in the child. Parents fear that the school environment might be impersonal. They want to find out if the teachers also care for their child. The questions, however, were in code: "Do they expect you to have taught your child numbers? She can already count to ten. Maybe I shouldn't have taught her." "He didn't go to nursery school. He'll be lost with 30 other children." "How do they hang clothes up? I mean, if two children have white sweaters, how do they know which is Jane's?"

Mother-teacher competition took the form of considering the teacher perfect, or else being very critical that the teacher might be imperfect and mishandle something.

"I don't see why he acts so good at school."

"Why doesn't she know how to handle him? She's the teacher."

"I was a terrible stutterer; my nephew, too. The school never did anything about it until fifth grade."

"I told my child to push him back if a child pushes him at school."

"What if she doesn't like school?" (What if she likes it better than home?)

"I don't want the teacher to punish *my* child."

"He waited until he got home to go to the toilet; he's used to me unsnapping his pants."

"I'm afraid I'll sound stupid if I ask the teacher what the 'visual phonic' system is."

Some parents eventually expressed the insight that the teacher might be fallible, and that it might be permissible to let the child know that they themselves do not know everything.

The teachers were there at the meetings to tell the parents explicitly that they did care about the child and that they also cared about the parents. ("I hope you'll call me if he doesn't get home okay." "I'll call you if he doesn't get to school." "I'd like to have parents come to the classroom.")

The teachers' presence alone showed that they were willing to give an evening to listen and talk with parents. The teachers gave strong messages that they welcomed a note from parents if there were home crises or if the child had nightmares. The parents were also told that they did not have to handle everything themselves anymore. If the child did not adjust after a reasonable time, parents could use the teacher or counselor as a "sounding board" or a "back stop." Parents reinforced the teachers' communications by telling them how much it meant to them that teachers send home notes or their children's paintings. Some parents said they had always been afraid of talking to teachers, but were no longer afraid.

To further the bond between parent and teacher, they were both alerted to children's projections of fears. For example, if a child reported, "The teacher said I was stupid," the parent need not join the child's attack on the teacher or reassure him immediately. Rather, the parent could listen to the child's feelings of anger or hurt to determine what he is really concerned about.

Other issues between parents and teachers emerged. For some parents, the enriched school environment, with its array of toys and advanced educational offerings, brought forth their feelings of inadequacy. Would the school offer their child more than they offered him at home? Again, such deep concerns were not brought out openly. The group leader did not have a mandate to confront parents with such feelings. Rather, they were expressed and shared when they emerged naturally.

Re-establishing Communication with the Child. Some of the discussion inevitably concerned how parents might communicate with their child if he became less communicative. "How do you draw a child out? How do you go about it?" "He doesn't say anything." The small groups offered specific suggestions on how to restore the link with the child. Group leaders gave some guidance on the clinical skill of listening to a child without being too intrusive or too reticent. One mother told how she ignores her child when she is busy, but when she does listen she bends over so he can look into her eyes. It was explained that an important thing that they can do to help their child through this crisis is to listen to how he feels and comfort him. They might also join in his feelings of mastery. They could lessen his confusion by helping him discriminate spheres of authority, backing up the teacher's authority at school while insisting on their own at home.

In summary, interventions in the small groups were at a minimum. It was not necessary to intervene. It was enough to help the parents share their mutual fears and feelings, and realize that they were not alone or unique; and a few specific suggestions on how to help their child through the crisis usually emerged from the group.

RESULTS

Population

Parents who came to the meetings were overwhelmingly those with a first child or a last child going to kindergarten. It was expected that parents of first children would come, since they would be most anxious about their parenting. However, many mothers of last children also came. Mothers of last children expected to be relieved to see their last child leave, but were surprised to find it such a loss. Some suddenly felt "older," which might be interpreted as losing their mothering role. Others felt the added impact of reactivated separations that they had experienced with older children leaving home. There was some tendency (not statistically significant) for more parents of boys than girls to come, which can be interpreted as more concern for boys' social and academic adjustment.

Effect

Since kindergarten entry is not considered a traumatic stress, it is not appropriate to think in terms of preventing mental health casualties. However, the program can be considered in public health terms as generally upgrading mental health education in the community, immunizing people against this and other separations.

General observations about effectiveness can be made. Parents often continued discussing separation beyond the scheduled program, until 11 o'clock at night, which suggested that they found it worthwhile. Mothers volunteered to phone other parents to let them know about subsequent meetings. Often parents who attended subsequent meetings repeated with strong conviction what they had learned earlier about anticipatory guidance and crisis reactions. Mothers spontaneously told of giving other kindergarten mothers anticipatory guidance.

"My next door neighbor called me last night. She was all upset that Jimmy hadn't eaten supper on his first night of kindergarten or last night either. I told her it was just the normal anxiety reaction and it was okay and that she should come to our second meeting tonight."

One expectation was that school personnel and parents might be more aware of crises and especially separation anxiety. In the middle of the year a teacher left. Mothers in the neighborhood who had been in the program were able to help children with this crisis. Then, in the spring, some parents requested that the mental health staff discuss the crisis of first-grade entry.

The main effect of the program appeared to be realized. Parents seemed to find this intervention worthwhile, reporting to teachers that it helped them and volunteering to help others the same way in subsequent years by being group leaders.

By-products

Teachers reported much better relationships with parents. "This year we used parents a lot in the classroom. Any time we needed extra hands we had them." "This has been a very good year, one of our best. One of the reasons must have been the kindergarten entry program." "There was much more general parent contact, especially sending notes back and forth."

There were some unexpected consequences for the school system, apparently modeled in part after kindergarten entry with parents, such as a principal beginning a series of discussions at parents' homes. Other by-products included referrals from counselors and kindergarten instructors for parent education courses,[5] and the development of regular consultation to all kindergarten teachers in the district.

One long-range goal was to train mental health staff and trainees with a brief success experience in crisis intervention and mental health education. Since their participation in the program, the staff seems more alert to separation anxiety in conducting therapy or consultations in the community. A second long-range goal was to train parents as mental health resources. Parents have supported the program, some having been with the program for several years and reaching proficiency as group leaders, others being new every year.

Note on More Recent Developments

In the years since this program began, it has gradually moved more and more into the province of the school's responsibility. The kindergarten curriculum consultant for the school district and a mental health professional meet with the teachers in spring, discussing the crisis of entering kindergarten and role-playing responses to parents. Kindergarten teachers who had met with parents the previous year tell new teachers what to expect and how they handled certain situations. Kindergarten teachers now conduct the meetings with parents by themselves, usually during the daytime at school or in a house near school.

REFERENCES

1. BERLIN, I. N. Mental health consultation in schools as a means of communicating mental health principles. *Journal of the American Academy of Child Psychiatry*, 1962, *1*, 671–679.
2. CAPLAN, G. *Principles of preventive psychiatry.* New York: Basic Books, 1964.
3. KLEIN, D. C., AND ROSS, A. Kindergarten entry: A study of role transition. In H. J. Parad, ed., *Crisis intervention: selected readings.* New York: Family Service Association, 1965.
4. SIGNELL, K. A. An interaction method of teaching consultation: Role-playing. *Community Mental Health Journal*, 1974, *10*, 205–215.
5. SIGNELL, K. A. Parent-child communication course: A mental health education model. *exChange*, April 1973, *1*, 50–53.
6. SIGNELL, K. A., AND SCOTT, P. A. Mental health consultation: An interaction model. *Community Mental Health Journal*, 1971, *7*, 288–302.
7. SIGNELL, K. A., AND SCOTT, P. A. Training in consultation: A crisis of role transition. *Community Mental Health Journal*, 1972, *8*, 149–160.

Article 4 Parental Divorce in Early Childhood

JOHN F. MC DERMOTT, JR.

Since the divorce rate is highest in the first few years of marriage, its victims tend to be predominantly younger children, who are the most vulnerable to disharmony in the family and fear more for their own basic security when parental disharmony becomes pronounced and persistent. A number of authors have discussed the nature of the potential effects of parental divorce upon children.[1,4,6,7,9,10,11]

The legal event is undoubtedly much less traumatic to all involved than the emotional divorce which has inevitably preceded it. We recognize that many "intact" homes involve much more discordance than homes broken by divorce. But while the divorce may clear the air and reduce strain on the child in the long run, it also operates as an acute stress in itself since it is a major change for the young child. During the divorce period the child experiences a disruption of regular experiences with two parents, and may be uncertain of what will become of him in spite of the best reassurances. What happens to the young child during this period? Obviously, there is a wide range of divorce experiences which will have all types and degrees of effects on children.

It is difficult to separate the effects of divorce from those of the prolonged trauma and strain preceding it. The child's reactions also depend upon such factors as his or her age, sex, extent and nature of family disharmony prior to divorce, each parent's personality and previous relationship with the child, the child's relationships with siblings, as well as the emotional availability of all of these important people to him during the divorce period, and his or her own personality strengths and capacities to adjust to stresses such as separation in the

JOHN F. McDERMOTT, JR., M.D., is now Professor and Chairman of Psychiatry, University of Hawaii, School of Medicine. At the time of this study Dr. McDermott was Associate Professor of Psychiatry, University of Michigan School of Medicine, and Director of Inpatient Services, Children's Psychiatric Hospital, University of Michigan Medical Center.

Source: The American Journal of Psychiatry, Vol. 124 (1968), 1424–32. Copyright the American Psychiatric Association.

past. Furthermore, in any study of children's reactions to divorce, it is important to recognize the considerable difficulty in differentiating the impact of several factors: (1) direct impact on the child of the strife around the divorce; (2) immediate reactions of the child to the loss of a parent; (3) the impact of the divorce on the remaining parent, reverberating in the child; and (4) the impact, probably some time later, of the loss of a parental model.

Our research group at the Children's Psychiatric Hospital has begun a long-range study of the nature, degree, and duration of the impact of divorce on children. This report is limited to one small dimension and will open the way for the study of larger groups. It is an attempt to offer some observations of the changes in behavior of a series of 16 children, age three to five, around the time their parents were being separated and divorced. The objectives were: (1) to see how often the divorce period is indeed a time of significant stress for children of this age; (2) to see which conflicts and anxieties seem to rise to the surface and how they are manifested; (3) to identify the resources upon which the child draws to adapt himself; and (4) to learn how the school may assist him in coping successfully during this period of stress.

SETTING AND METHOD

The study took place at Children's Play School, a private nursery school for normal children. It employs a psychiatric consultant, who spends four hours per week at the school working with individual teachers regarding any children in their group whose behavior seems to be a problem. He observes and interacts with the children in their nursery setting, but does not evaluate or treat them individually. He is also available to any parents who may wish to confer with him regarding their children, and generally meets with two or three of the parents each month. The school is located in a university community, and its population comes mostly from white, middle-class families.

The teachers maintain ongoing anecdotal records on all their children at weekly intervals (observations of personality characteristics, play, and relationships) and any significant changes are recorded as they happen. The records of 16 children age three to five, ten boys and six girls, whose parents were separated and divorced during their nursery experience (1956–1966) were examined. During the period of several months surrounding the final separations and legal decrees, changes in children's behavior from previous levels and from those of the group were noted by the teacher and recorded. Two of those in the series also had direct contact with mental health clinics for psychiatric evaluation during the one- to two-year period the group was followed, and one of these children was treated in outpatient psychotherapy.

The limitations of this study should be emphasized. (1) The data were collected unsystematically with no attempt to evaluate home behavior or individual family conflicts and there was no long-range follow-up. The only cases in which details of family relationships were obtained were those three cases in which the psychiatric consultant's help was actively sought by the family. (2) The sample is small and narrow. (3) Similar changes may occur in children whose parents were not divorced; furthermore, the changes described in the children of divorced parents might be seen at other periods of their lives and as reactions to other stresses.

However, the special contribution of this study should also be emphasized. (1) The population is a nonclinical one. Many previous studies have focused on children of divorce who were patients of mental health clinics, with chronic changes and scars from previous difficulties being difficult to separate from reactions to the acute trauma. (2) It is based on direct observation rather than on parental reporting. (3) It consists of immediate observations before, during, and after the crisis period rather than retrospective data obtained years later. (4) It focuses on a specific age range. In all but the two most recent cases, the data were gathered retrospectively from records routinely kept by the school without thought of the study.

FINDINGS

Reactions and their expression were, of course, highly individual. However, acute behavioral changes presenting management problems were noted in ten of the 16 children studied, or 62 percent. An additional three cases, or 19 percent, while not showing acute symptoms, demonstrated a consolidation of previously noted personality traits which could be considered to represent hidden problems and reactions. These findings contrast sharply with Goode's[4] series in which only 18 percent of mothers reported significant management problems in their children around the time of final separation and divorce. There may, of course, be considerable differences according to the reporter of disturbance, who was the mother herself in Goode's series, and the teacher in this one. More will be said about the mother's reaction later.

The children in this study are divided into four groups according to the clustering of behavior; for the convenience of examination each group will be discussed separately.

The Unchanged Group

First of all we will consider the three children whose behavior did not appear to the school personnel to change during the divorce period—two boys and one girl. This is not to say that they were not unaffected by the divorce, but their adaptive techniques seemed more highly developed. It may very well be that the problems were handled so satisfactorily at home—that guilt, blame, and anger had been anticipated, handled, and worked through there in such a fashion—that behavioral derivatives did not appear at school. The children's relationship with each parent seemed to be a solid but distinct one—they were on good terms with each parent following the divorce. They actively worked out the problem of separation instead of passively and helplessly experiencing it.

One boy insisted on tackling the difficult and time-consuming job of making two entirely different Christmas gifts for his mother and father, while the other children were making a joint present or replicas for them. His parents seemed very friendly and took turns driving him to and from school during and after the divorce. Another child seemed able to articulate concerns exceptionally well. Once, while the group was discussing different kinds of puppy dogs, she announced a great discovery: "Do you know that not all married people love each other?" Subtle readjustments in personal identity were noted in the boy who no longer wanted to be called by his nickname, a nursery abbreviation, insisting "Please call me Donald because that is my whole name!"

These three children had previously been described as well-adjusted, well-liked, spontaneous, and imaginative in art and play. A good example was seen in the boy who, during the period of his parents' divorce, demonstrated an exaggeration of his already acknowledged creativity in role play. Who he was during the day was determined by the color of the shirt he wore to school in the morning. If it was blue, he would announce "I am a policeman," and if it was red, "I am a fireman," and he would play that role throughout the morning, weaving it in and out of the activities. Often he would ask the other children to pretend that they were stranded or trapped so that he could actively rescue them. While he usually seemed to deal with his anxieties by assuming the identity of a powerful rescuing person instead of an aggressor, there were occasional regressive ones as well; "Today I am a puppy dog" was followed by a morning filled with clownish behavior on his hands and knees and playing on the floor.

Play and fantasy are of particular interest as diagnostic aids to determine the severity of difficulties children are having, as it is well recognized[10] that both defects and normal problems of development regularly come to light in the way the child plays. Many authors[2,5,12,13,14] have described the function of play in mastery of problems. For example, play is a method of assimilating piecemeal an experience which was too large to be assimilated instantly in one swoop. Divorce in some instances would seem to fit the category of a traumatic event in which there is an onslaught of more events in a relatively brief interval of time than the immature ego can endure. In contrast to the above group, our three other groups manifested varying degrees of disturbances in play, which in turn seemed to prevent them from coping, adapting, and restituting.

The Sad, Angry Children

This group (eight children, seven boys and one girl), the largest and perhaps most typical, initially met their trauma with shock, anger, depression, and defenses of denial and regression, blaming others for their problems. Changes in these children's behavior at school most commonly seemed to be manifestations of diffusely and vaguely experienced anger and feelings of grief, loss, and emptiness. Typically, they became possessive, noisy, restless and pushing, kicking, hitting, and occasionally biting peers in contrast to previous behavior.

One child would hit others, become startled by his own actions and begin to cry, saying that he didn't know why he did it. Another child began finding angle worms to kill in the backyard. Several of the more popular of these youngsters would exclude a former close friend from a favorite game and enjoy seeing the friend devastated by it.

One child, a four-year-old boy, who was previously regarded as well socialized, was in many respects more or less typical of the children in this group. He began knocking down blocks and throwing other children's toys, dishes, and puzzles, which seemed to displace anger and assign to other children the role of the helpless sufferer whose house of blocks was coming down on top of him. Needless to say, whenever he noted two or three other children playing house, they became his immediate target.

His bullying and destructive play, in which frank identification with the aggressor could be seen, only seemed to repeat the anxiety-laden situation and not to master it. Heightened fear of body injury and of new activities were noted at this stage of development when concern over the integrity of the body and the self is natural anyway. In addition to counterphobic defenses there was a play regression to an earlier age level when the entire body had been used as a toy, with sensorimotor enjoyment of smashing. However, he still "pretended" that he was something or somebody else, not just acting as the one- and two-year-old who runs into and through things without caring.

Previously noted to be an imaginative and creative child, he still played roles, even though using his body in a more primitive fashion than before. When he hurled himself around the playroom, he shouted, "I'm Bill the steamshovel, I can knock over houses!" or "Bombs away, I'm a B-29 bomber!" (Some youngsters would use large toys—bicycles or trains—as extensions of their own bodies to be propelled and to propel them, to make a great deal of noise, to run over the other children's toys, and slyly knock over block buildings as part of the machine's powerful function in reaching its destination at all costs.)

Rest period and transition times between activities in school were particularly difficult for this boy and his group to tolerate. He became hypertalkative and restless when it seemed that inner thoughts were pushing themselves into painful prominence, yet could not be fully or directly discharged. He would break into tears for no apparent cause in the middle of a game or on the merry-go-round. He would cry excessively, out of proportion to a real difficulty encountered, and state, "It's just not my day."

Often he would sit on the periphery of the group of which he had been a member, looking tired, sad, gloomy, and lost. Formerly the school extrovert, and now unable to function in the group or in twos or threes, he moodily sat on a two-seated swing and would have stayed there for hours if left alone. When he deliberately hurled his juice at the table and was asked why, he said in a puzzled fashion, "I don't know, I'm just not happy any time any more."

He seemed to be silently and *secretly* mourning the lost, disparaged father, even though the reason for his depression was obvious to everyone else. Perhaps because his mother's feelings toward the father caused a dilemma for the child it was difficult to speak of him and still to retain her favor. He may have feared that since Mommy had stopped loving Daddy it could happen to him too, and feared losing her favor by missing his father. However, while he could not identify the problem, he could identify his depressive affect which could not be successfully warded off through his aggressive behavior and he thus opened the way for others to help him with these feelings.

Two children in this group demonstrated a depression which disrupted their play to an even more marked degree. They were characterized by giving up and not trying, certainly red flags to signal depression. In these children the loss of the capacity to play creatively seemed part of a vicious cycle preventing the child from mastering this depression which itself was interfering with his capacity to play. Extreme boredom and inhibition were typically seen during "free play" periods when the activity was left up to the child. During free periods their comments were typically, "What can I do, I can't think of anything to draw."

While they seemed able to externalize feelings in structured play periods, there was a stereotyped quality to their work which reflected their dilemma in its content and associations. Pictures drawn were typically a boy looking for his parents or his lost kite, or a girl who had lost her way home. In fact, the art room was avoided by the more seriously affected children for a period of several weeks or even months, a place where things were left unfinished and only later during restitution became attractive again.

Typically these children could no longer sit passively and enjoy listening to stories as they once had, particularly those involving family relationships such as "Goldilocks and the Three Bears." There would be a good deal of chatting and sometimes the need to interrupt and identify characters as bad and evil and to announce to everyone why the story was "wicked."

They also showed regression from house and human family play to animals. One child fled crying from playing house when asked to be a family member. One boy assumed a father role. but then would plead to the little "girl-mother" for food and to be allowed to sleep in the house, sucking his thumb, and able to do nothing but eat one imaginary meal after another.

No conclusion could be drawn as to when the time of crisis occurred for this group of children during the divorce period. In some cases the impact followed the separation, in others when the divorce decree became final some months later. Under ideal conditions, the impact occurred early and resolution followed. In this group of children signs of restitution and readjustment generally began to become apparent after a period of six to eight weeks.

The Lost, Detached Children

The third group of children, in which regression and disorganization were more severe, consisted of two youngsters. It is probable that these children were reacting to the divorce not as a stress in itself but as a trigger to long-standing, preceding disturbance in ego development.

The most severely affected of these children tended to lose his personal belongings and to wander about aimlessly, crying, bored, and detached. He forgot the location of his locker, his personal landmark at school. He became unwilling or unable to dress and undress himself any more. He repeatedly provoked rejections and hurts and made insatiable demands for affection and approval.

This child, who had been previously noted as unusual and evaluated at the local child guidance clinic, demonstrated a greatly exaggerated need for reassurance that he was liked and was almost unable to distinguish emotionally the teacher from his mother. When disciplined he would cry, "You don't love me— I know you like me but you don't love me. I know you don't want me in this school, you want me to go home, don't you?" He would hit other children and sob to the teacher, "They don't love me."

The other child's detachment from everyday events and self-absorption and preoccupation was somewhat different—more striking by its acuteness. She grew increasingly irritable, tearful, tense, and bossy, lost interest in imaginative play altogether, soiled and wet at home and in school on occasion, chewed on the stuffed animals that she brought to school, began sucking her thumb, chewing her hair, and asking the teacher to readjust her clothing and to retie her shoes many times a day so that they would be tighter and tighter, apparently to heighten sensation and awareness of her body. She would also pull up her underpants as high as she could, severely irritating the skin of her perineal region. She appeared to be masturbating in this way and eventually began masturbating openly with a dreamy, far-away look in her eyes.

Both of these children were referred to mental health clinics for psychiatric evaluation and treatment. The girl, whose regression was most dramatic, responded with apparent restitution to her previous adjustment level in 6 to 12 weeks; the boy did not demonstrate as much improvement.

Hidden Problems: The Pseudo-Adult Girls

In general, changes in behavior seemed more dramatic and acute in boys than in the girls, who tended to be more sulky and petulant. The girls' changes may not be so easily seen as a "problem," yet may be even more serious signs of potential life disturbance than the previously described acute changes. In the three girls who constituted our last group, it seemed that personality constriction and quarrelsome attitudes, bossiness, and pseudo-mature mannerisms were a way of dealing with conflict.

For example, one of these girls seemed quite popular with her peers, but on close study her relationships were quite superficial, with her frantic search for a real friend always being unsuccessful. She looked and acted like a caricature of Shirley Temple, was described as an excellent model for children's clothes, a "public relations child." She was a fussy dresser who wore clothes too elegant for nursery school and often demanded compliments from the teachers. At the

Halloween party she dressed as a "beautiful lady" and was in her glory. She used grown-up expressions, at lunch time exclaiming, "This is absolutely delicious!"

These girls showed no great behavioral changes during the time of divorce and continued to play with other children but tended to become even more pseudo-adult and bossy, scolding and lecturing their peers with comments about their health and manners as well as the rules of the games. The teacher said of one of them, "It is sometimes difficult to remember that she is the youngest child in the group as her behavior seems so adult-like."

One explanation of these characteristics is to view them as a kind of identification with a real or fantasied part of the mother that has heightened meaning for the children at the time of divorce—features often considered characteristic of the hysterical personality in adults. It may represent a premature, sudden distorted freezing of personality traits with which they had been experimenting, or identification with a caricature of the mother whose husband could not find genuine warmth in her, the "superior, nagging wife" who is always right. It sug-gests an identification with the "wife of the husband who leaves home" rather than with the more positive qualities in the mother expressed in other ways and seen at other times.

These girls seemed to be reacting not so much to the divorce as to life at home, and the persistence of this bond between them and their mothers signalled future adjustment difficulties. For some girls this may serve as a way to ensure the mother's love during a period of intense emotional strain. The crucial question with the girls, of course, is how fixed and permanent such a picture becomes, that is, whether the traits are internal as well as external manifestations.

The Boys: Disruption of Masculine Identification

Such defensive identification seemed less clearly evident in the typical behavior of the boys who comprised the previously described angry, sad group. Yet it is possible that their behavior too represented the part of the father perceived or fantasied with heightened intensity at the time, the blustering pseudo-potent figure or the passive pleading father playing at house. Boys of this age are in the process of switching their object of identification from the primary one, the mother, to the father, and we would expect that his loss at this time would be a matter of immediate critical importance in addition to potential long-range effects on personality development.

There seemed to be an acute and violent disruption of the process of masculine identification formation at the very least in several of the boys, as contrasted with what appeared to be a consolidation of a particular form of identification in the girls. This phenomenon might, of course, involve several possible contributing factors. Guilt over the secret satisfaction of having won over a rival of the same sex is well known. But there may also be a sudden upheaval because of the loss of the antagonist upon whom aggression had been safely focused at this stage of life, and who had served to keep it at a controlled level, and it may even be that some reactions reflected the children's frightening fantasy that the father was banished from the home by the mother as a punishment for masculine aggression.[3]

DISCUSSION

The exquisite sensitivity of these young children to the ways in which parents were feeling toward each other is in contrast to the parents' tendency to consider them as immune, not old enough to participate in the process of working out problems of the family disruption. Significantly perhaps, the news of family disruption was remarkably slow in arriving at the school from the parents of almost all the children studied, in spite of frequent contact between them and their children's teachers. Sometimes the news was blurted out by the child himself, who announced, "My father doesn't live with us anymore—he has an apartment" or "I spent Christmas with my father." More typically, the school personnel first became concerned and puzzled about changes in the youngster's behavior and upon inquiring about it with the parent, were told of the separation and divorce. One teacher's retrospective notes state that before the news of the father's absence was out, "We knew something was bothering him, he was just not the boy we knew, and we suspected a problem at home."

At this point it may be appropriate to speculate further about the discrepancy between Goode's[4] mothers, who reported only 18 percent of their children as management problems during the period of divorce, and our direct school observations in which the majority of children appeared affected. While it is possible that some children behaved no differently at home and found release from their conflictual feelings at school, it seems more likely that in many cases acute changes in children at this time remain hidden to mothers. There may be a tendency toward witting or unwitting avoidance of the issue of divorce by the mother because of her uncertainty in suddenly having to deal with the child by herself.

It was surprising to hear several mothers say that their child had not asked any questions about the father's absence. Of course, the child's denial and acting as if it had not happened at all may reflect his wish that this were the case, and perhaps the silence and unspoken feelings constitute a secret and special bond between mother and child with all its implications for the future. If the mother either ignores or denies the child's reaction and is able to pretend that the husband did not really exist as a member of the family, then she need not feel a responsibility to face the painful problems of readjustment for the child and herself without him.

Needless to say, teachers experienced considerable relief when they knew of the divorce and seemed to feel intuitively that the school could offer some extra security so essential for the child during this period. That is, behavior could then be managed along certain lines, neither excusing and ignoring it out of sympathy nor simply attempting to suppress it. Behavior could be seen as a way of telling teachers something which could then be put into words. It was used as a means of unveiling both angry and depressive feelings so that adult support could be given during this period of family and personal upheaval. Redirection of behavior through suggestion and structured activity with adult participation were the most frequently employed techniques.

During the divorce period, absences from school were significantly more frequent than in the past; in one case more than half the school days were missed during a semester. The reasons given for these absences often seemed to suggest where the difficulty lay—with a harried mother when the explanation given was that of transportation problems or with a depressed child when there were many frequent colds, stomachaches, puzzling periods of apathy and listlessness, or just feeling "tired." Exquisite ambivalence about separation from each other was noted in both the mothers and their children. The youngster might say, "I like to stay in school better than home" one moment, and then become unhappy and cry for home the next. One mother was persistently late in picking up her child after school, kept him home frequently for not very substantial reasons, yet said, "It ruins my day when he stays home," seeming to sense mutual hostility between herself and her son.

In many instances there seemed to be almost a moratorium for mothering, as it had been performed in the past, during the divorce period. When a mother finds her husband no longer available at home to struggle with, it may be a relief for her in one sense but it may also provide a void in which to ponder her mounting feeling of inadequacy about having failed as a wife. Frequently she must face this crisis alone and may have little to give to the children because of her own preoccupation and reaction to loss. A rush to other interests characterized many mothers. Several immediately thrust themselves into work or resumed their schooling, perhaps to ward off feelings of grief and worthlessness as well as to become self-sufficient and capable of supporting themselves and the children.

In some instances, children were perceived as a handicap to a new career or to ideas of future remarriage. Two mothers openly stated that the children reminded them of the unhappy past relationship from which they were trying to free themselves. One casually remarked to the teachers how much the boy reminded her of her absent husband ("full of tensions like his father") at this particular time, and it is probable that this was an unwitting process in many others. Such impressions are in agreement with Lehrman,[8] who stated that the mother often identifies herself in the role of the oppressed in a struggle over the need for love and thus may temporarily be unable to function in the normal, giving role of a mother to her children.

SUMMARY AND CONCLUSIONS

Although the individual nursery school child's reactions to the impact of family disruption by divorce varies considerably, some general considerations apply in most cases and may tend to shape the role of school personnel.

1. To the majority of children of this age divorce has a significant impact and represents a major crisis. There is often an initial period of shock and acute depressive reactions. Clinically observed regressive phenomena were followed by restoration of previous skills and subsequent resolution and mastery both in play and verbally.

Sex differences were noted, with boys demonstrating more dramatic changes in behavior, characterized by the abrupt release of aggressive and destructive feelings. Boys seemed more vulnerable to gross disruption of identifications already in process than the girls. Some of the children, principally girls, seemed to show a tendency to identify with selected pathological features of the parent of the same sex.

These findings suggest that the well-known readiness of the three- to five-year-old child to form identifications, when coupled with family upheaval, may produce sudden appearance or further consolidation of pictures which are characterized by pathological fragments of the parents' personalities. In the most seriously affected children in which individual psychotherapeutic intervention may seem indicated, the reaction to divorce presents a mixed picture superimposed on more long-standing disturbance.

2. Since the immediate effects of divorce are manifested by the impairment of the capacity to master anxiety and depression through play, it is important to identify and work with these children for preventive purposes. It may even be argued that the school has an obligation to intervene at this time in order to prevent reactions from going underground and thus to prevent future disorder, since key relationships in the family are often temporarily disrupted because of geographic inaccessibility of the father and emotional unavailability of the mother in her usual role. Thus, the teacher may find herself even more forcibly thrust into the role of an interim parent substitute.

3. Finally, the mental health consultant may be able to pick up signs and symptoms of the behavioral changes from the teachers, to assess the severity of these, and to advise the teacher. He may also be available to discuss with the parents how to cope with the effects on the child via explanation and to emphasize the importance of school for the child during this period. The children should not stay home because they are bored or diffusely irritable. He may also coordinate the preventive work with the child between the school and home in order to help the child resume the normal tasks of maturation.

REFERENCES
1. DESPERT, L. *Children of Divorce*. Garden City, N. Y.: Dolphin Books (Doubleday and Co.), 1962.
2. EKSTEIN, R., AND FRIEDMAN, S. "On the Meaning of Play in Childhood Psychosis." in Jessner, L., and Pavenstedt, E., eds.: Dynamic Psychopathology of Childhood. New York: Grune & Stratton, 1959.
3. FREUD, A. *Normality and Pathology in Childhood*. New York: International Universities Press, 1965.

4. GOODE, W. *After Divorce*. Glencoe, Ill.: Free Press, 1956.

5. GREENACRE, P. Play in Relation to Creative Imagination, Psychoanal. Stud. Child 14:61–80, 1959.

6. JACOBSON, P. Differentials in Divorce by Duration of Marriage and Size of Family, Amer. Sociol. Rev. 15:235–244, 1950.

7. KELLY, R., NORTH, J., AND ZINGLE, H. The Relation of the Broken Homes to Subsequent School Behaviors, Alberta Journal of Educational Research 11:215–219, 1965.

8. LEHRMAN, P. R. Psychopathological Aspects of Emotional Divorce, Psychoanal. Rev. 26:1–10, 1939.

9. MAHLER, M., AND RABINOVITCH, R. "The Effects of Marital Conflict on Child Development," in Eisenstein, V., ed.: Neurotic Interaction in Marriage. New York: Basic Books, 1956.

10. MOHR, G. The Threat of Divorce, Child Study 25:7–9, 1947.

11. NEUBAUER, P. The One Parent Child and His Oedipal Development, Psychoanal. Stud. Child 15:286–309, 1960.

12. OMWAKE, E. "The Child's Estate," in Solnit, A., and Provence, S., eds.: Modern Perspectives in Child Development. New York: International Universities Press, 1963.

13. PELLER, L. E. Models of Children's Play, Ment. Hyg. 36:66–83, 1952.

14. WAELDER, R. The Psychoanalytic Theory of Play, Psychoanal. Quart. 2:208–224, 1933.

Article 5 Reversibility of Trauma

A Follow-up Study of Children Adopted When Older

ALFRED KADUSHIN

A common supposition, which has been elevated to the status of an axiom, is that children emotionally damaged in early childhood are likely to be scarred for life. In order to determine whether reversibility of damage is possible, a follow-up study was conducted of a group of ninety-one children who were placed for adoption when they were at least 5 years of age or older and who had suffered considerable trauma before adoptive placement. At the time of the study most of the children were in early adolescence, the average age of the group at follow-up being 13 years and 9 months.

The group of children included forty-nine boys and forty-two girls. They were white, of average intelligence, and without physical handicap. They had been removed from their own homes at an average age of about 3½ years, had been placed for adoption at the average age of 7 years and 2 months, and had experienced, on an average, 2.3 changes of homes prior to adoptive placement.

EARLY DEVELOPMENTAL DATA

In all instances, the children in this study group became available for adoption as a result of court action to terminate parental rights because of neglect and/or abuse. The families from which they came were atypically large, 52 percent having five or more children. They had lived, during their infancy, in socially deprived circumstances, in substandard housing, with families whose incomes were

ALFRED KADUSHIN, PH.D., is Professor, School of Social Work, University of Wisconsin, Madison, Wisconsin. This paper was presented at the Sixth International Congress of Child Psychiatry, July 1966, Edinburgh, Scotland. The research on which the report is based was conducted with the support of a grant from the Children's Bureau, U.S. Department of Health, Education, and Welfare.

Source: Reprinted with permission of the National Association of Social Workers, from Social Work, Vol. 12, No. 4 (October 1967), 22–33.

most often below the poverty line. Natural parents had limited education, only 2 percent of the fathers having completed high school. When employed, they worked in unskilled or semiskilled occupational categories. The marital situation in the natural home was, in almost all instances, conflicted. In addition, the natural parents presented a picture of considerable personal pathology compounded of promiscuity, mental deficiency, alcoholism, imprisonment, and psychosis. The average number of specific social and personal pathologies exhibited by each of the natural families from which these children came was 5.7.

The relationship between the natural parents and these children was most frequently characterized by physical neglect, although 31 percent of the group were described as having experienced an emotional relationship that was "normally warm and accepting." Physical abuse was encountered with only 4 percent of the mothers and 10 percent of the fathers. In the largest percentage of cases physical neglect was accompanied by emotional indifference or emotional neglect and, less frequently, by emotional rejection. In 70 percent of the cases the child was physically neglected by the mother, in 40 percent the child was emotionally neglected (mother indifferent, showed no affection, recognition, encouragement, or approval), and in an additional 7 percent the mother was emotionally punitive (active disparagement of child, overt expressions of hostility or rejection).

In about 48 percent of the cases there was some relationship with parental surrogates prior to the first foster placement. In 23 percent of the cases this relationship was with grandparents; in 11 percent it was with older siblings.

Data on the backgrounds of these children were derived from the records of the agency that placed them for adoption, the Division for Children and Youth of the Wisconsin State Department of Public Welfare. In each instance there was a separate voluminous record available on the child, the natural parents, and the adoptive parents.

Categorization of the record data was done by a graduate social worker. A 10 percent sample of the records was independently rechecked by another trained worker. As a result of this procedure reliability of categorization for items requiring some subjective interpretation was satisfactorily determined.

Of principal concern to this particular report is the confirmation of the fact that these children lived at an early age during a prolonged period under conditions of serious deprivation. The worker's categorization of the record data is confirmed by the court action.

The courts in the United States, as elsewhere, are reluctant to terminate parental rights and do so only when it is clearly established to the satisfaction of the court that the living situation presents a real danger to the child. Often, even when such danger is clear and immediate, the courts tend to temporize by maintaining the child in the home and assigning a social agency to supervise the family in the hope that they can be helped to change. It is generally only in the most egregious circumstances that a child is removed from the home and parental rights terminated. The fact that all of these children were removed from the home by the courts and parental rights terminated corroborates the social worker's negative characterization, based on record data, of their early living situation.

CRITERIA OF OUTCOME

Follow-up interviews were conducted jointly with the adoptive father and mother, using a semistructured interview form. The interviews were conducted by graduate social workers, were tape-recorded, and lasted 2–2½ hours on the average.

The focus of the interview centered on the parents' satisfactions and dissatisfactions with the adoption, the problems they encountered, and the adaptations they made. Consequently, the child was not contacted and no attempt was made at a direct assessment of the child's functioning. The presumption was that parental satisfaction is related to the child's functioning. A child who is showing poor achievement in school and demonstrating behavioral difficulties in the home and with his peer group is not likely to have parents who express satisfaction in the relationship. A high correlation between parental satisfactions and outcome measures that include the child's adjustment has been established empirically in other adoptive studies.[8,31]

All of the taped interviews were transcribed, the typescripts averaging about fifty double-spaced pages of type. The researchers had available a total of some 4,300 pages of interview transcription. Each typescript was read and scored independently by three readers, all of whom were graduates of a school of social work and two of whom were working, at the time they engaged in this task, in the adoption unit of a public agency. Two of the three readers had no knowledge of the background material with regard to the family or the child whose interview they were reading.

Since the principal concern was with adoptive parent satisfaction and dissatisfaction in the adoptive experience, the primary outcome criteria of adoption "success" derive from this focus. Two measures of this were obtained. One was a ratio of satisfactions to dissatisfactions articulated by the parents during the course of the interview. Independently, the three typescript readers checked off on a prepared form that listed eighty different satisfactions and dissatisfactions the particular satisfactions they perceived the parents as having expressed explicitly or implicitly. An item was tabulated only when the same item for the same parent was checked by at least two of the three readers.

A ratio was computed for each case based on the total number of satisfaction-dissatisfaction items that appeared in the interview and were checked by two or more of the readers. Table 1 shows the distribution of such ratios. The ratio was computed to the nearest whole number so that a ratio of 1–1.4 appears as a ratio of 1–1 and a ratio of 1–1.6 appears as a ratio of 1–2.

The second outcome criterion measured was a composite score of parents' over-all satisfaction in the adoptive experience with the child. This was derived

Table 1. Ratio of Expressed Satisfactions to Dissatisfactions

Ratio	Number	Percentage
5 or more to 1	60	66
4–1	2	2
3–1	4	4
2–1	1	1
1–1	8	9
1–2	4	4
1–3	5	6
1–4	1	1
1–5 or more	6	7
Total	91	100

from a checklist completed independently by each of the parents at the end of the interview and from judgments made independently by four graduate social workers—the worker who interviewed the parents and the three typescript readers. The method of scoring applied the corrective of the social workers objectivity to the parents' more subjective judgments. Once again a preponderance of responses fell in the upper end of the scale, some 87 percent of the parents being judged to be either "extremely satisfied" or "more satisfied than dissatisfied" in the adoptive experience with the child.

The two criteria of outcome were correlated with each other at a very high level ($r = .89$, $P < .001$). The two different kinds of criteria give similar kinds of outcome data—the composite ranking criterion indicates 78 percent of the adoptions as successful, 13 percent unsuccessful, and 9 percent that might be regarded as neither successful nor unsuccessful. The ratio of satisfactions to dissatisfactions criterion for these categories indicates 73 percent, 18 percent, and 9 percent, respectively. A comparison of these results with other adoptive outcome studies, the subjects of most of which were children adopted in infancy, indicates that the outcome of the group studied here was only slightly less successful.

The general conclusion indicating a considerable level of success is somewhat unexpected and raises a question of great interest. The data on the background factors and developmental history of these children indicate that almost all of them lived, during early childhood, under conditions of great social and emotional deprivation. The families into which they were born and from which they were removed by court action after the community had recognized the dangers of such an environment for healthy child development were characterized by considerable social and interpersonal pathology. Given the conditions under which these children lived during their most impressionable years—in poverty, inadequately housed, with alcoholic, promiscuous parents who frequently neglected them, sometimes abused them, and only rarely offered them the loving care that is the prerequisite for wholesome emotional development—how can one explain the generally favorable outcome of these adoptive placements? A successful adoptive experience might have been anticipated for the 31 percent of the children who had a good relationship with mothering figures, but the general success level is closer to 82–85 percent.

It would be easy to dismiss the conclusion out of hand by suggesting that the interview material presented by the parents had little relationship to the actual experience. The social workers who conducted the interviews rated each of the parents on the basis of their involvement, defensiveness, sincerity, and the like in the interview. The overwhelming impression was that most often most of the parents were being honest. Whenever possible information obtained from parents was checked with earlier record material; this check revealed few discrepancies. The material was also consistent with a projective technique used at the end of the interview, a sentence completion form filled out by each of the parents. It seems that the question raised needs to be answered on its own merits.

OTHER STUDIES

Before attempting an explanation, however, it might be well to point out that other studies have shown the same unexpected results with similar expressions of surprise on the part of the researchers. In each instance the children studied turned out to be more "normal," less "maladjusted," than they had any right to be, given the traumata and insults to psyche that they had experienced during early childhood.

Theis, in summarizing her impression derived from one of the first large-scale follow-up studies of foster children, expresses gratification at the adjustment of the group, 80 percent of whom came from "bad backgrounds." She says:

> Our study of the group as a whole, in so far as the subjects have demonstrated their ability to develop and to adjust themselves to good standards of living, and perhaps even more strikingly, our study of individual members of it, leave us with a distinct impresson that there exists in individuals an immense power of growth and adaptation. (p. 163)[26]

Roe and Burks did a follow-up study of thirty-six young adults who had, as children, been removed from their homes and placed in foster care because their own parents were chronic alcoholics. Other types of deviant behavior were associated with alcoholism; 81 percent of the fathers and 44 percent of the mothers of these children "were guilty of mistreatment or neglect of their children" (p. 38).[23] The authors note that since most of the children

> became dependent as a result of court action—this means that the first few years of life of these children were spent in a home situation which left much to be desired—and that they were probably subjected to traumatic experiences during the early years of their lives. (pp. 382–383)[22]

As adults, at the time of the follow-up interview, "most of these subjects have

established reasonably satisfactory lives, including adequate personal and community relationships and most of them are married" (p. 388).[22] The authors
are prompted to ask:

> How did it happen that these children turned out as well as they did? How
> did it happen that in spite of these [adverse] factors many of them have be-
> come not only useful citizens but reasonably contented persons working
> adequately, with pleasant family lives and sufficient friends? No one who has
> read the records of some of these lives and pondered on them can escape a
> profound sense of awe at the biological toughness of the human species.
> (p. 391)[22]

Maas conducted a follow-up study of twenty children who had been removed
from home during infancy and early childhood and placed for at least a year in
a residential nursery. He reviewed agency records, interviewed parents, and saw
the children themselves, who, at the time of the follow-up study some twenty
years later, were young adults. Maas reports:

> Although these twenty young adults may have been seriously damaged by
> their early childhood separation and residential nursery experiences, most
> of them gave no evidence in young adulthood of extreme aberrant reactions.
> ... To this extent the data support assumptions about the resiliency, plas-
> ticity and modifiability of the human organism rather than those about the
> irreversibility of the effects of early experience. (pp. 66–67)[16]

In a follow-up study of independent adoptions conducted by Witmer, a
group of fifty-six children were identified as having lived under "possibly traumatizing conditions" prior to adoption. These children had lived under adverse
physical conditions and the psychological situation "was even more pathetic." At
follow-up there was little difference in the adjustment ratings achieved by this
group of children as contrasted with other adoptees placed at a similar age and in
similar adoptive homes but who had not experienced such "possibly traumatizing
conditions." (pp. 286–287)[31]

Meier completed a follow-up study of sixty-one young adults who had grown
up in foster care. All of the group had experienced five years or more in foster care
and none had returned to his own family. About half of the group had been removed from their own homes before the age of 5. Between their first foster placement and their discharge from foster care at 18, these children had experienced
an average of 5.6 living arrangements. Most of them had been removed by the
courts from their own homes "in which they had experienced inadequate care"
(p. 197).[18]

Based on lengthy interviews with the group—now young adults—Meier concluded:

> The vast majority of the subjects have found places for themselves in the
> community. They are indistinguishable from their neighbors as self-support-

ing individuals, living in attractive homes, taking care of their children adequately, worrying about them, and making some mistakes in parenting, sharing the activities of the neighborhood and finding pleasure in their association with others. (p. 206)[18]

In a later report Meier says:

Child welfare workers are continuously baffled, as well as heartened, by the fact that over and over again they see children removed from impossibly depriving circumstances who, by all the rules "ought" to be irreparably harmed, who, nevertheless, thrive and grow and learn to accept love and affection and respond to it. (p. 12)[17]

Lewis followed up 240 children who had been studied at a children's reception center following removal from their own homes. Seventy-one of these children came from problem families "characterized by gross neglect and squalor" (p. 84).[15] On follow-up two years later 64 percent were doing "good" or "fair," 36 percent were assessed as "poor" (Table 71, p. 120).[15]

Rathbun reported on a follow-up of thirty-three foreign children who, after having suffered considerable deprivation in their own countries, were placed for adoption in the United States. Interviews by caseworkers with the adoptive parents six years after placement, supplemented by contact with the schools, showed that "the adjustment of the majority was judged adequate and in some cases notably superior" (p. 6).[20] The report concludes by noting:

The consistency of the ratings for all categories in which assets outweigh liabilities, points in the direction of a considerable degree of reversibility of the effects of early psychic damage. (p. 131)[20]

Another research group studied twenty-two Greek children institutionalized during the first two years of life and subsequently placed for adoption in the United States. Testing and interviews five years after adoption rated only two of the children as "poorly adjusted." The researchers concluded that "despite early deprivation the children have done remarkably well" (p. 19).[9]

Welter studied seventy-two children placed for adoption in the United States when older than 5 years of age. Thirty-six of the group were children born outside the country and transferred to it for adoption. Some 85 percent of the children were judged to be showing "good" to "excellent" adaptation on follow-up (Table 40, p. 126).[29] Welter notes, in summary:

Perhaps the single most important implication may be drawn from the fact that according to the social workers [responsible for working with the children] both of these groups of older adoptive children . . . , despite extended exposure to massive deprivation, have indicated a degree of responsiveness to a restitutive environment and a reversibility of early psychic damage which seems to exceed even the most optimistic assessments of the studies on maternal deprivation and separation we have seen thus far. (p. 164)[29]

A somewhat different series of studies helps to support the general contention we are trying to substantiate here—the contention that childhood trauma can be overcome and an initially adverse developmental history does not, necessarily, inevitably, lead to subsequent incapacity for social functioning.

Victor and Mildred Goertzel, both psychologists, raise this question in their study of the developmental history of eminent people. Using published biographies of four hundred prominent people, all of whom "lived into the 20th century," they codified material available on the nature of the childhood experiences of these people. They established that the greatest majority of the group came from homes that demonstrated considerable pathology and that the childhood experiences of many of the group were replete with potentially psychogenic factors. "Only 58 of the 400 can be said to have experienced what is the stereotyped picture of the supportive, warm, relatively untroubled home" (p. 131).[11]

Renaud and Estess conducted a typical clinical history interview with "100 men whose functioning could be described as distinctly above normal by all ordinary standards" (p. 786).[21] The findings of the detailed interviews indicated that this group of men who

> . . . functioned at above average levels and who were substantially free of psychoneurotic and psychosomatic symptomatology, reported childhood histories containing, seemingly, as many "traumatic events" or "pathogenic factors" as we ordinarily elicit in history taking interviews with psychiatric patients who are in varying degrees disabled by their symptoms. (p. 796)[21]

Having been exposed to such "traumatic events" and "pathogenic factors," these men, who never would have come to the attention of mental health clinicians, have been able to reverse the expected developmental direction.

Similar findings are noted when one seeks to differentiate, on the basis of background factors, an essentially normal group from a group manifesting emotional pathology. Schofield and Balian matched a group of 178 schizophrenic patients with a group of 150 people who had never manifested any psychiatric disturbance. Matching was in terms of age, sex, and marital status. A comprehensive clinical life history interview was conducted with the normal group and a comparative item analysis was made. In general, there was a surprisingly great degree of "overlap of the normal subjects and schizophrenic patients in the distribution of the personal life history variables" (p. 222).[24] While it is true that about 35 percent of the schizophrenic patients had a pathological relationship with their mothers as against some 19 percent of the normals, the question is raised as to why the 65 percent of the schizophrenics who had an essentially normal relationship with their mothers broke down and why the 19 percent of the normals who had pathological relationships with their mothers did not. The question as it stands, of course, puts too much of a strain for causation on one isolated factor, but the question can be repeated for a variety of supposedly significant explanatory items and it can be repeated for configuration of items that supposedly "explain" emotional maladjustment. The authors conclude by stating:

> The finding of "traumatic" histories in nearly a fourth of the normal subjects suggests the operation of "suppressor" experience or psychological processes of immunization. (p. 225)[24]

Similar kinds of studies by Munro[19] and Brill and Liston[5] show considerable trauma in the childhood backgrounds of psychiatrically normal populations.

In the studies reviewed by Bowlby[3] in his widely disseminated and influential report on the dramatic negative effects of early trauma, it might be significant to call attention, in this context, to the fact that, as Yarrow reminds us (p. 20),[32] a sizable proportion of the children in each of the studies listed did not show the predicted negative reactions to separation and deprivation.

One might, for comprehensiveness, cite Bowlby's later modification of emphasis on an irreversibility in his study of sixty children who had experienced prolonged institutionalization in a tuberculosis sanitarium early in life.[4] Contrasted on follow-up with a control group of schoolmates who had not experienced early separation, there seemed to be only a 20 percent greater incidence of maladjustment in the sanitarium group as compared with the control group. For most of the sanitarium group, then, the effects of early trauma seem to have been reversed.

The work of Clarke and Clarke showing cognitive recovery from severe early deprivation is also relevant here.[6] Adolescents and young adults coming from backgrounds in which they had experienced cruelty, neglect, separation, and long periods of institutionalization were able, in a benign environment, to achieve significant increases in IQ test scores. Clinical reports also support the contention of reversibility of trauma in response to special psychotherapeutic effort.[1, 2, 12]

FACTORS RELATING TO REVERSIBILITY

Does all of the above then imply a contradiction of the most important tenets of child rearing—namely, that continuous contact with one set of loving, accepting, understanding parents providing the proper emotional as well as physical environment is the best basis for healthy biopsychosocial development? Actually not. The writer is not suggesting that neglect, abuse, and physical deprivation are not harmful. As a matter of fact, each of the studies cited shows that a more detailed contrast within the follow-up group invariably favors those subsets of children who were provided with a more benign environment prior to separation. For instance, while most of the children in the Maas study and in the study by Roe turned out reasonably well adjusted, those who came from less pathological backgrounds did better than those who had been subjected to greater trauma (p. 67; Table 1, pp. 381–382).[16,22] In the present study of adoption of older children outcome is positively related to natural mother's acceptance. The data do not argue for a rejection of the generally accepted tenets but rather for recognition of at least the partial reversibility of the effects of deprivation.

Explanation for reversibility lies with biological, social, and psychological factors. There is much empirical evidence indicating that constitutional factors play a part in differing levels of vulnerability to deprivation and resiliency in recovering from deprivation.

Studies of differences among children immediately after birth, before one can postulate differences owing to the effects of variations in their surroundings, indicate that children differ in many ways that have significance for the result of interaction between themselves and their environment. Thus, Thomas studied 130 children from the first months of life onward. From the moment of birth children differed in terms of activity, adaptability, distractability, persistence, mood, intensity of reaction to sensory stimuli, threshold of responsiveness, and so on. He notes:

> Given constancy of environmental factors the reaction will vary with the characteristic of the child upon whom the relatively constant stimulus is brought to bear. . . . This holds true for all aspects of the child's functioning . . . including reactions to situations of special stress, such as illness, radical change in living conditions or abrupt shifts in geographic environment. . . . This view of the child stands in contrast to the assumption that environmental influences as such have determinative effects. . . . Our findings suggest that exclusive emphasis on the role of environment in child development tells only part of the story and that responses to any regimen will vary in accordance with primary patterns of reactivity. (pp. 84–85)[27]

These conclusions are supported by additional research that points to the same fact—that children are different at birth in ways that are crucial for personality development.[7,10,30]

If biological determination of individuality, as contrasted with environmental or experiential determination, is given greater consideration, the resiliency of the children in this study group may be more explicable. One child's trauma, which makes future positive adjustment very improbable, may be another child's inconvenience, the effects of which, given reasonable opportunity, may be reversed.

There may, however, be another general line of explanation somewhat more sociological in nature. These children adopted when older made two important shifts in moving from their own home to the adoptive home. They made the change referred to above, from a home that offered little in the way of meeting their needs in terms of affection, acceptance, support, understanding, and/or encouragement to the adoptive home, which offered some measure of these essential psychic supplies. They also made a change from a lower-class, multiproblem, generally disreputable family living in a slum-ridden area in the community to a middle-class, reputable home in an area of the community that had some status.

A child's self-concept is developed as a result of his experience in the intense relationship with significant others within the intimacy of the group, particularly in relation to the most significant of all others, the parents. A child who perceives himself as acceptable to the parents perceives himself as acceptable to himself. But however important this factor is in building the child's self-concept, it ignores the impact of the wider world, which soon begins to transmit messages to the child relevant to his conception of self that he cannot ignore. The Negro child has to have a much stronger positive self-image, initially, to withstand the corrosion on this image of the thousands of overt and covert cues that come from the predominantly white environment, all of which say black is bad, white is better, black is down, white is up, black is subordinate, white is supraordinate.

This problem operates along class lines as well as along color lines. The same kinds of messages are received by subsets within the white community. The standard-bearers of the white community behave toward the lower-class, multiproblem, disreputable white family in such a way as to transmit the message, loudly and clearly, that the family is unacceptable. The child identified with such a family, carrying its name and associating with its members, inevitably begins to be affected by this pervasive negative labeling.

The child is then removed from this family, placed in a decent-appearing, middle-class home in a nice-looking neighborhood, identified with a well-organized family of father and mother who act in a responsible, respectable manner. He now receives messages that proclaim his acceptability, that support, reinforce, and strengthen whatever components of self-acceptance—however limited—he has been able to develop within the family. Burks and Roe, in attempting to explain the better-than-anticipated outcome in foster care of children removed at the median age of 5 from homes in which the parents were alcoholic and/or psychotic, point to this as a factor that needs consideration:

> Had these children remained with their own outcast families, they, too, would have been, in a sense, outcasts. They could react only by identifying with their families and rejecting the community and all its customs, or by rejecting their families and striving ceaselessly somehow to achieve membership in the group which had despised them. (p. 116)[23]

The children having been removed from such homes and placed in acceptable ones, the authors note:

> It seems very probable that residence in a home which is a respected part of
> the community, and the child's acceptance as a member of that community,
> make possible the formation of an organized ideal derived from the attitudes
> and forms of behavior of the community which can function as an inte-
> grating force. . . . (p. 116)[23]

Srole, in his report of a large-scale effort to assess the level of "mental health"
of people in midtown Manhattan, reports that mental impairment is related to
socioeconomic level. More of the "poor" in his study were apt to be "mentally
impaired." In attempting to explain this, the report points to the factor of
community rejection of the lower-class child, which intensifies tendencies toward
maladjustment.

> In many areas of his experience the lower class child encounters the con-
> tempt, implicit but palpable, in the non-verbal behavior of others who think
> of him in the symbolism of such words as rubbish, scum, dregs, riff-raff and
> trash. These devastating judgments inevitably force their way into his own
> self evaluating processes. (p. 198)[25]

Thus, in moving the child from a lower-class home and social environment to a
middle-class adoptive home and social environment, the agency has "rescued"
him from a situation that intensifies the problem of adjustment to one that
assists the child in making a positive adjustment.

Supporting these two factors that dispose toward reversibility of trauma for
children adopted when older—the biological factor of constitutional resiliency
and the sociological factor of upward displacement, which reinforces self-acceptance—is the more important factor of making a therapeutic milieu available to
such children. Adoption is not psychotherapy, but its psychotherapeutic potential is like a good marriage, a true friendship, a new and satisfying job, an enjoy-
able vacation. It can help to repair old hurts.

What is involved here are different approaches to an explanation of what is
psychotherapeutic. The rationale that is basic to the traditional approach to
therapy (specific remedial measures) suggests that the behavior itself is not of
primary importance since the behavior is merely a manifestation of some underlying intrapsychic conflict. Since behavior is merely symptomatic of underlying
disturbances, these inner causes, rather than the behavior itself, should be the
focus of attention. The behavior itself is purposive and beyond rational control
or simple re-education, exhortation, persuasion, or the like, since the individual
is motivated to act in this way in response to a conflict that he cannot resolve
because its nature is not fully available to conscious awareness. Changes in behavior may be achieved, but unless the basic conflict is resolved, other, equally
disabling, symptoms may be substituted for the symptom that is no longer manifested.

It would consequently be futile merely to seek to change behavior and/or
relieve symptoms without attempting to trace and resolve the conflict, the "real"

problem from which the symptoms originate. It follows then that effective therapy is directed not toward changing behavior but toward achieving understanding, toward an "expansion of consciousness" so that it includes the hidden sense of conflict.

> If the client can be helped to understand why he behaves as he does or to recognize and understand the origin of his neurotic tactics that continually defeat him, he will gradually abandon the inappropriate behavior and substitute therefor more rational tactics in the management of his life. (p. 474)[13]

The promotion of self-understanding, of insight, is the most effective approach toward helping people with their problems and all strategies—reassurance, universalization, desensitization, catharsis, clarification, interpretation—are valued because they free emotional energy or change the balance of intrapsychic force in such a way as to maximize the possibility for self-understanding. Even environmental manipulation—reducing, modifying, or mitigating external stress infringing on the client—is regarded as a desirable tactic primarily because this, too, frees ego energy (previously devoted to struggling with the environment) for dealing more effectively with the basic intrapsychic conflict. These postulates are fundamental to one view of what is psychotherapeutic deriving primarily from psychoanalytic psychology.

Another view of therapy derives from the learning and conditioning psychologies. Here the primary concern is with the behavior itself without concern for underlying "causes." The behavior is viewed as the result of some unfortunate learning, conditioning experiences that have taught undesirable, unadaptive approaches to interpersonal relationships. The concept of symptoms as a response to an underlying conflict, in this view, is unnecessary and superfluous (p. 8).[28]

The therapy is primarily focused on the behavior itself and is concerned with providing an opportunity for unlearning the unadaptive behavior and learning new, more adaptive modes of behavior. The therapy seeks to identify "unsuitable stimulus-response connections," to dissolve them and to teach more desirable ones. It seeks to identify the specific environmental conditions through which the undesirable behavior is controlled and sustained and to change these. The stress is on immediate experience and specific behavior.

The view expressed here regarding what is therapeutic about adoptions is closer to the second rationale for therapy outlined, the learning-conditioning rationale, than it is to the first, the psychoanalytic psychology rationale. The child's previous living experiences may have "taught" a view of parents and parental surrogates that resulted in neurotic, unadaptive behavior. Defenses were developed and behavior manifested that was in response to the nature of the situation to which the child was subjected. Moving into the adoptive home meant moving into an environment that was set up to condition the child to a change in behavior. Previously learned, now inappropriate, behaviors went unrewarded or were actively discouraged; new, more appropriate, more adaptive behaviors were rewarded and actively encouraged. Without any explicit effort to resolve whatever underlying intrapsychic conflict may or may not have been present, without any explicit effort to have the child develop insight or self-understanding into his distorted perception of himself in the parent-child relationship and/or his distorted expectations with reference to the parents' behavior toward him, the living experience provides the corrective in day-to-day learning. The living experience teaches new ways of relating to people and new ways of perceiving oneself. In this sense, the adoptive home is a therapeutic milieu—a healthy restitutive living experience. It acts as a large-scale conditioning matrix that stimulates and supports changes in the child's feeling and behavior.

Psychotherapy is, in effect, a condensed, systematic attempt to imitate curative, real-life situations and assure the availability of such a curative configuration to the patient. Environmental therapy—therapy that actually affords the child the opportunity to live in a healthier family situation and experience the possibility of successful interaction with parents—is, as Josselyn notes, "the least artificial form of therapy" (p. 120).[14] And, like all therapies, it is grounded in the supposition that experiences in the present can free us from the past and that the effects of such earlier experiences are reversible.

REFERENCES

1. ALPERT, AUGUSTA. "Reversibility of Pathological Fixations Associated with Maternal Deprivation in Infancy," in *Psychoanalytic Study of the Child*, Vol. 14. New York: International Universities Press, 1959. Pp. 169–185.
2. APPELL, GENEVIEVE, AND DAVID MYRIAM. "Case Notes on Monique," in B. M. Foss, ed., *Determinants of Infant Behavior*. New York: John Wiley & Co., 1963. pp. 101–112.
3. BOWLBY, JOHN. *Maternal Care and Mental Health*. Geneva, Switzerland: World Health Organization, 1951.
4. BOWLBY, JOHN, AND AINSWORTH, MARY. "The Effects of Mother-Child Separation: A Follow-up Study," *British Journal of Medical Psychology*, Vol. 29, Part 3 (1956), pp. 211–244.
5. BRILL, NORMAN, AND LISTON, EDWARD. "Parental Loss in Adults with Emotional Disorders," *Archives of General Psychiatry*, Vol. 14, No. 3 (March 1966).
6. CLARKE, A. D. B., AND CLARKE, ANN. "Recovery from the Effects of Deprivation." *Acta Psycholi*, Vol. 16, No. 2 (1959), pp. 137–144.
7. ESCALONA, SIBYLLE, AND HEIDER, GRACE. *Prediction and Outcome: A Study in Child Development*. New York: Basic Books, 1959.
8. FANSHEL, DAVID, AND JAFFEE, BENSON. "A Follow-Up Study of Adoption—Preliminary Report." Unpublished paper, New York, N.Y., 1965. Mimeographed.
9. "Final Report to Children's Bureau on Study of Adoption of Greek Children by American Foster Parents." New York: Research Institute for the Study of Man, November 1964. Mimeographed.
10. FRIES, MARGARET E., AND WOOLF, PAUL Y. "Some Hypotheses on the Role

of the Congenital Activity Type in Personality Development," in *Psychoanalytic Study of the Child*, Vol. 8. New York: International Universities Press, 1953. Pp. 48–62.

11. GOERTZEL, VICTOR, AND GOERTZEL, MILDRED GEORGE. *Cradles of Eminence*. Boston: Little, Brown & Co., 1962.

12. HELLMAN, ILSE. "Hampstead Nursery Follow-up Studies: 1. Sudden Separation and Its Effect Followed Over Twenty Years," in *Psychoanalytic Study of the Child*, Vol. 17. New York: International Universities Press, 1962. Pp. 159–174.

13. HOBBS, NICHOLAS. "Sources of Gain in Psychotherapy," in Warren G. Bennis, Edgar H. Schein, David E. Berlew, and Fred I. Steele, eds., *Interpersonal Dynamics*. Homewood, Ill.: Dorsby Press, 1964. Pp. 474–485.

14. JOSSELYN, IRENE M. *Psychosocial Development of Children*. New York: Family Service Association of America, 1948.

15. LEWIS, HILDA. *Deprived Children*. London, Eng.: Oxford University Press, 1954.

16. MAAS, HENRY. "The Young Adult Adjustment of Twenty Wartime Residential Nursery Children," *Child Welfare*, Vol. 42, No. 2 (February 1963), pp. 57–72.

17. MEIER, ELIZABETH G. "Current Circumstances of Former Foster Children," *Child Welfare*, Vol. 44, No. 4 (April 1965), pp. 196–206.

18. MEIER, ELIZABETH G. "Former Foster Children as Adult Citizens." Unpublished doctoral dissertation, Columbia University School of Social Work, 1962.

19. MUNRO, ALISTAIR. "Childhood Parent-Loss in a Psychiatrically Normal Population," *British Journal of Preventive and Social Medicine*, Vol. 19, No. 2 (April 1965), pp. 69–79.

20. RATHBUN, CONSTANCE, McLAUGHLIN, HELEN, BENNETT, CHESTER, AND GARLAND, JAMES. "Later Adjustment of Children Following Radical Separation from Family and Culture." Paper presented at the annual meeting of the American Orthopsychiatric Association, Chicago, Ill., 1964. Mimeographed.

21. RENAUD, HAROLD, AND ESTESS, FLOYD. "Life History Interviews with One Hundred Normal American Males: 'Pathogenicity' of Childhood," *American Journal of Orthopsychiatry*, Vol. 31, No. 4 (October 1961), pp. 786–802.

22. ROE, ANNE. "The Adult Adjustment of Children of Alcoholic Parents Raised in Foster Homes," *Quarterly Journal of Studies on Alcohol*, Vol. 5 (March 1945), pp. 378–393.

23. ROE, ANNE, AND BURKS, BARBARA. "Adult Adjustment of Foster Children of Alcoholic and Psychotic Parentage and the Influence of the Foster Home," *Memoirs of the Section on Alcohol Studies*, No. 3. New Haven: Yale University Press, 1945.

24. SCHOFIELD, WILLIAM, AND BALIAN, LUCY. "A Comparative Study of the Personal Histories of Schizophrenic and Nonpsychiatric Patients," *Journal of Abnormal and Social Psychology*, Vol. 59, No. 2 (September 1959), pp. 216–225.

25. SROLE, LEO, et al. *Mental Health in the Metropolis: The Midtown Manhattan Study*. New York: McGraw-Hill Book Co., 1962.

26. THEIS, SOPHIE VAN SENDEN. *How Foster Children Turn Out*, Publication No. 165. New York: State Charities Aid Association, 1924.
27. THOMAS, ALEXANDER, BIRCH, HERBERT, CHESS, STELLA, HERTZIG, MARGARET, AND KORN, SAM. *Behavioral Individuality in Early Childhood*. New York: New York University Press, 1963.
28. THOMAS, EDWIN J., AND GOODMAN, ESTHER. *Socio-Biological Theory and Interpersonal Helping in Social Work*. Ann Arbor: University of Michigan, School of Social Work, 1965.
29. WELTER, MARIANNE. *Adopted Older Foreign and American Children*. New York: International Social Service, 1965.
30. WITKIN, H. A., *et al. Psychological Differentiation*. New York: John Wiley & Sons, 1962.
31. WITMER, HELEN, HERZOG, ELIZABETH, WEINSTEIN, EUGENE, AND SULLIVAN, MARY. *Independent Adoptions—A Follow-up Study*. New York: Russell Sage Foundation, 1963.
32. YARROW, LEON J. "Maternal Deprivation—Toward an Empirical and Conceptual Reevaluation," *Psychological Bulletin*, Vol. 58, No. 6 (November 1961), pp. 459–490. Reprinted in *Maternal Deprivation*. New York: Child Welfare League of America, January 1962. Pp. 3–41.

Developmental Life Transitions: Bereavement in Childhood

Learning about death is an essential part of growing-up that can be a crisis experience for a child if it involves the death of someone very close. We have just learned in the previous section that prolonged separation from one's parents or family can create feelings of anger, confusion, anxiety, and guilt, but that children are usually able (with some help) to come to terms with these feelings and reestablish some equilibrium in their lives. Death is absolute and final; it is the ultimate separation. Young children cannot fully understand death or distance themselves emotionally from it. It is therefore especially difficult for them to deal with the major upheaval a death can cause. Yet the majority of children do manage to handle such a crisis and even to grow stronger because of it.

A child's first experience with death is usually with someone outside his immediate family, most commonly an older adult such as a grandparent. The impact is limited because there is often some distance (emotional and physical) between the child and the dead person and because the death does not affect the child's daily life directly. The child must deal with his feelings of loss, however, and in doing so begins to build up his own conception of death and its meaning.

In the first article Saul Harrison, Charles Davenport, and John McDermott had planned a study of the reactions of children in a psychiatric hospital to the assassination of President Kennedy, who was a well-known person but somewhat distant from the children's lives. The focus of the study shifted as it became clear that the staff's observations were distorted by the adults' own grief and their inappropriate expectations of the children. Despite differences in age and development among the children, staff members expected a uniform "adult-like" soberness and grief from them and often blamed those who did not respond this way. Many of the staff members, who were usually very sensitive to the children's thoughts and feelings, chose to avoid or ignore the children's grief reactions.

Koocher,[4] interviewing school-age children, found that they were able to discuss the subject of death without anxiety. He concluded that it is adults who impart the sense of taboo to the topic. Because both intellectual and emotional problems are involved, adults have a crucial role to play in helping children understand, and adjust to, a death, a role made doubly difficult by the "forbidden" nature of the subject and by the fact that the adults involved are often grieving for their own loss.

When an immediate member of the family dies, there is no longer the safe distance necessary to cushion the shock. The very core of a child's life, his family structure, is threatened. In the second article Albert Cain, Irene Fast, and Mary Erickson examine children's reactions to the death of a brother or sister. Children's initial reactions to a death include loss of appetite, incessant talking about death, regressive behavior like thumb-sucking, antisocial behavior and other expressions of severe anxiety.[6] As a more long-term reaction, younger children often develop distorted views of illness and other causes of death (for example, that coughs or bruises or whatever symptom their dead sibling might have had can lead to death).

The sense of responsibility or culpability, another facet of causality, is also a serious problem to a bereaved child. Cain, Fast, and Erickson found that over half of the fifty-eight children they studied felt very guilty. In trying to determine "what" caused the death, a young child may end up thinking in terms of "who" caused the death,[1] blaming himself, a parent, or even the doctor or hospital. Some children also develop death phobias, fears that they will die at the same age or from the same cause as their dead sibling. Binger[2] found many of these same reactions in his study of children whose sibling had died of leukemia—guilt, vulnerability, resentment of parents for "allowing" it to happen, preoccupation with death.

In addition to the struggle to understand the concept of death, the surviving children must face a readjustment of family relationships. Roles within the family change with the absence of one member. The dead child may have been a protective older brother, a younger sibling who could be dominated, or perhaps the family scapegoat. Parents sometimes compound this problem by imposing the identity or characteristics of their dead child on a surviving sibling; for example, some parents even changed the name of a surviving child to that of his dead sibling. Even without such adult prompting, children may try to preserve some tie with the dead person (thereby easing the sense of loss) by taking on some of his attitudes or interests.

Many of these same problems face the child whose parent has died. The loss of the parent in his or her role as provider, protector, and adult model also brings its own additional difficult adjustments. The child's first reaction and defense is usually denial, a refusal to accept the reality and the permanence of death. Children under the age of seven or eight especially have problems comprehending the finality of death.[5] Children may interpret adult explanations of heaven and guardian angels as indications that the missing person is alive in some far-away place. For an adult who is also grieving it can be painful to correct a child's misconceptions. Lynn Caine writes of her six-year-old daughter's efforts to understand her father's death even after a year.

> "Mamma," she asked, "does my daddy know I'm in the first grade now?"
> How do you answer a question like that? I didn't know how to handle it. What should I say? The question hung there. Why does she have to ask me first thing in the morning when my head is filled with cotton?
> The answer was obvious. If Buffy's daddy knew that she was in the first grade, he wouldn't be dead. If he had any consciousness at all, he would not be dead. There is no consciousness past death. I couldn't let Buffy think that.
> "No, Buffy," I said. "Daddy is dead. He doesn't know you are in the first grade now. But if he did know, he'd be very pleased." (pp. 202–3)[3]

In the third article Diane Becker and Faith Margolin focus on the special problems of young children (under seven years) in experiencing the death of a parent, and on how surviving parents can aid their children's efforts to cope. They found that the children who could not accept the death right away either spoke of the parent as if alive or else avoided the subject altogether (see also McConville, Boag, and Purohit).[5] A common adult tendency to try to insulate children from death (not having them attend the funeral, for example) may encourage the use of denial and avoidance, delaying the children's real acceptance of the death and thus the beginning of a healthy readjustment. Becker and Margolin suggest that formal religious or cultural observances like memorial services or cemetery visits can help parents and children share their feelings openly and stimulate the kind of discussion in which children's misconceptions, fantasies and fears can be dealt with by parents.

Children can make successful adjustments, even to the death of a significant person in their lives, if the normal grieving process is allowed to "run its course." This process is often complicated by the problems of the grieving adult on whom the child is greatly dependent. The coping mechanisms that an adult uses (for example, the initial emotional "numbness" that enables a widow to make funeral arrangements and attend to other necessary "business") may actually raise additional problems for the children (such as encouraging them to deny the death or their sad feelings). Parents (and health professionals who are in contact with bereaved families) can generally prevent the development of a serious disturbance in a bereaved child by being aware of adult attitudes and behaviors that hinder a child's normal grieving. After the initial defensive use of denial, children usually need to ask questions about death, to seek reassurance about its possible consequences (for example, that the remaining parent will continue to take care of the child), and to share their memories and feelings about the dead person. In these ways children are able to attain a realistic understanding of death and to grow because of their experience with it.

REFERENCES

1. ARTHUR, B., AND KEMME, M. L. Bereavement in childhood. *Journal of Child Psychology and Psychiatry*, 1964, 5, 37–49.
2. BINGER, C. M. Childhood leukemia—emotional impact on siblings. In E. J. Anthony and C. Koupernik, eds., *The child in his family: The impact of disease and death*, Vol. 2. New York: Wiley, 1973.
3. CAINE, L. *Widow*. New York: William Morrow, 1974.
4. KOOCHER, G. P. Talking with children about death. *American Journal of Orthopsychiatry*, 1974, 44, 404–11.
5. McCONVILLE, B J.; BOAG, L. C.; AND PUROHIT, A. P. Mourning processes in children of varying ages. *Canadian Psychiatric Association Journal*, 1970, 15, 253–55.
6. McDONALD, M. A study of the reactions of nursery school children to the death of a child's mother. *Psychoanalytic Study of the Child*, 1964, 19, 358–76.

Article 6 Children's Reactions to Bereavement

Adult Confusions and Misperceptions

SAUL I. HARRISON

CHARLES W. DAVENPORT

JOHN F. MC DERMOTT, JR.

This report was conceived during the course of a study of the response to President Kennedy's assassination involving children hospitalized at Children's Psychiatric Hospital, University of Michigan Medical Center, Ann Arbor, Mich. The staff's conflicting perceptions of the children's reactions to the tragic event were strikingly evident. Torn between their own needs and those of the children, the staff was confused about the therapeutically appropriate means of exposing the children to the details of the assassination, the shooting of the accused assassin, and the funeral. Even experienced personnel, ordinarily intuitively responsive to the children's needs, were markedly uncertain about the handling of children at that time. Thus, the focus of investigation shifted from study of the reactions of the children to the exploration of conflicting staff perceptions and attitudes. This report of the staff's reactions is intended as a limited contribution to the accumulation of knowledge[2,15,17,23,28] concerning the capacities, perceptions, and behavior of interested adults dealing with bereaved children.

METHOD

Two types of observational data were studied. The first was the routine written reports of psychotherapists, nurses, child-care workers, teachers, occupational therapists, and recreational therapists. These were recorded at the time as a regu-

From the Department of Psychiatry, University of Michigan, Ann Arbor.

SAUL I., HARRISON, M.D., is Professor of Psychiatry, University of Michigan, Ann Arbor.

CHARLES W. DAVENPORT, M.D., is Clinical Assistant Professor of Psychiatry, University of Michigan.

JOHN F. McDERMOTT, M.D., is Professor and Chairman of Psychiatry, University of Hawaii, Honolulu.

Source: *Archives of General Psychiatry*, Vol. 17 (1967), 593–97. Copyright 1967, American Medical Association.

lar part of hospital procedure, and thus were in no way influenced by the present study. The second type of data was gathered retrospectively in the form of written material and group discussion, with staff representing various disciplines. These data were focused on the children's reactions and the patient-staff interaction, and were collected after this study was undertaken.

OBSERVATIONS

Staff Conceptions of How Children Should React

There was considerable disagreement between staff members over what the children's appropriate response should be. Typically, staff members perceive and react to the children in a highly differentiated and understanding manner. However, in this instance their covert and, at times, overt expectations seemed to call for a uniform and "adult-like" reaction of soberness, if not downright grief. Indeed, the staff responded to children's jocular behavior and lack of reverence with moral indignation. A striking example was provided by a 12-year-old boy with ulcerative colitis who laughed when the bugler hit a wrong note while playing taps at the late President's funeral. Understandably, the staff's immediate reaction was one of distaste; but subsequently, over the course of the next two months, there appeared to be increasing difficulty in the interaction between this boy and ward staff. It should be noted that this connection had not been recognized until it was brought up in one of our retrospective meetings. On another occasion, the staff became indignant when some of the youngsters protested and expressed resentment about a party being cancelled on the evening of the President's assassination. It seemed as though the proper adult mourning attitude was expected of the children, regardless of their developmental stages or degree and kind of pathology.

Management of Children's Reactions

There was uncertainty among staff members about how to respond to the children's reactions. In some instances, children's grief reactions appeared to be avoided by the staff. Several recorded observations noted that children failed to comment about the death, and staff members presumed the children were not reacting to it, despite noting a distinct behavioral change. For instance, an aggressive borderline psychotic 13-year-old girl, who suffered periods of depression, slept a great deal on the day of the President's assassination. This had been her usual behavior when depressed in the past. She mentioned dreaming about being killed, but reportedly "dropped the subject." Apparently the depressive reaction in this girl was either consciously or unconsciously ignored. Some of the staff seemed to focus on verbal reactions and responded to misbehavior with immediate firm controls. Other staff members perceived misbehavior as a means of mastering anxiety and, therefore, considered such behavior helpful. Also, assassination games, such as shooting or being shot, were tolerated—as if such identifications and repetitions were means of constructively mastering anxiety.

Degree of Exposure to the Details of the Tragedy

In keeping with Children's Psychiatric Hospital's typical mode of individualizing each child's treatment, there were no general administrative directions as to how much the children should be exposed to details of the tragedy. Each ward was free to handle their patient's reactions individually. This resulted in a remarkable lack of agreement among ward staff about the proper amount of exposure for the children to the details of President Kennedy's assassination. Some staff members were convinced that television provided a vehicle for working through grief reactions to President Kennedy's death; others disagreed, and were concerned about how to allow the children to watch "history in the making" while staying within the limits of what would not be emotionally harmful. As might be expected, these two opposing attitudes led to conflict and confusion about when and how long the children should watch the continuous television coverage of the events. Frequently, attempts were made to divert the children's attention from the repetitious broadcast. At times this seemed to be in response to the children's request; yet at other times, the children were frankly disinterested in diversions, and angry when denied the right to watch television.

In our intramural school, some of the teachers attempted to cover the subject of the President's assassination as a structured formal learning experience in the current events part of the class on the day following the funeral. Other teachers, however, assiduously avoided the subject; some despite the fact that the discussion of current events was scheduled for that day.

The Ability of Grieving Adults to Observe Objectively

Retrospectively, there appeared to be numerous distortions in the staff's perceptions of children's reactions, apparently as consequences of the participant observers' own grief. For example, there was an isolated report claiming that an autistic 9-year-old girl demonstrated "an appropriate affect" in response to the assassination, contradicting several other reports describing no change in her usual bizarre behavior.

In another instance, during one of the retrospective discussions, a nurse was unable to remember the presence of a particularly omnipotent, bizarre boy with whom she generally worked closely. In fact, she doubted that he had been admitted to the hospital by the time of the assassination. She felt that if he had been admitted, he must have gone home early that weekend. In contradiction to her conviction, there were many notes describing his usual bizarre behavior during the entire weekend.

Often it was impossible to decide whether a given instance represented misperception on the part of the staff, adaptation by the children to the behavior of bereaved adults, or manifest mourning on the part of the children.

Thus, the ward for the youngest and generally most regressed children reported an amazing unanimity of "mature" behavior and response. Far less fighting and bizarre behavior than usual was reported that week; the staff perceived this as expressive of consideration by the children for the adults' feelings. The

accuracy of this surprising observation and the validity of the motivational explanation cannot be ascertained, and both shall have to remain open to question.
If, however, we assume the observation is undistorted, there remains doubt as to
whether the children's unusual behavior was a reaction to the tragic events, imi-
tation of the adults, or adaptation to the emotional unavailability, and perhaps
irritability, of the staff. Such difficulties in obtaining accurate behavioral observations, much less valid interpretations, was particularly evident in assessing the
retrospective data.

The Effect of Adult Grief on Children's Reactions

There were some instances in which it was clear that the child was responding not only to the President's assassination, but also to the reaction of the
bereaved adults. For example, a 10-year-old boy with severe behavior problems
watched adults carefully for cues in how to react, and tended to mirror what he
observed. In the hospital's intramural school, there was a marked similarity between the staff's reaction and the children's reported reactions. The teachers
tended to describe similar reactions from all their students, seeing them as a
group rather than in terms of individual differences. It was difficult to avoid
the feeling that these reports depended far more upon the observer than on the
children observed.

COMMENT

Study of the observations and the reactions of children made by clinical child-
care workers during a period of mass-mourning[26,29] demonstrated a striking degree
of estrangement from the children, in marked contrast to the emphatic relationship these same adults generally enjoyed with their charges. Furthermore, it
should be noted that the observers were experienced in giving written and verbal
reports about these children. At other times, and in relation to other issues, this
was invariably accomplished without evidence of the massive misperceptions and
distortions cited above.

Concurrently it was clear that the therapeutic staff were vague about children's concepts of death.[1,3,8,18,19,21] There was much uncertainty and disagreement on what children knew about death, how much information and emotional
impact they could tolerate, what they should be exposed to, and in what fashion.
Underlying some of the manifest attitudes appeared to be an assumption that
children are incapable of possessing a reasonable concept of death. Paradoxically,
this was often in juxtaposition with the expectation of an adult-like reaction from
the children. At times, this was expressed in a tendency to dictate moralistically
about how the children should behave and "properly grieve." Children who did
not conform to this stereotype tended to be blamed, overtly or covertly, for alleged shortcomings. Stated briefly: "children don't know about death; yet they
should mourn as adults do." This is remindful of Durkheim's[7] assertion that

mourning is more than an expression of individual feelings, and also reflects a duty imposed by a group that forces one to weep.

There were surprisingly many instances in which otherwise sensitive and intuitive people thought the children were unaffected by the tragic events, despite their own observations of unequivocal behavioral shifts. One cannot help but wonder how much this reflects the ubiquitous fear of death[27] that results in a widespread denial of death's inevitability and finality—in essence, a societal evasion of the entire issue of death. Assisted by technological progress and urbanization, American culture seems involved in a conspiracy to remove death from conscious awareness. Bereavement has become an individual matter, with an attendant deritualization of mourning.[12,13,16] As institutional patterns diminish the bereaved are left increasingly to fend for themselves, with concomitant growth of uncertainty. Such an avoidant attitude on the part of our staff was in evidence; for example, in the failure of the administrative staff to give direction, and in the presence of the emotion-denying "history in the making" viewpoint.

It should be emphasized, however, that our society has never had much in the way of identifiable guidelines to follow in dealing with children's confrontations with death. Although the tendency of children to equate death with a long journey is no longer encouraged as much as in the past, by adults "protectively" telling children that a recently deceased relative has taken a long trip, we lack a ritualized prescription of what to tell children, what to do with them, what to have them do, etc. Anthropological reports occasionally mention the role of the "children of the deceased" in the death rituals of other cultures, but indications as to the age of the participating offspring are rarely found.

Mandelbaum[16] mentions the well-defined special ceremonial roles for the children of the recently deceased in the Kota (India) "dry funeral," a communal event celebrated every year or two. Stewart[25] describes that, following death until the funeral feast, the children of a deceased Negrito (Philippines) are expected to dream repeatedly of their parent and be willing to do whatever request is made in the dream. In neither of these reports, however, are there any hints as to how young the "children" are. Fortes,[9] on the other hand, describes the ritual of headshaving practiced by children, as young as 10 years, of deceased Tallensi (Africa). Furthermore, the youngest grandchildren of the deceased Tallensi are assigned a ritualized role in the funeral.

Does our staff's display of confusion relate also to some aspects of the controversies[6,10,11,13,14,20,22,24] about the incidence and duration of children's manifest depressive reactions stimulated by object loss, as well as their capacity to work through the mourning process adequately? Certainly, our experiences suggest that one should question the validity of accepting the descriptions of children's bereavement reactions given by mourning adults. In our data it was impossible to distinguish between adult misperceptions and confusions, the children's reaction to the tragedy, and the children's reaction to the changes in the adults.

When a young parent dies, the children are often immediately shunted off to the homes of relatives or friends. This is frequently accomplished without the active participation of the surviving parent, whose profound grief and shock may render him incapable of providing even the most basic and minimal care for his children. Furthermore, the helpful relatives and friends who bring the children to their homes typically leave them in the care of older cousins or sitters as they return to the surviving parent, whose manifest need for help is often so much more obvious than the children's. Thus, retrospective investigators[5] may be forced to rely on reports from people whose perceptive apparatus were blurred by overwhelming emotion. Clearly, there is a great need for continuing "objective" observations, despite the fact that objectivity is impossible for a feeling person to attain when dealing with a small child whose parent has just died.

SUMMARY

Following the assassination of President Kennedy, the staff of Children's Psychiatric Hospital, torn between their own grief and the needs of the children, manifested a significantly atypical amount of confusion and insensitivity in their handling of the children. This highlighted the lack of societal guidelines or rituals for dealing with children's confrontations with death in our culture. Furthermore, the staff's descriptions of the children's reactions were so riddled with disagreements and distortions that one cannot help but raise questions about the validity of accounts of children's bereavement reactions given by mourning adults —just such accounts as figure prominently in our growing literature on childhood bereavement.

REFERENCES

1. ANTHONY, S. *The Child's Discovery of Death*, New York: Harcourt, Brace, and World, 1940.
2. BECKER, D., AND MARGOLIN, F. Parental Attitudes Towards the Young Child's Adaptation to the Crisis of Loss, read before the 43rd annual meeting of the American Orthopsychiatric Association, San Francisco, April 1966.
3. BENDER, L. *Aggression, Hostility, and Anxiety in Children*, Springfield, Ill.: Charles C. Thomas Publishers, 1913.
4. BOWLBY, J. Grief and Mourning in Infancy and Early Childhood, *Psychoanal. Stud. Child.* 15:9–52, 1960.
5. CAIN, A.; FAST, I.; AND ERICKSON, M. E. Children's Disturbed Reactions to the Death of a Sibling, *Amer. J. Orthopsychiat* 34:741–752, 1964.
6. DEUTSCH, H. Absence of Grief, *Psychoanal. Quart.* 6:12–22, 1937.
7. DURKHEIM, E. *The Elementary Forms of the Religious Life*, London: George Allen & Unwin Ltd., 1915.
8. FAST, I., AND CAIN, A. C. Children's Distorted Concepts of Death, read before the 40th annual meeting of the American Orthopsychiatric Association, Washington, March 1963.
9. FORTES, M. *The Web of Kinship Among the Tallensi*, London: Oxford University Press, 1944.
10. FREUD, A. Discussion of Bowlby, *Psychoanal. Stud. Child.* 15:53–62, 1960.

11. FREUD, A., AND BURLINGHAM, D. *Infants Without Families*, New York: International Universities Press, 1944.
12. KRUPP, G. R. The Bereavement Reaction: A Special Case of Separation Anxiety—Sociocultural Considerations, *Psychoanal. Stud. Soc.* **2**:42–74, 1962.
13. KRUPP, G. R., AND KLIGFIELD, B. The Bereavement Reaction: A Cross-cultural Evaluation, *J. Religion Health* **1**:223–246, 1946.
14. MAGLER, M. On Sadness and Grief in Infancy and Childhood, *Psychoanal. Stud. Child.* **16**:332–351, 1961.
15. MAHLER, M. Helping Children to Accept Death, *Child Study* **27**(4):98–99, 119–120 (Fall) 1950.
16. MANDELBAUM, D. G. "Social Uses of Funeral Rites," in Feifel, H. (ed.): *The Meaning of Death*, New York: McGraw-Hill Book Co., Inc., 1959.
17. MOHR, G. *When Children Face Crises*, Chicago: Science Research Associates, 1952.
18. NAGY, M. "The Child's View of Death," in Feifel, H. (ed.): *The Meaning of Death*, New York: McGraw-Hill Book Co., Inc., 1959.
19. PIEPER, W. S. How Children Conceive of Death, read before the Area IV divisional meeting of the American Psychiatric Association, St. Louis, October 1965.
20. ROCHLIN, G. The Loss Complex, *J. Amer. Psychoanal. Assoc.* **7**:299–315, 1959.
21. SCHILDER, P., AND WECHSLER, D. The Attitudes of Children Toward Death, *J. Genet. Psychol.* **45**:406–451, 1934.
22. SCHUR, M. Discussion of Bowlby, *Psychoanal. Stud. Child.* **15**:53–84, 1960.
23. SHERRILL, L. J., AND SHERRILL, H. H. Interpreting Death to Children, *Int. J. Religious Educ.* **27**:4–6 (Oct) 1951.
24. SPITZ, R. Discussion of Bowlby, *Psychoanal. Stud. Child.* **15**:85–94, 1960.
25. STEWART, K. *Pygmies and Dream Giants*, New York: W. W. Norton & Co., 1954.
26. VOLKART, B. "Bereavement and Mental Health," in Leighton, A., et al. (eds.): *Explorations in Social Psychiatry*, New York: Basic Books, Inc., 1957.
27. WAHL, C. W. "The Fear of Death," in Feifel, H. (ed.): *The Meaning of Death*, New York: McGraw-Hill Book Co., Inc., 1959.
28. WOLF, A. W. M. *Helping Your Child to Understand Death*, New York: Child Study Association of America, 1958.
29. WOLFENSTEIN, M., AND KLIMAN, G. *Children's Reactions to the Death of the President*, New York: Doubleday and Co., 1965.

Article 7 Children's Disturbed Reactions to the Death of a Sibling

ALBERT C. CAIN

IRENE FAST

MARY E. ERICKSON

The purpose of this paper is to explore one portion of a relatively neglected territory in the domain of child development and psychopathology, that of children's reactions to death. Our culture's avoidant attitudes toward the realities of death[17,37,40] have until recently been too fully reflected in the sparsity of scientific investigations into attitudes toward and reactions to death. But in the last decade the study of death and dying[4,16,39] and particularly of children's profound reactions to the death of a parent, have received increasing empirical and theoretical attention.[9,10,18,19,28,31,34,35] By contrast, the investigation of children's reactions to the death of *siblings* remains in an early stage. We have barely progressed beyond the time when lengthy, intensive psychiatric case studies could note in a passing sentence that "one of the patient's siblings died when he was four" and omit any further reference to the event's meanings to the patient. Similarly, current studies showing a sharp awareness of the complex reactions of parents to the death of a child may omit any mention of the impact of the death upon other family members.

But there are now brief mentions of the effects of a sibling's death in a number of specific cases in the literature (to be noted later); three studies assessing the comparative incidence of early sibling deaths among a variety of clinical and control groups[8,28,33]; and a recent case study partly focused on a sibling loss.[21] Stimulated by some striking case material and tangential findings of a previous

ALBERT C. CAIN, PH.D., is with the Department of Psychology, University of Michigan, Ann Arbor.

IRENE FAST, PH.D., is now at San Fernando Valley Child Guidance Clinic, Sherman Oaks, California.

MARY E. ERICKSON, M.S.W., is at Children's Psychiatric Hospital, Ann Arbor, Michigan.

Source: *American Journal of Orthopsychiatry*, Vol. 34 (1964), 741–52. Coyright © 1964 the American Orthopsychiatric Association, Inc. Reproduced by permission.

investigation,[13] this study was undertaken in part to fill in the outline gradually sketched by these papers. Our growing case material soon served to demonstrate the limitations of perhaps the one notion of any currency about the import of sibling death, namely, the concept that the primary if not exclusive pathological impact of a sibling's death upon the surviving child is one of guilt over rivalry-bred hostile wishes which, through the early omnipotence of thought, are seen as having been fulfilled by and responsible for the sibling's death.[32,33]

The importance of multiple-based sibling tensions, rivalries and hostilities are visible to the most naive clinical observation, emerging vividly even in research investigations employing such minimal disguise as single standardized interviews with children asked rather directly about sibling attitudes and grievances.[20] The import such dynamics have for personality development cannot be overemphasized. But as the following case material indicates, we must proceed far beyond the perception—which would not do justice even to the psychoanalysis of the 1920s—of a child's siblings as almost exclusively representing rivals, sexual temptations or objects for Oedipal displacements, and a sibling's death as primarily, if not exclusively, producing guilt reactions.

After attempting to expand such conceptions, we will briefly note the preventive implications of some of our findings. And in a later study the data on children's reactions to a sibling's death will be reviewed in terms of prevalent theories of object loss and mourning. For, as Pollock[27] has noted, and as our data appear to indicate, the process of bereavement in such a context may show striking differences from other contexts of loss and bereavement (for example, the loss of a parent or a spouse, or parents' loss of a child), thus broadening and differentiating current concepts of mourning.

What follows, then, is a listing of varied forms of disturbed reactions by children to the death of a sibling.* The children whose cases are noted here were all psychiatric patients, and in each there was clear evidence that some—though by no means all—of the major symptoms and character distortions were substantially related to the death of the sibling. The individual forms of reactions are by no means mutually exclusive; in most of the children studied, a number of such reactions were intertwined. It is only for the sake of clarity and simplicity of presentation that the material sequentially focuses upon single dimensions of the surviving child's response. Obviously the cases are far more complex, the sources of disturbance numerous and difficult to disentangle, and each child's reactions in some ways unique. Nevertheless, we have cited only cases where reac-

---

* Given the nature of the data—collected in good part from closed files of materials ranging from outpatient evaluations to years of intensive inpatient treatment, with cases seen in various clinical settings each with its own emphases and orientation—it was often not comparable from case to case. As such, quantification in a number of instances would be futile if not misleading. Instead, where the data permitted, we have simply noted approximate percentages of the children in the sample who evinced a particular, disturbed reaction.

tions crystallized along fairly clear dimensions of personality development and disturbance. Omitted are such observations of more unsettled, immediate responses to a sibling's death as Cobb[15] succinctly described: appetite loss, dazed states, incessant talk about the death. Also omitted are many enduring symptoms we encountered that were originally precipitated by the sibling death, for example, nightmares, speech disturbances, enuresis, antisocial acting out, severe anxiety states.

Some exclusions are to be noted: We have reserved for discussion elsewhere children's often deep, if not immediately evident, reactions to their mother's stillbirths and miscarriages.[14] Nor will we deal with the even more complicated cases of infanticide in which one or more children survived a parent's attack while other siblings were killed. Initially we had also thought to exclude cases where the dead sibling died *before* the sibling we studied was born, but some of these cases share in so many ways the specific problems seen in our primary group of children that they are included.*

THE DISTURBED REACTIONS

Guilt Reactions and Their Vicissitudes

We found, of course, a heavy accent on guilt-laden reactions. In approximately half our cases, guilt was rawly, directly present. So, too, was trembling, crying and sadness upon mention of the sibling's death, with the guilt still consciously active five years or more after the sibling's death. Such children felt responsible for the death, sporadically insisted it was all their fault, felt they should have died too, or should have died *instead* of the dead sibling. They insisted they should enjoy nothing, and deserved only the worst. Some had suicidal thoughts and impulses, said they deserved to die, wanted to die—this also being motivated by a wish to join the dead sibling. They mulled over and over the nasty things they had thought, felt, said or done to the dead sibling, and became all the guiltier. They also tried to recall the good things they had done, the ways they had protected the dead sibling, and so on. The guilt was variously handled by each child in accord with his unique personality structure, with reactions including depressive withdrawal, accident-prone behavior, punishment-seeking, constant provocative testing, exhibitionistic use of guilt and grief, massive projection of superego accusations and many forms of acting out. The consequent deterioration in these children's functioning, especially in school, provided them further grounds for insisting they were rotten and worthless.

---

* The death of a child's close friend, neighbor, cousin, or the like, was found capable of producing disturbances in important ways similar to those reported here. "Loss" of siblings due to war-induced separations, splitting up of siblings when transferred to foster homes, and long-term institutionalization of a mentally defective or psychotic sibling also obviously have great emotional significance for a child, and demand active working through.

The sibling deaths among the cases studied ranged widely from chronic or sudden illnesses—leukemia, asphyxiation, heart conditions, severe infections—through car accidents, drownings, burning to death, accidental shootings, severe beatings and murder. The remaining siblings' actual involvement in or "responsibility" for the death of their siblings similarly ranged widely. In at least one-third of those cases in which one or more remaining siblings had, in a sense, been significantly responsible for the death, the findings were striking. The child's parents often clearly would *not allow* the remaining sibling to talk about the event. They rushed in with heavy repeated reassurances, quickly labeled it all an accident, and cut off any possibility of the child's telling them what he told (in some instances pathetically, eagerly told) us later—that it wasn't a total accident, that they had been mad at the sib at the time, partly wanted him hurt or "dead," and intentionally did what they did. The suppression of the child's need to confess his role in the incident appeared multidetermined. Parents understandably, if wrongly, felt or were professionally advised that allowing the child to talk about the incident would only upset him more and make it linger on in his and everyone's mind. The parents were fearful of being swamped by even further affects beyond those already overwhelming them, fearing not only their own intense grief but their repressed or suppressed rage at the child. Lastly, they needed desperately to avoid the open assessment of blame, for each parent was struggling with his own self-accusations. It was particularly this latter element that consistently kept the specific details of events surrounding the death remarkably vague in even the most important respects. One cannot help but conclude from these records that the clinicians later involved in these cases too often joined in this preconscious pact not to explore even the simplest details about the death.

The child's death typically stimulates an avalanche of superego accusations and overt blaming, which in turn undergo marked transformations. Generally, the remaining children and each parent, in some way and to some degree, blame themselves. Alternatively they blame each other, and often, needing to maintain positive images of and relationships with each other, they try defensively *not* to blame each other. Frequently a major solution is found that permits the anger and blame assignment so vital to object-loss reactions to occur without disrupting much-needed family relationships. This solution consists of placing the blame well outside the family circle onto neighbors, car drivers, lifeguards, doctors, or "the hospital." While this worked well for some families, with about 15 percent of our group the oscillating superego introjections and projections quickly wreaked havoc: Mothers sought divorces; a father denounced his wife for her responsibility in the death and abandoned his family forever; a girl ran away from home insisting her mother was a murderess; a boy killed his father whom he held at fault in his little brother's death (the autopsy had established otherwise). Nor does the blaming end there. Grandparents and other relatives also pointed accusing fingers, and in three cases neighbors called in the police much as has been occasionally reported in the pediatric literature on sudden, unexpected death in infants.[2]

Where either the realities or the aggressive fantasies surrounding the death left the remaining sibling struggling with intense guilt, the child typically grew extremely fearful of losing control of his anger and experienced himself as a monster and potential killer. He attempted various identifications which all cried out that he was in no way aggressive or capable of such behavior, generally withdrew and perceived situations in which he might do *anything* wrong as reverberating of the past wrongdoing, the "killing." In all but one of the instances in which the child had actually been an aggressive participant in his sibling's death, the child's parents, siblings or peers reinforced this reaction. They viewed the child as always latently dangerous, shrank back from him and never fully trusted him. As if this were not tragic burden enough for a child to carry, he often took upon himself even further guilt stemming from additional family reactions to his sibling's death, for example, his father's desertion, his mother's miscarriage or hospitalization.

Beyond immediate blaming reactions, in some cases one or both parents continued to maintain explicit or unconscious attitudes of blame toward a surviving child, with constant hostility and guilt-inducement toward the child and minimal love for him. The child was never conceded the possibility of "making it up." As one parent put it, "I guess we bore him a grudge."

In slightly less than one-quarter of our cases, guilt regarding the sibling's death was essentially *imposed* by the parents, but not by blaming a child for his sibling's death. Rather, it was that the child had shown no regret, no sadness, no grief at the loss of his sibling, perhaps blithely going off to play or eagerly using the dramatic news of the death to grab his playmates' interest. His parents, distressed at the child's lack of "proper sentiment," sharply rebuked the child and made him feel guilty over his apparent lack of grief. This was even more intense in a few instances in which the parents were horrified, then enraged by a child's outspoken happiness at being rid of his sibling. We originally considered such reports to be mostly retrospective distortions; we assumed this alleged "guilt over lack of grief" was primarily a displacement of guilts, guilts basically derived from the child's fantasied responsibility for his sibling's death. It appears to be that in some cases, but such guilt-induction over lack of grief clearly can exist powerfully in its own right.

Far rarer forms of guilt imposition were also encountered, for instance, in the referral of a three-year-old girl by her mother, who stated that the girl had become severely disturbed, tortured with self-accusations since her involvement in a farm accident in which she was held responsible for her sister's death. Clinical observation revealed, rather, an essentially normal, well-functioning three-year-old—in no way corresponding to her mother's description of a nightmare-ridden, nail-biting, depressed, agitated, self-accusing girl. The evaluation revealed, instead, a mother who had entirely projected her own guilty, tormented reaction to her child's death onto this child, then vicariously sought treatment for her.

Distorted Concepts of Illness and Death

Apart from such vicissitudes of guilt reactions, the siblings of the dead child often had quite confused, distorted concepts of illness, death, and the relationship between illness and death. The normal child's developmentally evolving concepts of death have received attention elsewhere,[3,23] as have some pathological distortions of such concepts.[13] These children, who had long been treated to conventional saws about people not dying until they were very old, struggled with the clear contradiction to this of their sibling's death, and its corresponding undermining of their confidence in adults' pronouncements. A few simply solved it temporarily by saying "you can't die 'til you're at least nine" (the dead sibling's age). Others swung back and forth, wanting to continue to believe they couldn't die until very old but having to deny the sharpest of realities to do so. They emerged with such confusions as, you die because you're small, you die young only at night, only girls die and the like. The lesson taken by almost one-third of the children was that growing up, growing older, meant you would die: Partial or total defensive regressions toward passive-dependent infantilism followed. A sample of one preschooler's concerns: His dead sister was in heaven, heaven was up in the sky, birds flew in the sky, would the birds eat his sister? And one hardly need elaborate upon how utterly distorted a young child's concept of his sibling's death was when he observed (or heard fragments of) attempts at emergency respiration, tracheotomies and so on. Here, the surviving child's ghastly fantasies centered not around illness but upon adult murderers killing a struggling child.

Children's distorted concepts of illness, still too little recognized by parents and pediatricians, were heavily present in those children whose siblings died due to illness. The children lived with frighteningly concrete disease notions, for example, that coughs, colds, "high temperatures" and bruises led to death. Thus, death was constantly imminent. Notions combining old parental urgings and cautionary tales about sleep, food, wearing galoshes, running about barefoot and the like, were elaborated into causes of illness and death (for example, "he didn't eat his vegetables"), and other more primary process fusions of concepts of food, germs and death were also present.

Disturbed Attitudes Toward Doctors, Hospitals and Religion

In the majority of such cases, the surviving children's fears of doctors were greatly heightened. Doctors were perceived as impotent in the face of illness and closely associated with illness and death, if not themselves seen as somehow responsible. Hospitals were even more terrifying, and going to a hospital was equated with death. This response by the children was not an isolated one: Often the parents distrusted and blamed doctors or hospitals, and later some remained reluctant to let their other children be hospitalized for necessary medical procedures.

Almost as strong was the child's confusion about God's portrayal as benevolent. Many of the children simply remained puzzled as to how and why their loved and loving God would have killed or at best "taken away" their sibling.

Some needed constant reassurance that God didn't really go around hurting people. A few others, awed by the concrete inches-away demonstration of their Lord's power, spoke of fearing or even hating God as their sibling's "murderer."

Death Phobias

Looming large in virtually all of the children's responses was an intense fear of death. A few children responded with such omnipotent attitudes and statements as "I can't die," and, "I couldn't be killed," but their defensiveness was transparent and their fears constantly broke through. Most prominent in these fears were talion fantasies and identifications with the dead sib: The children were often convinced not only that they, too, would die, but that they would die either at precisely the same age or from the same cause or under the same circumstances as the dead sibling. The children, of course, had a generally heightened awareness and fear of death, feeling it could strike at any moment, and at his other siblings or parents as well as himself. Notions of their parents' invulnerability, all-powerfulness and especially of their parents' strength as protectors, came crashing.

But it should be clear, the children's death phobias were not solely the products of talion fantasies and identification with the dead sibling. As noted by others,[5,13,22,26] a typical parental response to a child's death was fearful overprotectiveness of the remaining children. Normal parental concerns were intensified to extreme proportions, leading to restrictiveness, overprotection and infantilization. The parents' phobic vigilance, and the extremely dependent, phobia-breeding relationship into which the remaining child was often pressed, tended to heighten further the child's death phobia. Thus restricted from so many basic growth experiences, these children were generally immature, passive-dependent and fearful, feeling small, inadequate and vulnerable in an ever dangerous world. These parental attitudes sometimes focused primarily on just one child, for example, "the only girl we have left," and often were combined with open favoritism toward and indulgence of that child. The child was rarely punished or properly controlled; the other siblings were not permitted to interfere or even hold their own with this child, as the parents quite consciously tried to make up for past guilt-festered behavior toward the dead child through this living child. The child became quite aware of his special role and used it to the hilt.

Comparisons, Identification and "Mis-identification"

Previous mention was made of these children's identifications with the dead child, in the sense that they believed they would die much as their siblings had. In addition, in approximately 40 percent of the cases there were either immediate, prolonged or "anniversary" hysterical identifications with the dead child's prominent symptoms. These included hysterical pains, convulsivelike states, severe asthmatic attacks (the first occurring immediately after a sibling died in such an attack), and apparent almost total motor paralysis, which indeed nearly did lead to death. In such cases, awareness of the identification element in the symptoms was crucial to the medical diagnosis.

Identification, or as John Benjamin wisely calls it, "mis-identification," played another major role with roughly one-fifth of the children in our group. As some children are specifically conceived to replace a dead child,[13] these children were unconsciously mis-identified with the dead sibling, becoming the parents' open "substitute" for him. In a few stunning cases, parents even changed the *living* child's name to that of the dead child, while in other cases newly born children were given the dead child's exact name or his name slightly changed. The dead child's identity was further imposed upon the sibling by the parents' (and at times the community's) continual open identification of the two—their looks, posture, way of talking—and by the parents' expectations and demands upon the replacement being based on the idealized image of the dead child. All of this severely warped the "replacement's" identity formation. The parents' relationship with the substitute child, was, in brief, almost totally dominated by their image of the dead child. The children, of course, could not possibly replace the dead child for their griefstricken parents. They found this strange task of being yet not being the dead child hopeless, resented it, were resented by their parents for not fulfilling it, and were aware of their parents' basic wish that *they*, not the dead sibling, had died.

Consistently unfavorable comparisons between the surviving children and their dead siblings occurred in almost one-half of our cases, both in these replacement cases and in many others.[7,25] The comparisons were relentless, often quite open, even in parents who were aware of how damaging this was. Comparisons could extend over any and all areas of behavior, but were particularly focused upon school performance. The surviving child initially wondered if he could measure up. He tried to compete with his dead rival, but even his best efforts went for naught, given the hyperidealization of his dead sibling. Soon such children gave up, at times joining in the idealization of the dead sibling, at times unconsciously engaging in vengeful school failures.

Disturbances in Cognitive Functioning

In a manner distinctly different from the poor school performance in some of the above cases, other children, 15 percent of all cases, revealed major distortions of cognitive functioning related to the death of their sibling. These otherwise intellectually intact children displayed profound cognitive distortions in which occurred unbelievable encapsulated "ignorance," fluctuating pseudo stupidity, seeming lack of knowledge of the child's own age, reversals and distortions of concepts of old and young. General "not knowing" or particular areas of ignorance, and especially specific disability in concepts of time and causality seemed unconsciously vital to these children: these deficits appeared traceable to denial mechanisms surrounding the sibling's death (its causes, its relation to age, and so on).

Impact of Changes in the Family Structure

A child's death also had both direct and indirect effects upon his remaining siblings by its disruption of internal balances of interrelated roles and functions in the family structure—in brief, by shifts required in the "family dynamics." [1,38] The remaining siblings directly lost, for instance, a playmate, a companion, an older brother who "ran interference" for his younger sib, a protector, a scapegoat, a baby whom a four-year-old girl needed to mother, an ally against a fierce border-line psychotic mother, a younger sister who could be actively dominated and controlled in order to master a mother's sadistic controlling behavior, and the like. But the indirect effects of the loss could be even more encompassing. Some fathers who strongly needed a son and lost their only son began (and succeeded in) masculinizing their daughters. Needs to infantilize a child after a baby's death interfered with a surviving preadolescent son's struggle for independence; a mother's need to act out vicariously through a daughter's tempestuous rebelliousness, blocked by the daughter's death, sought expression through a remaining daughter, previously the family's "good girl." A family of four that had existed contentedly but precariously on a preconscious arrangement in which each parent virtually owned one of their two children, burst asunder upon the death of one child and its shattering of the bargain. Similarly, when veiled counter-Oedipal involvements of a father with his only daughter were ended by her death, he turned in an unmistakably erotic manner to his youngest son. In these and far more complicated ways did a child's death rebound upon his siblings through realignments of the family dynamics.

Impact of Parental Mourning

In at least one-fourth of the cases, the primary impact of the sibling death seemed less specific in nature and more diffuse in its effects. Here the primary impact consisted of the parents' profound grief reactions and prolonged mourning. Thus, the mothers, a few of whom required brief hospitalization, were generally severely withdrawn, preoccupied and depressed. They unendingly reworked the details of the child's death, blamed themselves as well as others for the death, alternately assured themselves they loved the child and guiltily accused themselves of past harshness with the dead child and all manner of shortcomings as mothers. Findings such as severe anxiety states, insomnia, nightmares, incessant talk about the death and the dead child, auditory hallucinations of the dead child and rage-filled agitation were not uncommon. Amidst these reactions, mothers were often completely incapable of providing any love for, or even attention to, the remaining siblings, and could barely stumble through the simplest household chores. Our lesser data upon the father's reactions indicated that for the most part, in accord with our culture's role prescriptions for men, the fathers were more overtly calm, stoic and effectively functioning. But they also tended to let themselves cry when no one could witness it, were dazed, preoccupied and heartsick, all the while overtly "intact," hearing little of what was said to them, constantly forgetting things, muted to automaton behavior.

Another damaging element may be added when the child's death occurs after prolonged illness and hospitalizations. In such cases the heightened disturbance in the face of sudden death is avoided, and gradual anticipatory mourning may take place, but other strains upon the family members are exaggerated. Major contributions to mental health have been made by those who have vividly aided our understanding of the impact of hospitalization upon children: the separation from parents, the strange new surroundings and people, completely different routines, traumatic medical procedures and so on.[11,30] As a consequence of such influences, some pediatric wards have extended visiting hours, allowed a mother to move right into the hospital with her child, or worked out programs of active parental participation in hospital care of the ill child.[6,24] While this is of immense help to the hospitalized child, and probably vitally necessary to hold down otherwise spiraling guilt in the mothers of dying children, we found as did Cobb[15] that it can also do striking damage to the dying child's siblings. During such times they become essentially motherless, their mother not only being emotionally drained by her ordeal, but often having little actual time left for the other children (the more so if the hospital is at any distance). In her absence, the children are often left with neighbors, or even distributed among relatives for prolonged periods. The remaining children often lose not just their mother's love and concern but her very physical presence. When available at all she is rushed, burdened, sapped and irritable, and, should the children complain, they are likely to draw a hasty, guilt-inducing retort. Illness, too, may come to be seen as the only route to the mother—but far more crucial is her physical or psychological absence and almost inevitably disturbed state. And as the children then turn to their father they find little emotional support, for he is bereft of his wife, has heavily increased responsibilities, and is struggling to suppress his own anguish.

CONCLUSIONS

Reviewing the clinical data, the determinants of children's response to the death of a sibling were found to include: the nature of the death; the age and characteristics of the child who died; the child's degree of actual involvement in his sibling's death; the child's pre-existing relationship to the dead sibling; the immediate impact of the death upon the parents; the parents' handling of the initial reactions of the surviving child; the reactions of the community; the death's impact upon the family structure; the availability to the child and the parents of various "substitutes"; the parents' enduring reactions to the child's death; major concurrent stresses upon the child and his family; and the developmental level of the surviving child at the time of the death, including not only psychosexual development, but ego development with particular emphasis upon cognitive capacity to understand death. The effects upon the child obviously are not static, undergoing constant developmental transformation and evolution.

Clearly, these data require related investigations into the factors differentiating relatively successful integrations of sibling losses versus pathological consequences akin to those depicted here. Such investigations, complementing further clinical study of the psychopathology of "sibling death" patients, will be of particular preventive value. We are, of course, committec to the preventive application of the findings from these sibling death cases, and are much encouraged, even excited, by the preventive opportunities therein. For in all child deaths, professional people (namely, pediatricians and ministers) of potentially immense preventive-therapeutic assistance are almost automatically involved—an impressive contrast with many other equally needy but relatively inaccessible subclinical or "in crisis" groups.

Our optimism, though, remains restrained by this study's demonstration that the problems involved are *not* simply those of a surviving child's unrealistic, inappropriate guilt or fear of death. Rather, as seen in the cases described here, the complex pathological distortions involved in children's disturbed reactions to the death of a sibling include such areas as affect, cognition, belief systems, superego functioning and object relationships. The distortions are not merely intrapsychic, they are inevitably intertwined with and partially products of the dynamics and structure of the family. In fact, they may significantly involve the extended family or the child's peers or even his general neighborhood and community. They include not only profound immediate reactions, the least of which often are physical and psychological symptoms, but tendencies toward enduring symptom formation and distortions of character structure.*

Accordingly, the preventive task is a large one, and preventive efforts in such cases must necessarily be comprehensive. Recently, pediatric workers have wisely insisted that in child deaths the physician's responsibility is not alone to the dying child but to the entire family unit.[12,24] Others have reminded us that "family unit" includes the dead child's *siblings* as well as his parents.[29,36] A full preventive-therapeutic approach to the dead child's siblings, integrated with assistance to the grieving parents, remains to be carefully spelled out. But recognition of the need for such efforts represents a major step toward preventing what we elsewhere called the senseless arithmetic of adding newly warped lives to the one already tragically ended.

REFERENCES

1. ACKERMAN, N. W. 1958. The Psychodynamics of Family Life. Basic Books, Inc. New York, N.Y.
2. ADELSON, L. AND E. R. KINNEY. 1956. Sudden and unexpected death in infancy and childhood. Pediatrics 17(5):663–699.
3. ANTHONY, S. 1940. The Child's Discovery of Death. Kegan Paul. London, Eng.

---

* A small clinical study of adult patients delineating some enduring characterological manifestations of early sibling death reactions has recently been initiated.

4. Aronson, G. J. 1959. Treatment of the dying person. *In* The Meaning of Death, H. Feifel, Ed. McGraw-Hill Book Co. New York, N.Y.
5. Bakwin, H. 1948. Pure maternal overprotection: maternal overaffection. J. Pediatrics. 33(6):788–794.
6. Bierman, H. R. 1956. Parent participation program in pediatrics oncology: a preliminary report. J. Chronic Dis. 3(6):632–639.
7. Blanchard, P. 1957. Psychoanalytic contributions to the problems of reading disabilities. *In* The Psychoanalytic Study of the Child, Vol. 12. International Universities Press. New York, N.Y.
8. Blum, G. S. and S. Rosenzweig. 1944. The incidence of sibling and parental deaths in the anamnesis of female schizophrenics. J. Gen. Psychol. 31 (January, 1):3–13.
9. Bowlby, J. 1961. Childhood mourning and its implications for psychiatry. Amer. J. Psychiat. 118(6):481–498.
10. Bowlby, J. 1961. Processes of mourning. Int. J. Psychoanal. 42(4–5):317–340.
11. Bowlby, J., J. Robertson and D. Rosenbluth. 1952. A two-year-old goes to hospital. *In* The Psychoanalytic Study of the Child, Vol. 7. International Universities Press. New York, N.Y.
12. Bozeman, M. F., C. E. Orbach and A. M. Sutherland. 1955. Psychological impact of cancer and its treatment: III. The adaptation of mothers to the threatened loss of their children through leukemia: Part 1. Cancer 8(1):1–19.
13. Cain, A. C. and B. S. Cain. On replacing a child. J. Amer. Acad. Child Psychiat. In press.
14. Cain, A. C., M. E. Erickson, I. Fast and R. A. Vaughan. Children's disturbed reactions to their mother's miscarriage. Psychosom. Med. In press.
15. Cobb, B. 1956. Psychological impact of long illness and death of a child on the family circle. J. Pediat. 49(6):746–751.
16. Eissler, K. 1955. The Psychiatrist and the Dying Patient. International Universities Press. New York, N.Y.
17. Feifel, H., Ed. 1959. The Meaning of Death. McGraw-Hill Book Co. New York, N.Y.
18. Fleming, J., S. Altschul, V. Zielinski and M. Forman. The influence of parent loss on personality development and ego structure. Unpublished paper.
19. Hilgard, J. R. 1960. Strength of adult ego following childhood bereavement. Amer. J. Orthopsychiat. 30(4):788–798.
20. Koch, H. L. 1960. The relation of certain formal attributes of siblings to attitudes held toward each other and toward their parents. Monogr. 78. Soc. Res. Child Devel. 25(4).
21. Kolansky, H. 1960. Treatment of a three-year-old girl's severe infantile neurosis: stammering and insect phobia. *In* The Psychoanalytic Study of the Child, Vol. 15. International Universities Press. New York, N.Y.
22. Levy, D. M. 1943. Maternal Overprotection. Columbia University Press. New York, N.Y.
23. Nagy, M. I. 1948. The child's theory concerning death. J. Genet. Psychol. 73:3–26.

24. NATTERSON, J. M. AND A. J. KNUDSON, JR. 1960. Observations concerning fear of death in fatally ill children and their mothers. Psychosom. Med. 22(6):456–465.

25. ORBACH, C. E. 1959. The multiple meanings of the loss of a child. Amer. J. Psychother. 13(4):906–915.

26. PEARSON, G. H. J. 1949. Emotional Disorders of Children: A Case Book of Child Psychiatry. W. W. Norton & Co. New York, N.Y.

27. POLLOCK, G. H. 1961. Mourning and adaptation. Int. J. Psychoanal. 42(4–5):341–361.

28. POLLOCK, G. H. 1962. Childhood parent and sibling loss in adult patients. Arch. Gen. Psychiat. 7(4):295–306.

29. RICHMOND, J. B. AND H. A. WAISMAN. 1955. Psychologic aspects of management of children with malignant diseases. Amer. J. Dis. Child. 89(1):42–47.

30. ROBERTSON, J. 1958. Young Children in Hospitals. Basic Books, Inc. New York, N.Y.

31. ROCHLIN, G. 1953. Loss and restitution. In The Psychoanalytic Study of the Child, Vol. 8. International Universities Press. New York, N.Y.

32. ROSENZWEIG, S. 1943. Sibling death as a psychological experience with special references to schizophrenia. Psychoanalytic Review. 30(2):177–186.

33. ROSENZWEIG, S. AND D. BRAY. 1943. Sibling deaths in anamnesis of schizophrenic patients. A.M.A. Arch. Neurol. Psychiat. 49(1):71–92.

34. SCHARL, A. E. 1961. Regression and restitution in object-loss: clinical observations. In The Psychoanalytic Study of the Child, Vol. 16. International Universities Press. New York, N.Y.

35. SHAMBAUGH, B. 1961. A study of loss reactions in a seven-year-old. In The Psychoanalytic Study of the Child, Vol. 16. International Universities Press. New York, N.Y.

36. SOLNIT, A. J. AND M. GREEN. 1959. Psychological considerations in the management of deaths on pediatric hospital services: I. the doctor and the child's family. Pediatrics. 24(1):106–112.

37. WAHL, C. W. 1959. The fear of death. In The Meaning of Death. H. Feifel. Ed. McGraw-Hill Book Co. New York.

38. WATSON, A. S. 1963. The conjoint psychotherapy of marriage partners. Amer. J. Orthopsychiat. 33(5):912–922.

39. WEISMAN, A. D. AND T. P. HACKETT. 1961. Predilection to death. Psychosom. Med. 23(3):232–256.

40. WOLF, A. W. M. 1958. Helping Your Child to Understand Death. Child Study Association. New York, N.Y.

Article 8 How Surviving Parents Handled Their Young Children's Adaptation to the Crisis of Loss

DIANE BECKER

FAITH MARGOLIN

The influence of unmourned loss of a parent during early childhood on sub-sequent personality development has been recognized since Helene Deutsch's classic paper, "Absence of Grief." [1] For the past five years the James Jackson Putnam Children's Center has been studying the reactions of young children to the death of a parent. This paper presents a related area of interest, observations of how the surviving parents attempted to help their young children adapt to such a loss.

We will discuss the following areas: (1) how and when the parents informed their children of the death; (2) how parents dealt with their children's questions and theories about the death; and (3) how parents helped their children experience the feelings the loss produced.

The research sample consisted of nine children under seven years of age from seven families. All had experienced the death of a parent within six months prior to the first contact with the Center. A criterion for sample selection was that the children, on the basis of diagnostic evaluation, did not appear to be in need of psychotherapy. Of the seven deceased parents, one had been a mother, six had been fathers. Four of the fathers had had professions and three had been skilled workers. The three major religions were represented.

The seven parents were seen in weekly casework interviews over a one- to four-year period, while the nine study children were seen once weekly by psychiatrists. Research goals of work with the parents were to study the interaction between parent and child around the crisis of the loss and to gather information about the child's behavior and adjustment. Our therapeutic goals were to sup-

DIANE BECKER, M.S.S., is with the Wellesley Public School System, Wellesley, Mass.

Source: *American Journal of Orthopsychiatry*, Vol. 37 (1967), 753–57. Copyright © 1967 the American Orthopsychiatric Association, Inc. Reproduced by permission.

port the parents' working through of their grief and to guide them in dealing with their children's reactions to the loss. Recognizing that each person has his own style and pace of mourning, we intervened only when we observed a parent having unusual difficulty in his mourning or when the parent's behavior was hampering the child's adaptation to the loss. Because denial is recognized as a valuable and essential defense in the early stages of mourning, and since young children whose egos are too immature to bear the full impact of the death of a parent use denial to a greater exent than adults, we interfered with this defense only when parents were concealing facts from their children.

How to tell the children was a problem faced by all parents immediately following the death. Six of the seven parents told their children that the parent had "gone to" or was "taken to" heaven. Three of these six did not believe in heaven themselves. Six parents also waited before telling their younger children of the death. The older children in these six families were told within 24 hours; the younger, after a lag of a few days to a few months.

> The C. family was fairly typical. They were referred to the Children's Center by their minister six weeks following the death of father, age 29, from a heart attack. The four children ranged in age from 1 to 8 years. Mother told her two daughters, ages 8 and 6, the morning after the death that their father had been taken to heaven by God. She reminded them of their grandparents who had recently died and said that their father was now with them. In relating the news of the death, Mrs. C. struggled to maintain her composure, not wanting to cry in front of the children. Jane, the older girl, cried and couldn't eat her breakfast. The younger daughter, Karen, said she was going to forget and would not cry. Mother did not know whether to send them to school that day. She asked them if they wanted to go to a friend's house. They were quite eager to go and both cheered up, ate their breakfast, and left. The third child, a two-year-old boy, was told a day later that daddy was with God.

Generally it was difficult for the surviving parents to deal with the children's initial denial, as illustrated by the following cases:

> When Mrs. L., after a two-month delay finally told her daughter, Donna, three and a half years old, that daddy had not recovered from his illness, had died and gone to heaven and was not coming back, the child did not appear to react and continued speaking of daddy as if he were alive. Mother did not know what to say to her when Donna made comments about daddy fixing something or suggested saving food for him. She often avoided answering at all.
>
> During Mrs. D.'s five month, terminal illness her three-year-old daughter Judy would comfort her with words such as, "You will feel better, Mommy, and you will get well." Mr. D. waited a week before telling Judy of her mother's death as he felt he had to gain composure before he could talk to

her. He told her, "God took Mommy to heaven." and she answered, "Oh, just like my turtle," and continued with her play. Father assumed that she understood. However, for more than a month Judy continued talking about mother as if she were alive. She would say, "Let's buy something for Mommy," or she would make something for her. Mr. D. would repeat that Mommy wouldn't be back. He said it was very painful for him that she did not seem able to face the fact.

None of the nine children studied attended the funeral, and most were sent to the homes of friends and neighbors during the period immediately following the death and remained there until after the funeral. Only one mother expressed regret that her child had not been present at his father's funeral.

Children, adapting to loss, develop their own explanations and theories about death. The fantasy of the deceased parent being alive in heaven was common among our study children, and was sometimes encouraged by the surviving parent, as in the cases of John and Betsy.

John B., age 6, frequently talked about his daddy being in heaven. His mother assumed that he heard this from other children because she had never told him. However, she did not challenge the fantasy, believing it made John feel better.

Betsy L., six and a half years, asked her mother if she could send a Christmas present to her father. Mother answered, "No," and discouraged further discussion. Betsy repeatedly asked: "What do they do in heaven, what do they wear there, what do they eat?" Mrs. L., perplexed by these questions, replied that she didn't know but would find out, thus supporting Betsy's fantasy that her father was alive in heaven.

The parents' promotion of the children's fantasies or their inability to correct them sometimes reflected their own difficulty in gradually withdrawing libido from the deceased:

Despite the fact that Mr. D. did not believe in heaven and immortality himself, he told his daughter Judy that "Mother is an angel and angels watch over you," since he thought this fantasy would be comforting to her. Mr. D. himself had an ambivalent relationship with his wife. Not until three years after her death could he admit to his social worker that he had had negative feelings toward his wife. Only then could he begin to decrease visits to the cemetery from once every three weeks to occasionally.

Three parents delayed for as long as one to two years telling their children that the deceased parent's body was buried in the ground, justifying this delay by their belief that knowledge of burial would be too frightening or upsetting to them and should be avoided. In the D. family the children would visit the cemetery with their father, place flowers on the grave, and be witness to relatives crying without anyone mentioning their mother's death or burial. During these visits the children never asked why they were there. This suggests to us that the father's silence communicated to them a feeling of the taboo nature of the subject.

Some parents were able gradually to discuss the fact of burial with their children and take them on visits to the cemetery. The visits stimulated the children to ask for information that furthered their comprehension of death:

> Six months after his father's death, six-year-old John began asking his mother to take him to the cemetery where his father was buried. In discussing this at the Center, Mrs. B. told her social worker that she, herself, did not find comfort at the cemetery, but went only because it was her duty to tend to the flowers. She did not know whether she should take John and said she did not want to upset him too much. He had never been there and he had not gone to the funeral because she had wanted to protect him. Seven months later, John and his mother made their first visit to the cemetery. John asked many questions, "What happens to the body in the ground?" "Where is the head, where are the feet?" Once home he began to talk much more about death and about his father, and said that he wanted to be buried near him. He would talk about how daddy used to do things and said he wished daddy hadn't died. Mother found these discussions extremely painful but John persisted in talking and questioning and she was able to allow him to do so.

Although individual differences were observed in the parents' reactions to the loss and in their style of mourning, all seven parents showed a marked tendency to protect their young children from feeling the pain of loss. Although these parents had volunteered to share with us their own feelings and observations of their children's reactions to the death, they were reluctant and seemed unable to share feelings with the children. Eventually most parents were able to talk factually with their children about the death, share memories of the deceased with them and include them in memorial observances, but few could encourage them to express feelings about the loss and none could openly share his own feelings.

In conversation among family members, affect was often isolated or the subject of death and the deceased was avoided when the family obviously was feeling sad. Occasionally, references were ignored by parents. It is our impression that often such insulation of the young child is a displacement for components of the parents' own conflicts around painful issues of mourning. One mother described her need to avoid talking about her feelings for fear she would cry and not be able to stop; she believed this would be too upsetting for the children. She further saw, as did other parents, that she could not bear to face the intensity of her children's feelings and thus avoided talking with them of their father's death except in a matter-of-fact manner. This mother, however, did cry a great deal during her interviews and described crying for long hours at night after the children were asleep. Although she gradually became more comfortable in speaking of her own feelings, she could barely allow the children to express theirs.

We found that formal religious or cultural observances of the anniversary of the death helped the surviving parents and their children share feelings about the death which otherwise was difficult for them to do. This was particularly true of the L. family in which there was characteristically little talk of the father except in answer to direct questions from the children. Mrs. L. cried only when alone. The family could not share even mild painful feelings in reaction to ordinary daily stress, the lack of mutual support being quite striking:

> During the months following her husband's death, Mrs. L. had great difficulty dealing with six-and-a-half-year-old Betsy's sad and angry feelings. She responded to Betsy's increased whining with anger and often hit her. Several months after her father's death, Betsy asked for his picture. Mother gave Betsy the picture but avoided talking with her about her husband because she thought it would make Betsy sad.
>
> Mrs. L. procrastinated for several months in setting a date for the unveiling ceremony commemorating the first anniversary of her husband's death. The week of the actual anniversary she traveled alone across country to vacation with a distant and unfamiliar relative. Betsy and her sister, Donna, four and a half years old, remained at home with a baby-sitter. No mention had been made within the family of the anniversary of father's death. After returning from her trip, Mrs. L. struggled several months to plan the belated unveiling ceremony. One of her concerns was whether the children should attend as she feared the effect on them of seeing family members cry and wail. With our encouragement Mrs. L. did include the children in the ceremony. Although the children had rarely talked about their father for the several months prior to the unveiling, they were able to do so after the visit to the cemetery.
>
> Mrs. L. believed the experience of the unveiling was so valuable that she planned the commemoration of the second anniversary so that her daughters could be present. The memorial service consisted of the family lighting a candle and saying prayers in the home. Mrs. L. and Betsy cried during the ceremony. The younger daughter, Donna, asked why they were crying and Mrs. L. answered that they were crying about missing daddy, adding that it was natural to feel sad and cry.
>
> Since then Mrs. L. and the children have been able to speak even more freely about father and the meaning of death. Donna, now five and a half, asked, "Did Daddy used to pick me up?" and "Would a new Daddy build us a garage?" Betsy, eight and a half, has asked since the second anniversary, "Will the body always be in the casket?" and, "What goes to heaven, then?"

On the basis of this and other observances we concluded that when the children are given adequate preparation, the shared experiences of memorial observances and visits to the cemetery provide a unique and valuable aid to the parent faced with helping the young child adjust to loss. Such events work against avoidance, denial and repression of painful feelings and facilitate the process of adaptation.

CONCLUSION

The parents in our study showed a marked tendency to insulate their young children from the painful aspects of loss; they tended to promote avoidance and denial of the finality of death, and of feelings in relation to it. Their efforts to insulate their young children paralleled, in a striking way, community efforts to protect surviving parents from confrontation with painful feelings in reaction to death as described in an earlier paper by Gates, Roff and Stiver.[2] It is our impression that there is in our culture a taboo concerning death which is manifest in a general disability to cope with grief and loss. Although more permission is now given by adults to young children to inquire about their world, their relationships, and their bodily functions, this permission is rarely extended to the subject of death. Our experience has shown that children are very much aware of and do react to death and that a natural curiosity persists in the child who has lost a parent even when questions and references go unheeded.

We suggest that community caregivers such as pediatricians, teachers, clergymen, and social agencies be oriented to bring to the public knowledge that:

1. The young child *is* aware of death and reacts to it.
2. It is helpful to permit the young child to inquire about death, to share memories, and to observe and share feelings with the adults in reaction to the death of a significant person.
3. When adequate preparation is given young children, the shared experiences of memorial observances and visits to the cemetery further comprehension of death and facilitate the normal process of mourning.

REFERENCES

1. DEUTSCH, HELENE. 1937. Absence of grief. Psychoanalytic Quarterly 6:12–22.
2. GATES, P. H., C. ROFF, I. STIVER. Studies of the significance of death of a parent for young children. A preliminary communication. Paper read at 1965 annual meeting of the American Orthopsychiatric Association, New York, New York.

Developmental Life Transitions: The High School and College Years

IV Going to college is one of several experiences that are major steps in the transition from adolescence to adulthood. College entrance presents a number of specific challenges representative of the general problems entailed in growing up, such as increasing autonomy and responsibility. Such major challenges are not easily met, and many students find the transition difficult. For those with insufficient emotional preparation, unrealistic expectations, or particular cultural handicaps, it can represent a crisis in their lives. Despite this, most students are able to marshal their own coping resources, seeking outside help when necessary, in order to get through the difficult times and to learn from them.

Students entering college face significant testing of their skills in a number of areas—emotional, social, and intellectual. Separation from their families, seen as a major threat to personal security and development at a young age (see Part II), now becomes a welcome opportunity to develop independence and self-esteem. In a group discussion seminar with first-year nursing students Levin[3] found many who were upset by their separation from their families. These feelings are often the beginning of a vicious circle: the depression about being away from home causes difficulty in studying and poor performance; this in turn causes more depression and fear of failure and expulsion, and so on. The student is not the only one affected by the separation; for some families the readjustment is very difficult (for example, the home-centered mother facing an "empty nest") and puts added strains on the departing teenager.[2]

Developing a sense of autonomy, making and being responsible for one's own decisions, is a task closely related to that of leaving the family circle. Making new friendships, often with people from different backgrounds, becomes important as family and old friends are left behind. Questions of adult sexuality must be resolved in the framework of increasing independence and new peer relationships. To all these are added the significant intellectual challenges that college students encounter.

In the first article Earle Silber and his colleagues studied the ways in which competent high school seniors dealt with their anticipation of entering college. They found that those students considered by their teachers and fellow students to be academically, interpersonally, and socially competent had certain personality attributes in common. They reached out for new experiences. "It's going to be a change from my normal routine . . . I'm looking forward to it" was a typical comment. They took an active approach to new situations (for example, several wrote to their prospective roommates during the summer to get acquainted before school began). In general, they took pleasure in mastering difficulties. One student commented, "There's an enjoyment of solving difficult problems and learning new things. I feel it's also a challenge."

When the situation involves the anticipation of future difficult challenges, as with these students, building up a confident self-image becomes the immediate task. Silber describes several specific techniques used to accomplish this. Some students reminded themselves of analogous experiences in the past that they had managed well (for example, one girl who had moved to a new neighborhood and had to make new friends) and concluded that they would be able to do so again in the new circumstances. Putting college entrance and its attendant difficulties in a framework of the continual process of maturation, rather than as a complete break away from the known past, offered some reassurance. Some rehearsed their expected new roles ahead of time, like the boys who bought their own clothes knowing they would have to do this in college or those who took their first summer jobs in the adult world. Identifying themselves with a group having a record for doing well (such as graduates from their own high school), lowering their levels of aspiration (recognizing they might not be at the top of their class at college, for instance), and selectively perceiving encouraging aspects in the college experience (for example, assuming that colleges were probably adept at choosing compatible roommates) were other popular methods used by students to develop a sense of their competence to deal with the anticipated college experience.

Success in building up a self-image of competence alleviated some, but not all, of the anxiety that students felt about entering college; therefore, techniques of managing unavoidable distress were also examined. By telling themselves that they were not alone in their concerns ("everybody's in the same boat"), asserting that a little bit of worrying was probably beneficial, anticipating and acting on possible future problems, and using fantasy and imagination to rehearse solutions for anticipated difficulties, many students were able to keep their natural anxiety from incapacitating them.

In the second article George Coelho, David Hamburg, and Elizabeth Murphey followed these same students through their first year at college. Freshmen face many new social and academic tasks, such as heavier course loads and more responsibility for self-regulation of schedules, with the inevitable small failures and disappointments that such challenges entail. Those students who were able to keep their anxiety within reasonable limits had evolved techniques of maintaining their self-esteem (for example, by setting different levels of aspirations in different areas of commitment) and of handling the immediate disappointment of academic setbacks (by looking at long-range possibilities, by finding gratification elsewhere temporarily, or by reevaluating their expectations of themselves).

Coelho, Hamburg, and Murphey point out that another important aspect of coping, besides self-manipulation of feelings and attitudes, is the effective use of environmental resources, especially of human resources. Peer friendships are very important to students because of the emotional support and the opportunities for learning that they offer. Through informal group discussions, the pooling of information and experiences, and friends willing to serve as sounding boards or give support in times of crisis, students gain new intellectual and emotional insights and further the cumulative learning process that coping involves. Mechanic[4] found that graduate students undergoing the stress of doctoral examinations also used many of these coping patterns.

In the third article Elvin Mackey, Jr. discusses the special problems that black students on predominantly white campuses face in addition to the more common difficulties of this transition period discussed above. One student, who was from an all-black high school, felt he had been thrown into a "sea of white people." He began experiencing "white-outs" and periods of "altered consciousness" that he could not remember afterward; finally he had to leave school. Gibbs[1] identified three other modes of adaptation in addition to the withdrawal mode given in Mackey's example: separation (hostile movement against the dominant culture), assimilation (conformist movement toward the dominant culture), and affirmation (self-accepting movement with the dominant culture). Mackey found that black students were much less likely than white students to utilize health and counseling services offered by the college. He suggests that when a college makes available health professionals and counselors with whom these students can feel comfortable, they are able to overcome their reluctance and recognize and utilize the human resources presented.

The college experience presents older adolescents with a variety of challenges. Students must learn new emotional and social skills along with their academic training. The transition from the protective family circle, where generous allowances are made for childhood weaknesses, to the adult world where independence and accountability are prominent criteria of worth is a hazardous one. But, as Coelho, Hamburg, and Murphey suggest, successful coping with this "major developmental transition may build self-esteem that will support renewed efforts in meeting the difficult tasks of maturer years."

REFERENCES

1. Gibbs, J. T. Patterns of adaptation among black students at a predominantly white university: Selected case studies. *American Journal of Orthopsychiatry,* 1974, *44,* 728–40.

2. Langsley, D. G.; Fairbairn, R. H.; and DeYoung, C. D. Adolescence and family crisis. *Canadian Psychiatric Association Journal,* 1968, *13,* 125–33.

3. Levin, S. Some group observations on reactions to separation from home in first-year college students. *Journal of American Academy of Child Psychiatry,* 1967, *6,* 644–54.

4. Mechanic, D. *Students under stress: A study in the social psychology of adaptation.* New York: Free Press of Glencoe, 1962.

Article 9 Adaptive Behavior in Competent Adolescents

Coping with the Anticipation of College

EARLE SILBER

DAVID A. HAMBURG

GEORGE V. COELHO

ELIZABETH B. MURPHEY

MORRIS ROSENBERG

LEONARD I. PEARLIN

INTRODUCTION

The late adolescent–early adult period involves certain developmental tasks that derive from the emergence of biological sex maturity and the social requirements of adult life in a particular culture. These tasks may be met in many ways; different patterns of coping behavior may get the job done. The accomplishment of each task, and the way in which it is accomplished, may have consequences for the entire adult life of the individual and those close to him.

What are the tasks? A list of them may be long or short, depending on the level of specificity utilized. In contemporary American society—and indeed in many industrialized, urban, technologically complex cultures—certain tasks achieve prominence because they are requirements of the transition from being a child to being an adult. Adult roles, responsibilities, and opportunities are sufficiently different from those of childhood in such cultures that a major transition is involved here. During these late adolescent–early adult years, circumstances require a rather far-reaching change in one's way of life.

In the early psychoanalytic literature adolescence has been described mainly from the point of view of the resurgence of the infantile instinctual drives at this period of life.[5] Along with the developing interest in ego-psychology, more and more attention has been focused on the ego's efforts at mastery, in relationship to both instinctual drives and external stimuli; thus Anna Freud in *The Ego and the Mechanisms of Defense*[2] and in her more recent paper, "Adoles-

EARLE SILBER, M.D. is Supervising and Training Analyst, Washington Psychiatric Institute, and Clinical Professor of Psychiatry, Georgetown University School of Medicine, Washington, D.C.

Source: *Archives of General Psychiatry*, Vol. 5 (October 1961), 354–65. Copyright 1961, American Medical Association.

cence" [3] calls attention to some crucial defense mechanisms operating during this developmental period. Erikson's psychosocial theory of ego development [1] conceives of phase-specific developmental tasks which must be mastered in each segment of the life cycle, emphasizing the mutual transactions between the individual and his society. Erikson views adolescence as a sanctioned intermediary period between childhood and adulthood, during which the lasting pattern of inner identity is achieved.

The usual source material for describing adolescence comes from studies of clinically disturbed people, either from reconstructions through adult analysis or from analyses conducted during adolescence.[3] However, we wished to study a group of effective adolescents and to use a naturally occurring life situation as an extended laboratory through which the adolescent's adaptive behavior could also be studied. We were interested in exploring a life situation which offered an opportunity for examining adolescent behavior patterns in response to the potential challenges of new specific life tasks.

We therefore selected for study one of the commonly occurring experiences in our culture which may represent a challenge for personal growth in late adolescence: the transition from high school to the freshman year of college. If the shift from high school to college is taken as a revealing sample of the larger transition from adolescence into early adulthood, certain tasks become apparent. Prominent among these are the following: (1) separation from parents, siblings, and close friends; (2) greater autonomy in regard to making important decisions, assuming responsibility for oneself, and regulating one's own behavior; (3) establishing new friendships; (4) pressures (internal and external) toward greater intimacy and adult sexuality; (5) dealing with new intellectual challenges. Our aim was to learn how competent adolescents cope with the tasks presented by this particular life situation.

What do we mean when we say that an individual is coping with these tasks? We must decide what we are willing to call *coping behavior*. How will we recognize it when we see it? From a personality–social viewpoint, coping behavior has 2 broad components: (1) the effectiveness with which each task is accomplished; and (2) the cost to the individual of this effectiveness. Behavior may be considered to serve coping functions when it increases the likelihood (from a specified vantage point with respect to a specified time unit) that a task will be accomplished according to standards that are tolerable both to the individual and to the group in which he lives.

The time unit in this study was the 16-month period from the spring of the senior year of high school through the end of the first year of college. The project was naturalistic and exploratory in style. Relying on individual interviews with a small group of competent students who were in the process of changing from high school to college, we aimed at describing and understanding their adaptive behavior. This paper will focus on coping mechanisms these students utilized while still in high school in anticipation of the transitional experience.

CRITERIA USED IN THE SELECTION PROCESS

We recognized that there are a variety of patterns of competence and that any particular definition need not be regarded as all-inclusive. We chose to define our sample of students around certain criteria of competence which were used as the main bases for selecting them for study. We were interested in students who indicated a capacity to integrate competence in three major life areas: (1) in their academic work at school, (2) in their ability to maintain interpersonal closeness with a peer, and (3) in their ability to participate in social groups. This is not to say that this is the only way to measure or define competence, but for purposes of our study we wished to be explicit about the bases on which we did define it.

Certain steps in the selection process will clarify how these students were chosen and also define limiting factors which determine some unique characteristics of this particular sample of students. We wish to make it clear that we do not regard our sample as being in any way "representative" of the high school senior population.

These students also came from a rather similar socioeconomic background. The sample was drawn from the senior class of a high school which is located in a suburb of Washington, D.C. This suburb is a middle-class residential area, populated by a large percentage of professional people, managerial and supervisory governmental workers and a smaller percentage of less skilled workers. The per capita income level is high in comparison with the country at large.

After a talk to the senior class of the high school about the project, students could volunteer by returning a follow-up letter which was signed by themselves and their parents. From those who volunteered students were picked from the top half of the class who indicated that they were apt to be going to a college in the Eastern part of the country. This factor was introduced since we planned to visit our students at their respective colleges in a later part of the project. In the initial selection we preferred students who revealed a concentrated area of interests as reflected in their extracurricular activities. In addition, we used personality ratings of the students which were made by their teachers. For each student ratings were made by 8 teachers in regard to the following traits: motivation, industry, initiative, influence and leadership, concern for others, responsibility and emotional stability. We picked those students who had received the most favorable ratings by their teachers.

Finally, a number of students were interviewed and those who appeared overtly troubled, reported disturbing symptoms, or with whom the interviewer had difficulty in establishing comfortable rapport were excluded. Fifteen students (6 boys and 9 girls) were finally selected for a series of interviews on a weekly basis during the latter part of their senior year of high school. The selection was made chiefly on the basis of over-all impressions of the interviewer in relationship to the specified criteria of competence. This judgment was based upon the interviews plus reference to academic grades, school testing, and ratings. The findings to be presented represent a distillation of pertinent material from 6 or 7 interviews (spaced from April to September prior to the student's departure for college) with each of the 15 students selected for study.

DESCRIPTION OF THE STUDENTS IN RELATIONSHIP TO THE CRITERIA FOR COMPETENCE

These students showed satisfaction in their scholastic achievement and did not reveal any overt anxiety about their intellectual ability. Their class standing in the high school, based upon their grade records, seemed to us to reflect their intellectual endowment as measured by mental maturity tests administered by the high school. Not completely relying on these test scores, we included in the selection process an estimate by the interviewer of the extent to which he felt the student was making use of his intellectual potential.

These students showed a varied pattern of commitment to academic interests, ranging from those who stressed intellectual activities as a central value for them to those who included it as one important attribute in being "well-rounded." For the majority, effective work habits, coupled with their intellectual ability, accounted for their satisfactory academic standing.

In the area of interpersonal relationships we were interested in finding students who were competent in making and maintaining friendships with a peer, either of the same sex or the opposite sex. Our subjects were characterized by an absence of manifest anxiety in this area and by satisfaction with their close friends. We were attentive to such characteristics as the relationship having a feeling of mutual importance to both partners, a regard for the other person's welfare, the relationship not excluding other friendships, not being exploitative, and with a sense of shared interests and durability. We also looked for some experience in dating and relating to members of the opposite sex where a student indicated that his closest friend was a person of the same sex. The pattern of friendship included students with predominantly same-sex friendships, others more involved in dating, and another group who regarded their steady date as their best friend.

In their social relationships we looked for a satisfying membership in some social group, as revealed by actively participating in and feeling a part of such a group. We included in this either a companionship group or some participation in a service group at school or a group organized around a particular area of interest. These students were not necessarily the most "popular" of the senior class but they had found groups in which they could participate actively and feel accepted themselves. For a number of our students religious youth groups were important, both for their religious and social functions, with most putting greater emphasis on the latter. With about half of our students the friendship group included a circle of from 2 to 4 close friends. The remainder described a larger group of friends as constituting their predominant social group. Most of the girls in our sample belonged to more formalized social groups in the high school.

These students were very eager to cooperate in the study. In the interviews they were earnest and thoughtful in discussing their experiences. Their style of communication was characterized by a quality of frankness and minimal defensiveness. They expressed both positive and negative feelings with lively affect.

ANTICIPATORY COPING BEHAVIOR

In the latter part of the senior year of high school, the students having worked through the process of selecting and being selected by a college, were on the threshold of a new experience. In this situation, some adolescents may respond with increasing anxiety and disorganization. We had an opportunity to observe students who did not respond in this way and who were not overwhelmed by the anticipation of change. In the same way that one might seek to understand disturbance in functioning, we sought to understand the mechanisms operating in these students which maintained their psychological integrity and facilitated effective functioning in the face of challenge.

We will attempt to delineate the effective coping processes that were operating in the students. We recognize that the processes we will describe are interrelated and that the mere separating of them does an injustice to the concept of the total function of the personality. However, for purposes of description, we will consider them from different points of view, all of which are interrelated. First, we will describe general personality attributes which tend to facilitate involvement in and mastery of the new situation. Second, we will view ego operations which served to develop and maintain a self-image as a person adequate to the perceived requirements of a new situation. Third, we will consider those mechanisms which maintained distressful affective states within manageable limits. We will describe as many of the patterns as we were able to observe. This presentation, then, represents a composite of processes operating in our subjects; not all of these mechanisms were observed in all of the students.

I. GENERAL PERSONALITY ATTRIBUTES

The students' coping behavior will be described first in terms of general personality attributes which facilitated an active involvement in the new situation. Their behavior could be understood in terms of a motivation for actively seeking and dealing with manageable levels of stimulation and challenge in the environment.

1. Reaching Out for New Experience

We found the majority of our students expressed a very positive attitude toward newness. New experiences were not predominantly viewed as anxiety-laden and therefore to be avoided, but rather as desirable, exciting, and rewarding, and something to be welcomed. The tendency was not in the direction of maximal reduction of tension, but more positively in the direction of searching for stimulation and newness. The following are examples of such attitudes, from the interviews:

> I look forward to the people I'm going to meet, to the benefit I can get from talking to them and knowing new types of people.

> I'm looking forward to some sort of experience that's exciting.

> There is so much I haven't come in contact with which I want to.

> I'll be on my own. It's going to be a change from my normal routine. I enjoy new things. I'm not afraid of them. I'm looking forward to it.

The students saw in the new situation new opportunities and challenges which were viewed as opportunities for growth and change. For example, making new friends, being on one's own, separation from one's family and the new tasks involved in a more independent way of life were anticipated as desirable opportunities. We heard this expressed in these terms:

> It will be fun rooming with someone I don't know.

> I welcome the opportunity to take responsibility, and I will get plenty of chance.

Along with this desire for newness and change, students also felt that they did not wish to block themselves by remaining involved only in old relationships, as they went off to school. Thus, many of the students specifically said that they did not plan to room with high school friends or acquaintances. While they did not wish to give up these friendships, neither did they express a desire to cling to them as a way of avoiding new contacts with others. One girl, commenting on the fact that one of her closest friends is going to the same college, said:

> This will help, but I do not want to be with her all the time. I've known her for a long time and I'd rather try to make new friends. I intend to keep my old ones, but if you have someone like that you depend on too much, you don't go out of your way to meet others. I don't want to have one to depend on all the time like that.

2. Tendency for Activity
Another impressive characteristic of these students was their tendency to be active in facing the tasks of the transition. This reflected itself in the purposeful, highly autonomous way in which they assumed responsibilities for making preparations for going off to college. They were quite self-reliant in handling correspondence with the college, preparing their clothes, working out financial arrangements, etc. Two examples of the kinds of requirements that the students fulfilled

during this summer were as follows: One of the students shopped for her clothes, did her own packing, wrote to her future roommate, brought her correspondence up to date, planned her courses after studying the catalog, and made arrangements with her parents to open a bank account in the community where she was going to college.

Another student shopped for his clothes, after his mother had helped him make a list of what he might need, corresponded with his future roommate by means of which a collaborative relationship was already begun. He attended a dinner for freshmen going to his college from this area, arranged for medical and eye examinations, and made financial arrangements with his parents. He also corresponded with one of his future advisers at the college to work out plans for earning extra money by working at the school. He reviewed the courses offered in the catalog so that he had some idea about his preferences for the first year's program.

3. Pleasure in Mastery

Related to this optimistic attitude about new experiences and a hunger for them, in many cases there were indications of active enjoyment of problem-solving and pleasure derived from the process of figuring things out and in mastering them. When the students talked about their work at school there were frequent references to the sense of pleasure which came from tackling problems which were reasonably difficult, somewhat beyond the students' current level of attainment, but which could be solved. It seemed to us that our students enjoyed tasks that were challenging provided they were not overwhelmingly so. One student commented:

> There's an enjoyment of solving difficult problems and learning new things. I feel it's also a challenge.

These attributes of reaching out for new experience, a tendency toward activity, and a pleasure in mastery have been commented on by White in his developing the concept of *effectance motivation*. White regards effectance motivation as involving ". . . satisfaction—a feeling of efficacy—in transactions in which behavior has an exploratory, varying, experimental character and produces changes in the stimulus field." He views this kind of motivation as being ". . . moderate but persistent, and in this, too, we can discern a feature that is favorable for adaptation. Strong motivation reinforces learning in a narrow sphere, whereas moderate motivation is more conducive to an exploratory and experimental attitude which leads to competent interactions in general, without reference to an immediate pressing need." [7] We found this concept useful in understanding a quality in the personalities of our subjects. We viewed effectance motivation as providing a general press toward active involvement in dealing with the new situation. We will now consider other more specific mechanisms operating in our subjects.

II. DEVELOPING A SELF-IMAGE AS ADEQUATE TO THE NEW SITUATION

These students viewed themselves in a positive way. However, having an image of oneself as an adequate person is not a psychological state that exists outside of the pressing challenges of life experience. We view it as a dynamic process related to the requirements of external reality. During the period of anticipation of college we observed certain processes that served to develop and maintain a self-image as a person adequate to master the perceived requirements of the new situation. This represents a process of matching what one perceives as being required in the new situation with an image of oneself as being competent to deal with the new situation effectively. We observed a variety of behavior patterns that facilitated the development of such a self-image. These patterns will be described next.

1. By Referring to Analogous Past Experiences

In maintaining their sense of competency as to their readiness for the next step, our students often referred to relevant past experiences of their own which had been adequately mastered. In effect, the students seemed to identify elements in the new situation with some experience in their own past which they felt was analogous. By reference to their ability to handle this analogous experience in the past they could reassure themselves about their ability to handle these situations in the future.

These students, in talking about the change to college, often made reference to their experience in shifting from junior high school to high school. They recalled their ability to cope with this previous transition and made comparisons with the present situation as having features in common.

One girl, who has always had her own bedroom, reminded herself that when a girl friend came and stayed with her for a month, it worked out well and therefore she felt confident that she would be able to get along with her roommate at college. Another student reminded herself that she had had experiences with moving that had helped her to learn how to make friends and taught her to deal with strangers. These references presented for the student evidence of his ability to deal with similar situations in the past and thus provided a basis for confidence in dealing with them again in the future.

2. By Referring to Continuity with the Present Self-Image

These students had had a series of experiences which they regarded as relevant in preparing them for the situation at hand. They also had an image of themselves as young people ready for a new situation and able to meet the challenges ahead. They expressed in various ways the feeling that college represented a gradual part of a continuous process of growing and maturation. They tended to see a narrowing gap between their self-image and their image of college students. Long in advance of the actual departure for college a pattern of *anticipatory detachment* from parents supported the students' self-images as people ready for change. These students had a basis for confidence in their ability to live more in a world of their peers at college, since they already had had a wide experience in making friends and in establishing and maintaining meaningful relationships outside of their families while they were still in high school.

Our students expressed a feeling of readiness to grow away from their families and described a feeling of gradualness in this regard. They already had had previous experiences in being separated from their families and saw themselves in high school as young people moving in the direction of being more autonomous in relation to their parents. They felt ready to enter the next phase of their life experience. Thus change feels continuous with their present self-image. One student expressed this succinctly by saying:

> Going away is just another step on the ladder; it is not like jumping off a ladder.

For some students this anticipatory detachment extended to some of their existing ties with peers at high school. Thus, these students felt ready both for further separation from parents and for a shift in their peer relationships as well. We recognized that this preparatory behavior may have also served the function of warding off feelings of depression associated with loss.

3. By Learning About the New Situation in Advance

Another part of matching the self-image with the requirements of the new situation is to find out more in advance about what will be required in the new situation and what it will be like. Confidence tends to grow as the ambiguity of the new situation is reduced. Our students used a variety of channels for getting advance information about the new situation. They learned from correspondence with the college, through catalogs, talks with college friends, visiting college campuses, reading books about college, through counselors, teachers, friends, parents, and college representatives.

With those students having older siblings who had gone off to college, we heard references to learning about the experience through them and thus adding to a sense of preparation for the experience itself. For example, if there was some concern about how their parents would deal with the separation the experience of an older sibling leaving for college was brought to mind. One student recalled how her parents had been able to adjust to her brother's leaving for college. Siblings and friends served as important sources of information about college. Whether or not the ideas that the students gleaned were accurate, at least having this information served to reinforce a feeling that they were prepared for the new environment.

4. By Role Rehearsal

(*a*) As College Students: Most of our students seemed to be preparing themselves for college by rehearsing forms of behavior which they associated with college students. We could understand this as a process of attempting to deal with part of the experience in advance by rehearsing in the present the image of a "college student." This process is related to the pattern of *anticipatory socialization* as described by Merton and Kitt: "For the individual who adopts the values of a group to which he aspires but does not belong, this orientation may serve the twin functions of aiding his rise into that group and of easing his adjustment after he has become part of it." [6]

We saw numerous instances of this pattern of role rehearsal. Quite a number of the students reported that they began to read books that they felt were either required in college or would be useful for them in their college courses. A number of boys began buying their own clothes for the first time, saying they knew they would have to do this in college. One student who felt that scheduling would be an important part of the college experience tried to prepare herself for this by teaching herself to get things done in a minimum of time and not procrastinating or prolonging things.

Another boy, who viewed being collegiate in terms of becoming a gentleman, tried to act as he thought a college student would. He described to the interviewer his efforts at doing assignments, carrying out his responsibilities, and being a gentleman. For him, this also included letting adults through the door first, speaking when spoken to, and being more polite with girls.

Some of the special courses offered in high school were viewed as trial college experiences, thus simultaneously providing intellectual skills, bolstering the student's self-image, and coping with anxieties about intellectual abilities in this area. Students made references to these courses as "being on a college level." Some students referred to the experience of writing term papers as being an excellent preparation for college. For the most part they seemed eager for the kinds of experiences in high school which they felt would be relevant to what would be expected of them academically at college.

One of the students described in detail his experience in writing a term paper which he felt was like what he would be doing in college. He had the feeling that because of this experience and particularly because his teachers had told him that this was what students would be doing at college, he had a head start.

(*b*) As Adults: In addition to rehearsing college student behavior, there was also some rehearsing of functioning more like an adult before going off to college. This seemed to be related to getting more experience in mastering tasks in the adult world and strengthening the feeling of confidence to deal with adults on a more nearly equal level. We were very much impressed with the preparatory value of the summer job for these students. Those who worked felt that this experience helped them to develop confidence in dealing with adults. This went along with a shift in their own self-image with respect to the new adult role possibilities and helped to confirm their ability to handle new situations. For many students, this was their first experience in competing for jobs in the adult world. They were no longer working at jobs associated for them with adolescent status such as baby sitting, lawn mowing, etc. Many of them took competitive examinations for summer government positions. One girl said of her government job:

When you walk through that door you're on your own. If you don't make good, you're fired. It's not like school, where you're answerable to your parents through your teachers. I enjoy the experience of earning my own money and being on my own.

Another commented:

We were all equals, we were all working side by side. I was glad they called me by my last name. Otherwise it would have been like being a babysitter. It's the first time I've been almost looked on as a contemporary.

A student who planned to enter the ministry felt that his summer job was a broadening experience for him socially and would be an asset to him when he became a minister. He had previously worked at a gas station, but in the summer before going to college obtained a government job. He commented:

This gives me an opportunity to use my mind and associate with a different class. As a minister I will have to learn to associate with all kinds of people.

Thus, his job experience was regarded as a rehearsal opportunity in terms of future vocational goals as well.

Another student pointed out that in addition to earning money, she has learned that she is capable of meeting people and getting along with them easily. She enjoyed the feeling of responsibility. It was a novel and gratifying experience to have people give her jobs to do and leave her to her work with the understanding that she is capable of going ahead on her own.

5. By Group Identification

Another mechanism for the students' self-assurance that they would do well at college was to identify themselves as part of a group which shares a reputation for being adequately prepared for college. This was a view that our students had of their high school. These students, it is true, were drawn from a group of the more competent in academic performance and viewed themselves as being adequately prepared for college in terms of their own intellectual skills. In addition, however, they recognized they came from a high school which sent a high percentage of students to college. They expressed a feeling of pride in their academic preparation and revealed a feeling of confidence that attending this school had equipped them to deal adequately with the demands ahead. Many of them expressed the feeling that coming from their particular high school they were a jump ahead of students coming from most other high schools.

6. By Lowering the Level of Aspiration

Another mechanism had to do with students lowering their sights in terms of what they were going to expect of themselves in college. It was as if they were redefining an acceptable self-image that would permit them to maintain a feeling of satisfaction in their performance if it were at a lower level than in high school. This was most often expressed around concern about grades in college. This concern was fairly widespread and some of the students coped with it by lowering their expectations of their performance at college. This provided for them a cushion against possible future disappointment.

One student expressed rather well what several others felt, when she said that she wasn't really making firm plans about anything in college because

> If you set up certain standards for yourself, for instance belonging to a certain group, or making certain grades, or anything else, you're apt to be disappointed. It's better to be prepared to accept whatever comes your way.

We feel that this attitude did not reflect an abject resignation to the anticipated higher demands of college. Rather, these students conveyed to us that they would continue to try to maintain their academic standing. In the context of recognizing higher standards and the presence of other good students at college they prepared themselves for the possibility of a lower level of academic status.

7. By Selectively Perceiving Encouraging Elements in the New Situation

Our students tended to perceive college as a potentially friendly environment. This view of the unknown situation could facilitate a feeling that one would be adequate to mastery of the situation. Certain favorable characteristics were attributed to the college. Colleges were viewed as benign and wise institutions and particularly as places where support would be offered if needed. One student expressed the feeling that college would not be a sink-or-swim proposition, since he was sure he could always depend on the college advisers. He had the feeling that there would always be people to turn to if necessary.

Another attitude that was expressed was that college would be effective in placing students with compatible roommates. Some of the students spoke of their feeling that the college made efforts to place students with people who had similar interests. This view of the new situation helped sustain a feeling that there would be forces at work in the new environment which would facilitate one's being able to get along well.

III. MAINTAINING DISTRESS WITHIN MANAGEABLE LIMITS

In the preceding section we have described a variety of ways in which students reassured themselves that they were the kind of a person who would be able to manage the college experience well. It was not true, however, that our students approached this new experience without feelings of distress. We were impressed, however, that they were able to experience anxiety and contain it within manageable limits, so that it never became overwhelming to them. This capacity to contain anxiety has important consequences. A student who is able to experience anxiety without becoming overwhelmed by it is in a position to function more effectively and thus enhance the likelihood of using his assets in a more constructive way. For this reason we were interested in looking at those mechanisms which the students used in keeping anxiety within manageable limits.

1. The Supportive Function of Shared Experiences

Almost all of the students expressed an attitude that what they were experiencing in the way of anxiety was not unusual; rather, it was viewed as something that the majority of their friends in the same situation were also feeling. Thus, being aware of anxiety did not make the student feel uncomfortably different from others. Our students' close relationship with their peers made it possible for them to communicate their feelings to others and provided a situation in which they could be in touch with what others were feeling. In this way an awareness of anxiety did not trigger additional concerns about "What's wrong with me that I'm feeling this way?" The view of "We're all in the same boat" was one of the most common means of reducing anxiety. One student, in talking about the feelings of apprehension he felt just prior to leaving for college, said that he could dispell his apprehension with the thought:

> Everybody feels this. Everybody has to be a freshman once. Everybody's in the same boat that I am and there's nothing to really worry about.

In regard to concerns about making friends, these students felt that other freshmen at college would be just as concerned as they. They recognized this as a situation which would therefore be conducive to making friends. That is, others would have as much need for making friends as they themselves felt. This was illustrated in the following quotations:

> I feel confident in my ability to make friends. In college everyone's in the same boat. There's no reason to feel I won't make friends.

> There we'll all be in the same boat. We'll all be out to make friends.

> Everyone is in the same boat. I don't think making friends will be that difficult. Everyone will want to get to know people.

2. The Usefulness of Worry

The importance of the *work of worrying* in preparation for dealing adequately with a stressful reality situation has been pointed out by Janis in his studies of surgical patients.[4] Worrying can be viewed as adaptive when it contains a preparatory recognition of potentially difficult aspects of a life situation along with an opportunity to rehearse ways of dealing with the new situation. The students in our study did exhibit varying degrees of concern about the situation ahead; however, these concerns did not lead to a circular ruminative preoccupation.

Some of the students expressed an attitude that worrying about things had a useful function. For one student, certain magical implications were attached to worrying which then made the concern feel less threatening. That is, worrying about something somehow insured that things would work out well, as if one thereby extended his control over the uncertainties in the unknown situation. This seemed to be a way of allowing oneself to experience the distress without being overwhelmed by it. This quote from one of the interviewers illustrates this process:

> If you don't worry about something, then it will go wrong. As long as you worry a little about it, it will work out. . . . If you are overconfident and don't worry about some things, then I feel something's bound to go wrong, so that worries don't tend to bother me so much.

3. Present Activity in Anticipating Future Concerns

Some of our students' behavior seemed important, not only in dealing with elements in the present situation, but also in dealing with an anticipated future concern. For example, one student had some concern about whether his widowed mother could handle the household chores in his absence. He tried to get everything done around the house, including redecorating the basement and a number of little jobs that needed tending to before he left for college. Such behavior could be viewed as having future value in the new situation. It provided a reference point for handling anticipated guilt he might experience about leaving his mother while at college.

Two of our students, who had a close enough relationship during high school to consider marriage, made a decision to go to separate colleges; to date other people and "see what happens." As the girl expressed it, she felt even if they were to get married, if they hadn't had this experience, in the future she might have the thought "think of what would have happened if we had not been separated." She felt this would be helpful to their future relationship. To postpone a definite decision for a prescribed period of time provided a moratorium on making this decision immediately in college and thus freed her to deal with the new situation more actively. It also was a way of dealing with an anticipated future concern.

The value of the summer job experience has already been mentioned in an earlier section of this report. We felt that another useful aspect of the job experience was that it could provide some reference point in the future when there might be some concern about the element of dependency in his role as a student. That is, another value of the work experience might be to provide a proof of ability to function in a more independent and self-sufficient way before continuing in the role of a student.

4. Rehearsal in Fantasy of Future Behavior

Another way of dealing with some anxieties about the new situation was to rehearse how one might deal with some future contingency by imagining ahead of time what kind of behavior could deal with it. For example, one student had enjoyed her activity in a madrigal group in high school which was an important source of satisfaction for her. In discussing her anticipation of college, she spoke of hoping to find such a group in the new setting. In exploring this in fantasy, she said that if there were no such group available, that she could picture herself doing something about organizing one.

A more common area in which this mechanism was in operation was in regard to concerns about separation from home. One student imagined herself possibly needing help at college and she fantasied that she could turn to relatives not too far from the college she had chosen. Another student saw herself being more active about becoming more involved with her peers at college. She spoke of this in the context of her feelings that she will miss her family while at college and in the interview reassured herself by saying:

I suppose I'll become more familiar with people my own age.

Other students recognized that they would not actually be too far away from home and reassured themselves about any anticipated homesickness with the thought:

If I want to go home, I'll be able to.

SUMMARY

As an opportunity for exploring certain aspects of ego functioning in late adolescence, we selected for study a group of competent high school seniors who were anticipating the transition to college. During the spring and summer months preceding their departure for college, we observed a variety of coping behavior patterns in these students being mobilized in regard to the tasks ahead. These many processes served various ends simultaneously and tended to overlap and reinforce each other.

A general characteristic of our students was their tendency to reach out for new experience, a tendency to be active in dealing with challenge and an enjoyment in the sense of mastery. This pleasure in effectance served to facilitate an active involvement and pleasure in coping with the new situation.

In addition, more specific mechanisms were observed. Through a variety of ways, these students developed and maintained an image of themselves as adequate to the perceived requirements of the new situation. Students referred back to analogous situations in the past which had been adequately mastered, thus reassuring themselves about their ability to handle these situations in the future. They referred also to their present self-image as evolving gradually in desired directions, and through anticipatory detachment saw themselves as ready to establish new relationships at college. By seeking out information about the new situation, they reduced some of the ambiguity in it and, in so doing, felt better prepared to deal with it.

By role rehearsal they prepared for the new situation by rehearsing in advance forms of behavior which they associated with college students. In addition, they rehearsed behaving more like adults before going off to college. This summer job experience confirmed théir self-image as people adquate to master tasks in the adult world. It also strengthened their feelings that they could deal with adults on a more equal level. These students viewed themselves as members of a group which shares a reputation for being adequately prepared for college. They also dealt with concerns about performance at college by redefining an acceptable self-image that would permit them to maintain a feeling of satisfaction in their performance if it were at a somewhat lower level than in high school. In addition, they tended to perceive selectively encouraging elements in the new situation; by regarding college as a potentially friendly environment, they were sustained in feeling that one would be able to get along well there.

Other mechanisms seemed understood best as ways in which distress was contained within manageable limits. Our students were not without anxiety about college, but derived considerable support from an awareness that others were experiencing anxiety as well. "We're all in the same boat" was an attitude that was most common in maintaining anxiety from becoming overwhelming. Experiencing anxiety did not in turn signal a feeling of uncomfortable distance from others, but rather as something that was shared. Some students viewed their worrying as something useful, as if worrying extended control over the uncertainties in the unknown situation. Some of the students' behavior before going to college could be understood as providing a reference point for dealing with anticipated distress in the future. The knowledge that one was capable of earning money and holding down a job in the adult world could be a useful reference point in dealing with future concerns about the element of dependency in the role of student. Anxiety about some future contingencies could be dealt with also by rehearsing in fantasy how one would deal with that situation ahead of time.

In conclusion, during the period of anticipation of the new college experience, a variety of patterns are in process which serve to reinforce one another in a useful way. Attitudes of pleasure in dealing with newness, a process of matching the self-image with what is anticipated ahead, and mechanisms for containing distress reinforce one another to provide a general basis for confidence in facing the new situation. The active search for manageable levels of challenge in newness is more characteristic of the coping behavior of competent adolescents than a stabilized adaptation to the environment with maximal reduction of tension.

REFERENCES

1. ERIKSON, E. Identity and the Life Cycle, Selected Papers, Psychol. Issue, New York, International Universities Press, 1959, Vol. 1, No. 1, pp. 101–164.
2. FREUD, A. The Ego and The Mechanisms of Defense, New York, International Universities Press, Inc., 1946.

3. FREUD, A. Adolescence, Psychoanal. Study Child, 13:255–278, 1958.
4. JANIS, I. L. Psychological Stress, New York, John Wiley & Sons, Inc., 1958, pp. 374–388.
5. JONES, E. Some Problems of Adolescence, in Papers on Psychoanalysis, Baltimore, The Williams & Wilkins Company, 1948, pp. 389–406.
6. MERTON, R. K., AND KITT, A. S. Contributions to the Theory of Reference Group Behavior, in Continuities in Social Research, edited by R. K. Merton, and P. F. Lazarsfeld, Chicago, The Free Press of Glencoe, 1950, p. 87.
7. WHITE, R. W. Motivation Reconsidered: the Concept of Competence, Psychol. Rev. 66:329–330, 1959.

Article 10 Coping Strategies in a New Learning Environment

A Study of American College Freshmen

GEORGE V. COELHO

DAVID A. HAMBURG

ELIZABETH B. MURPHEY

INTRODUCTION

* * *

In this paper we focus on specific socioacademic tasks and various cognitive and interpersonal experiences of competent adolescents in their college freshman year. Our presentation contains three major sections. First, we outline academic problem-situations presented by the process of higher education at college; next, we illustrate various strategies for coping with these new intellectual and technical challenges; finally, we suggest some ways in which freshman coping behavior may stimulate developmental change and serve preparatory functions for adult social learning.

We hypothesize that ego-strengthening processes are facilitated by the mastery of socioacademic tasks of this transition. A coping strategy in our sense involves ego-processes in which two aspects are functionally interrelated: (1) maintaining a sense of worth as well as developing self-esteem as becoming and (2) managing emotional distress in the face of complex demands of the new college culture. Sometimes one aspect is more salient and dominant than the other. Coping functions involve not only self-manipulation of feelings in order to contain anxiety and maintain self-esteem, but also environmental management and realistic problem-solving. These functions, in general, tend to broaden the basis of self-esteem and self-expression in the adolescent's dealing with a wider social reality than he has known previously. Our data support the generalization that college-bound adolescents anticipate the transition from high school to college as a socially complex and intellectually demanding experience,[1,8,2] al-

GEORGE V. COELHO, PH.D., is at National Institute of Mental Health, Rockville, Md.

DAVID A. HAMBURG, M.D., is Reed-Hodgson Professor of Human Biology and Psychiatry, Stanford University School of Medicine, Stanford, California.

Source: *Archives of General Psychiatry*, Vol. 9 (November 1963), 433–43. Copyright 1963, American Medical Association.

though the nature of the academic challenge varies according to the institutional characteristics of different college environments.[4,6,7,9]

CRITERIA AND METHOD OF SELECTION OF THE STUDENT GROUP

Our qualitative data are based on intensive interviews with 14 volunteer subjects (nine female and five male), who were selected in their senior high-school year on their demonstrated competence in (*a*) academic work in school, (*b*) interpersonal closeness with a peer, and (*c*) participation in extracurricular activities and social groups.

*   *   *

These 14 students were enrolled in a wide range of college institutions differing in program emphasis, intellectual competitiveness, and size. Their college choices may be summarized as follows: (1) Five students enrolled in well-known four-year liberal arts colleges of the elite "Ivy League" variety; (2) Two students chose intellectually demanding prestige institutions with predominantly scientific and technical programs; (3) Three students went to small colleges which were much less prestigeful and less academically competitive than the elite colleges but typically strong in vocationally oriented programs of religious education; (4) Four girls enrolled in state universities—two in the local large university, one girl in an Eastern university, and the fourth in a Midwestern state university.

Our qualitative findings were derived from the whole set of 11 interviews conducted with each of the 14 subjects during an 18-month period. Four interviews were held in the Spring term of the senior high school year and three during the following summer. Four additional interviews were held throughout their college freshman year, namely, a campus visit about six weeks after college registration, two interviews when the student returned home for his Christmas and Easter vacations, respectively, and a terminal interview at the end of his freshman year.

The interviewers on the research staff, seven in all, each saw the same subject throughout. An interview guide was used and verbatim records were made during the interview. The focus was on a socially-oriented case history documentation of the individual student's life experiences, interests, activities, and relationships in his high school, family, peer group, and college environments. We were interested in specific problems he encountered in meeting various tasks of the transition and the means and resources he used for dealing with these problems. Our paper attempts to identify and analyze the major coping strategies of adolescents in the new environment of the college freshman year.

I. Socioacademic Tasks of the Transition from High School to College

The adolescent is exposed to complex new intellectual and technical challenges in college: (1) new subject matter in unfamiliar fields of knowledge; (2)

heavier course loads and more demanding intellectual work; (3) new ideas and techniques to be mastered under pressure of periodic examinations and deadlines; (4) assignments requiring greater initiative and organizing ability; (5) new fields of knowledge that have no immediate vocational application; (6) diverse college responsibilities requiring self-regulation in organizing time and activity; (7) the cumulative demands of campus life, curricular and extracurricular, requiring considerable autonomy in making decisions that often involve long-range and irreversible commitments.

These tasks may be met through diverse coping strategies which, in generic terms, involve not only the management of self-esteem and anxiety in the face of new standards of intellectual performance and academic competition with one's peers but also management of environmental resources. These tasks are critical in the sense that the adolescent confronts them while he is resolving the social ambiguities of living between two worlds—no longer a "school kid," and not yet the finished "college man." In a relatively pluralistic and open society, the adolescent is exposed to new possibilities of becoming—that is, developing desired characteristics consonant with his concept of the adult person he would like to be one day. During this period of role-transition and developmental change, the adolescent is also challenged to develop a broadened basis of self-esteem and self-expression in his dealings with strong and complex impulses on the one hand and an unfamiliar wider social reality on the other.

In identifying patterns of adolescent coping behavior in the college environment, we will illustrate the range and diversity of strategies used in meeting specific socioacademic tasks. We will suggest individual differences underlying various modes of competence, even though some tend to characterize the group as a whole. No individual studied manifested all the coping strategies described in this paper.

II. Maintaining Self-Esteem and Managing Anxiety in Coping with Socioacademic Tasks

A variety of cognitive and interpersonal strategies have been identified. We found it useful to list them under more general ego processes of anticipatory mobilization[3] and effectance motivation[10] to highlight the exploratory and prospective character of the individual's transactions with a novel and problematic environment. These strategies appear to be formed and employed at all levels of awareness and may be explicit or implicit in the individual's communication with others.

1. Projecting a Clear Self-Image As an Effective Doer. Most of these students, when interviewed in their senior high school year, saw themselves as effective doers capable of producing definite results through hard work. They were proud of their ability to be industrious and efficient. Many of these well-endowed students approached mathematics as a learning situation which prepared them for exacting standards of intellectual performance. Some of them expressed a "joy of achievement" and a "feeling of efficacy" in mastering difficult materials.

2. *Mobilizing New Combinations of Skills.* Some students relied on carrying over to the new learning environment efficient attitudes and organizational skills in budgeting time. Characteristically, however, most of them devised ways of dealing with new demands. The freshman year data indicate that they were able: (*a*) to look ahead and see clearly what was expected of them so that they would organize blocks of time; (*b*) to distinguish between primary and secondary demands on their time, subsidiating minor interests to major academic goals; (*c*) to study for long stretches of time without feeling bored, imposed upon, or resentful or to recover quickly from periods of negative affect, particularly through brief rewarding contacts with peers; (*d*) to concentrate under difficult conditions, or to rectify external conditions to improve their concentration efforts; (*e*) to diagnose the interests and attitudes of their professors; (*f*) to formulate intermediate goals that were attainable within a sequence of long-range work responsibilities. Not all students showed all these abilities in extraordinary measure, but each of them was clearly revealed in some students.

3. *Regulating Acceptable Risk in Different Areas of Commitment.* These students had considerable academic achievement in high school and relatively high levels of academic aspiration for college. Yet most of them were enrolled in colleges largely populated by students with similar academic records, so it was clear to them that some would have fallen below their level of high school performance. Moreover, most were eager to explore new territory and therefore did not permit a sharp deterioration in grades and other tangible signs of academic achievement.

Several students dealt with this problem by differentiating among levels-of-aspiration in different areas of commitment. For instance, one student set a "floor" under the acceptable level of performance in all course work and then pursued her extracurricular activities to the maximum extent compatible with this standard. Another student set a high level of aspiration in his "core" subjects (i.e., those he felt central to his career), while setting a much "lower floor" for other subjects. Such regulation tended to permit either broad experimentation with new experiences or exploration-in-depth of a core area of importance, while protecting against serious academic difficulty.

4. *Using Assets to Test New Images of Growth Potential.* Most students identified their assets not merely in terms of manifest past achievement but also in terms of their emergent interests and expectancies of self-change. They expressed confidence and zest in how they could learn to overcome previous limitations.

5. *Selecting Upperclassmen As Resource Persons.* Most students dealt with the perplexities of the new system by actively seeking information and taking cues from upperclassmen. Upperclassmen seem to help in providing emotional support, academic guidance and orientation, value reinforcement, and the confidante's touch of sympathy. In effect, upperclassmen helped in learning new academic skills or improving standards of performance so as to meet institutional demands and personal levels of aspiration. They also helped in pointing out

alternative pathways to personal fulfillment in the new situation and new bases for evaluating their potentialities. Upperclassmen in the dormitory may also expose the adolescent to new aspects of himself by suggesting alternative acceptable uses of time and energy. Some students were able to relax harsh demands on themselves by making use of suggestions from peers. By actively using upperclassmen, the adolescent is free to experiment with role-images of the developing freshman. Seeking out upperclassmen does not carry any adverse implications of a relationship of unequals but rather a solidarity of students "in the same boat." Forming contacts with upperclassmen also opens communication channels with student-peer groups in handling conflicts of opinion between student and faculty.

6. *Learning Through Part-Identifications with Faculty.* Most students were favorably oriented toward the faculty without seeking any close personal contact outside the classroom. These favorable images may have helped to realign the student's academic expectations with faculty demands. In the event of academic frustrations, this respectful attitude led some students to introspect and re-examine their own attitudes and abilities rather than to blame the "system." Some students felt concern about losing the confidence of their peer group if they fraternized too much with the faculty. Some need to demonstrate visibly their independence from their teachers in the same way as from the parents. Whatever the specific conditions, some learning that is incidental but quite consequential may occur through identifications with specific aspects of the faculty —their values, ways of relating, attitudes, personality traits. Accordingly, even without close contact with faculty, students may select identification models at a distance. In the final section we suggest some possible functions of these part-identifications with faculty or upperclassmen.

III. Short-Run Tactics for Handling Academic Disappointments

We were also interested in identifying short-run tactics for dealing with academic disappointments and dissatisfactions.

1. *Recentering Efforts Within a Long-Range Purpose.* One's academic efforts may reveal productive possibilities when framed within the perspective of emerging interests and values. A student planning to go into the ministry was doing poorly in advanced French. One day, however, he found inspiration in a speaker who asked for volunteers to come to Africa, which would require a knowledge of French. He was touched by a sense of mission to work harder and overcome his poor performance. He put forth extra effort and was able to bring his work up in French, using all the advantages of the tapes and language library in working out his problem. Follow-up data show that he spent the last quarter of his sophomore year in France.

2. *Accepting Alternative Gratifications Often Extracurricular.* A variety of substitute satisfactions may become available to students who are playfully alert to the informal aspects of their education. One of the students was initially disappointed in her academic program at an elite liberal arts college in the Midwest and thought her classes were rather dull because she had only one seminar.

Accordingly, she turned her energies into social and extracurricular activities, spending a lot of her time in the library, art theater, and museum. By the time of the Spring recess her satisfaction with college had increased.

3. *Remodeling Prefabricated Images of a Vocational Role.* The extension and differentiation of interests in this transition period may lead to explorations in role-modeling. A student who pushed very hard to be a predental major was doing poorly in chemistry, learned to ease the stereotyped expectations he had of himself. He was thus able to reestablish his self-confidence on the basis of new choices of role possibilities and control his doubts about his own worth as a person.

4. *Setting Intermediate Goals in Working Out Long-Term Plans.* Some students paced their steps toward future goals by marking out stages in attaining distant targets. Three students who were initially disappointed with their current academic program worked out practical bases for new intellectual stimulation in a novel cultural environment. They took courses to prepare them for study abroad (Molly in France, Bob in England, and Sarah in Germany).

5. *Referring to the High Academic Standards Set for College Admission.* Some students used their admission to a good school as a basis for evaluating their potential. One boy reassured himself that he had been admitted to a top-notch place famous for its distinguished scientists. Only outstanding students were admitted in the first place and belonging in such a group gave one a special significance, by definition; this feeling was important for him in handling anxiety about academic performance, which though clearly passing was below his extraordinary high school record.

6. *Projecting Optimistic Peer-Group Expectations.* In maintaining self-esteem and managing anxiety during the initial weeks at college the students' benign peer-group expectations played an important part.

The students' attitudes toward new friendships were reinforced by their parents' positive values. Most students, irrespective of their levels of academic potential, came to college valuing these social traits—friendliness, getting along with others, seeking out qualities in other people that one can like or respect. They also had substantial confidence in their own ability to make friends, recognizing these useful assets in adapting to college demands. This confidence was reflected in the student's self-image as a worthwhile person and his expectation that he would be so regarded by others. Their parents, too, who were interviewed a month before the student entered college, placed great emphasis on these values as basic assets and specifically as good preparation for college. This ability to form and maintain friendships seemed important for developing a broad and versatile repertory of coping skills in the socioacademic environment.

IV. Exploring and Using Their Interpersonal Environment

Coping functions are served not only by the individual's self-manipulation of attitudes and feelings but also by active exploration and use of human resources in the environment. The ability to make friends easily and to become significantly engaged in friendships has very important implications for dealing with academic crises and for learning skills in handling other developmental tasks. These students were selected for their ability to maintain interpersonal closeness with at least one peer. Throughout the freshman year we inquired as to what kinds of friendships these students formed in their freshman year and how these friendships help in managing anxiety over socioacademic tasks.

We have suggested in an unpublished report that the formation of friendship in the freshman year was a two-phase process.[5] Most of these students made friends rather indiscriminately in the early weeks at college. During this initial warming-up and reaching-out phase, it was important to have friends—any friends—and feel accepted by others. It was helpful to the freshman to know that there was some peer in the same class or dormitory—who was "in the same boat"—whom they could turn to and talk with without being uncomfortable about exposing to others their need for friendship. Friendships in the early period were formed usually on the basis of physical proximity. They were useful, especially to the freshman going to large universities, in preparing him to meet various new socioacademic demands and to learn his way around the campus maze. They helped to combat initial feelings of loneliness, to provide tension-relief as students talked about their concerns together, and to give orientation to classes, teachers, types of courses, through the impromptu bull sessions that arose in the dormitory unit. Later in the freshman year, there appeared a second phase of sifting out initial acquaintances and of working out deeper relationships based on shared interests and values. Students were able to use their peer groups and individual friendships in ways that facilitated acquiring skills for dealing with new socioacademic problems in this transition. The following illustrate some uses of friendship, though these are usually not the explicit goals in making friends:

1. *Clarifying New Self-Definitions and Career Possibilities.* Students were able to experiment with new alternatives in regard to career plans and choice of major by rehearsing these possibilities with others in informal discussions. A student can learn about himself in a differentiated way through the reactions of significant peers to his ideas and behavior.

New friendships provided the freshman with opportunities: (1) to start with a "clean slate" and to free himself of the need to reciprocate with stereotypical expectations on the part of others; (2) to "try on for size" new patterns of behavior in different kinds of relationships and new career plans that were more consonant with his emerging life style than the one he had earlier in school.

2. *Intellectual Stimulation Through Informal Discussion Groups.* Many students reported they experienced more intellectual stimulation from "bull sessions" than discussions centered in the classroom. These sessions provided opportunities for discussing class assignments, books they are reading, new subject matter and ideas, and in general, for expanding their intellectual horizons and developing a zest for learning.

3. *Learning Through Pooling of Information and Coping Skills.* A con-

siderable amount of informational exchange develops in the early phases of college life through informal group discussions which help in building up a predictable environment where certain standards and qualities in social and academic behavior are clarified.

4. *Learning Through Role Complementarity.* A common experience during college life is the subtle process of role-complementarity by which friends trade coping skills, help each other in their respective areas of strength, whether they are aware of doing so or not; for example, one student may help another by providing a model for the development of effective study habits and personal organization and perhaps get in return a model of effectiveness in the sphere of sex and dating behavior.

5. *Support in Time of Crisis.* In dealing with academic disappointments in one's grades and courses, or in general when the student experienced conflict and confusion about college experiences, both individual friends and friendship groups were supportive and gave reassurance about one's self-worth.

6. *Sounding-Board for Other Possible Points of View.* Even when friendships were not deep or close, peer group relationships provided opportunities for holding up a mirror to one's self in a friendly light. By meeting students of different regional and cultural backgrounds, a student recognizes the impact of his behavior on others and is likely to become aware of values other than his own. This may lead to some modification of his own values or to increased appreciation of other ways of life.

V. Possible Long-Term Developmental Consequences of Freshman Coping Behavior

Coping with specifically academic tasks not only helps to consolidate the student's self-esteem but may also have long-term consequences, though these may not be necessarily favorable by clinical criteria. We now suggest some of the directions in which these students moved. The freshman year opened up for them new cognitive and interpersonal resources through the stimulation of learning and peer group friendships. Many students felt encouraged to enjoy the intellectual process as a value in its own right. Several students spoke of changes in learning behavior and new modes of self-awareness. Some major dimensions of their novel intellectual and social experiences during their freshman year are as follows:

1. *Increased Intellectual Pleasure in Academic Work.* Most students, irrespective of school, reported an increasing sensitiveness to intellectual pleasures and a decreasing preoccupation about grades. One student, on finding she was much less concerned with grades, signed up for a heavier work load. Another found the work much harder but also more interesting than in high school. As she put it: "They require more work but it is much more intellectually exciting at college." Another student who was getting interested in learning, felt he did not want to stop learning in the summer and planned to study German on his own.

2. *Informal Group Discussions As a Learning Environment.* For many students, class discussions as well as informal bull sessions in the college dorm often provided new stimuli. Bob was studying seven hours a day, much more than in high school, but he seemed to enjoy it more because of the discussions. Louise was glad that she did not have to participate in class (as she did in high school) merely for the sake of grades.

3. *New Insights into the Higher Learning Process.* Another intellectual experience—which some students discovered in college—was the ability to think for one's self, to have one's own opinions, to use one's critical and analytical powers. Louise used the advanced English class as a setting to formulate her own opinions and present her own deductions. She was learning to depend on conceptualizing instead of memorizing as she had done in high school—she found that learning now was a help "to formulate answers to her own questions." Some typical extracts follow:

I can't deny my intellectual interests and now see there are other ways of life. I can express myself as being interested in intellectual things. I have been changed by the acceptance of intellectual interests . . . .

In high school the courses were never stimulating enough to get ideas. Here you are not afraid to do your own thinking and you want to bring out your ideas and discuss them.

Before I thought of God more as an omnipotent force in nature to account for what does exist . . . I feel I'm now in the process of analyzing these attitudes in myself.

In school I took what the teacher said. Now I'm not taking what the teachers say as gospel truth.

4. *Part-Identifications As Models for Becoming.* Identification with faculty members and even to some extent with highly respected peers may tend to serve corrective or augmenting functions in relation to earlier difficulties with parents. For example, a girl who has become increasingly dissatisfied with her mother, whom she perceives as devious and distrustful, identifies with a woman faculty member who is perceived by her and her peers as a straightforward and trusting person. Similarly, a boy whose father has been kind but ineffective identifies with a male professor who is remarkably similar to his father in a variety of attitudes and values, but who is clearly an effective person. Such identifications may be quite broad-scale, covering a variety of perceived attitudes—in effect, "I want to be the kind of person he is"—or may be quite selective part-identifications, linking oneself with a particular highly desired attribute of the respected person. In the latter instance, there is then the task of synthesis to be done, fitting this part-

identification with other part-identifications (past and current) in such a way as to facilitate a coherent sense of self as a distinct individual, worthy of respect and having a place in some highly valued reference group(s).

To Sum Up. We found no one single unitary or model pattern of strategies for coping with various intellectual-technical challenges. Indeed, we have been most impressed by the diversity of patterns of coping behavior in the new academic environment. To maintain a sense of worth and keep anxiety within non-interfering limits involves a readiness to mobilize inner personal resources to meet the new demands—especially the capacity for doing meaningful work, actively seeking out problem-solving opportunities, and working out diverse sources of intellectual gratification outside the normal academic curriculum. Our interview data suggested that the resolution of earlier disappointments (within a moderate range of severity) was helpful in coping with the disappointments encountered in transition from high school to college. By the same token, we suspect that effective resolution of these transitional disappointments would, on the whole, tend to prepare these young people for the inevitable disappointments of adult life. It seems reasonable to regard such coping behavior as involving complex skills acquired through long sequences of experiences with considerable transfer of learning from one stressful episode to another. One probable contributory factor is the enhancement of self-esteem that tends to result from ultimate mastery of a difficult distressing experience. The relevant kind of feeling might be expressed in such terms as "If I came through that earlier crisis quite well, I can surely handle this one." In general, we suggest that mastery of a stressful experience tends to contribute to a sense of strength, efficacy, or resourcefulness.

*   *   *

REFERENCES

1. EDDY, E. D., JR. The College Influence on Student Character, Washington, D.C.: American Council of Education, 1959.
2. FREEDMAN, M. B. Impact of College: New Dimensions in Higher Education, No. 4, Office of Education, U.S. Government Printing Office, Division of Public Documents, Washington 25, D.C., 1960.
3. HAMBURG, D. A. "Relevance of Recent Evolutionary Changes to Human Stress Biology," in Social Life of Early Man, edited by S. Washburn, New York: Viking Fund Publications, 1961.
4. PACE, R. C. Five College Environments, College Board Rev. 41:24–28, 1960.
5. SILBER, E.; COELHO, G. V.; MURPHEY, E. B.; HAMBURG, D. A.; AND GREENBERG, I. M. Formation and Functions of Friendship in the Freshman College Year Among Competent Students, Bethesda, Md.: National Institute of Mental Health, 1961, unpublished data.
6. STERN, G. G. Congruence and Dissonance in the Ecology of College Students, Student Med. 8:304–339, 1960.

7. STERN, G. G. "Continuity and Contrast in the Transition from High School to College," in Orientation to College Learning—A Reappraisal, edited by N. F. Brown, Washington, D.C.: American Council of Education, 1961.
8. SUSSMAN, L. Freshman Morale at M.I.T., Cambridge, Mass.: Massachusetts Institute of Technology, 1960.
9. THISTLETHWAITE, D. C. College Press and Student Achievement, J. Ed. Psychol. 50:185–191, 1959.
10. WHITE, R. The Concept of Competence: Motivation Reconsidered, Psychol. Rev. 66:297–333, 1959.

Article 11

Some Observations on Coping Styles of Black Students on White Campuses

ELVIN MACKEY, JR.

INTRODUCTION

An increasing amount of psychiatric literature is addressing itself to the special problems of newly arrived minority group students on predominantly white university campuses.[19] This particular population is one of high risk and thus becomes a target of appropriate concern for mental health and other professionals.[12] College can be a time of emotional stress for any student, regardless of ethnic or class status, as is evident by the already voluminous literature on this subject.[7] A particular problem posed by minority group students is their underutilization of available, though often inadequate, resource facilities.[21] The difficulties that traditional student health services might have in responding to culturally different student groups is not unlike the difficulties encountered by traditional health facilities outside of the university.[1] Even for the minority group specialist, reaching and gaining acceptance from this group can pose significant challenges.[2]

This paper will describe how one black psychiatrist attempted to get acquainted with and gather hypotheses regarding other newcomers to the university. It is felt that these experiences, limited as they are, have broader implications for mental health and other concerned professionals.

SOME RELATED LITERATURE

Edwards has described the origin, goals, and directions of the black student movement.[5] The content of the Yale Conference on cross-sectional views of black studies programs in universities has been presented by Robinson and associates.[18] Jones has presented an historical overview of the education of the black man in

ELVIN MACKEY, JR., M.D., is Assistant Professor, Department of Psychiatry and Human Behavior, California College of Medicine, University of California, Irvine.

Source: *Journal of the American College Health Association*, Vol. 21, No. 2 (December 1972), 126–30.

the United States and has discussed the controversial vicissitudes of black education.[10] Pierce has outlined problems that Negro adolescents might anticipate during the next decade and has offered potential solutions.[15] Erikson has discussed the dimensions of the problem of identity and related them to the emergence of national awareness of the Negro's position in the United States.[6] Coles and Hammond have described their contact, as white psychiatrists, with black students who come to university health centers.[3,9] Torrey and associates, have presented an overview of problems encountered by foreign students on university campuses and have hypothesized that certain students in this country might have similar experiences.[22]

ENTER THE NEW DOCTOR

Immediately after arriving at a university student health center, two years ago, as a trainee in child and adolescent psychiatry, I was confronted by several black administrators who had decided how I, the new black doctor, should be utilized. Minority group students were not coming to the health center, but were having intense adjustment problems on a strange new campus and, for some of them, a frightening new world. Having previously observed the rise and fall of a series of black scholars on California campuses, I hypothesized that a "black skin" was not enough to establish a working alliance with the new breed of black students. For myself, in particular, I had been removed considerably from minority group contact because of professional choices that I had made. I chose to get acquainted with the system—the students and the university—before making specific commitments. My observations on difficulties that black psychiatrists might encounter in leadership roles in the black ghetto, and challenges that black scholars have in working with the new breed of black students on white campuses have been previously described.[13]

   In choosing a role of participant-observer, to borrow Robert Coles's concepts, or research-therapist, as described by Hirsch and Keniston, I was aware of the nature of the "black backlash" to psychological research on black people and on the black community, as explained by Comer.[3] In the beginning, I saw students under informal circumstances and in informal settings such as faculty offices, dormitory rooms, coffee shops, picnics, and classrooms. To them I was a professional acquaintance and sometimes a friend but more often just another person who accepted them where they were. I was open and honest about my desire to learn what it was like for *them* being black on a predominately white campus. I identified myself as a member of the "establishment" and told them that I had been away from the mainstream of the black experience for a number of years. I audited courses in the Comparative Culture Program and conducted numerous informal and open-ended interviews with students and faculty in unstructured settings. Later, much later, I began to see a small number of black students on an informal referral basis from instructors. Eventually the students began to refer other students to me. These students taught me a lot and their pay was a listening ear with unconditional regard. I did not do formal psychotherapy or even give advice. I listened and asked questions, lots of questions, and I came to them.

THE OTHER NEWCOMERS

The students involved in this study, for the most part, were in the Educational Opportunity Program of the University. Generally, they did not meet the standard minimum requirements for admission to a university of four-year college, although there were exceptions. In all cases they needed the financial aid which was available through EOP. Some of them had attended predominately black secondary schools—in Watts, Los Angeles; West Oakland; Hunter's Point, San Francisco; Logan Heights, San Diego; and Santa Ana's "velvet" ghetto. Others, however, had just the opposite experience where they were one of few blacks in their secondary schools. This is not a statistical study and the examples that I will present are not necessarily representative of all of the black student newcomers whom I encountered. The *themes*, however, are prevalent enough to mention, as well as the hypotheses that they suggest.

CLINICAL ILLUSTRATIONS

(All names and identifying data have been changed to insure confidentiality, although each student gave me permission to use the reported clinical material.)

Case No. 1

One is immediately impressed by the *diversity* that black students recognize in themselves. One male student stated this as follows:

> There are four types of niggers here. There are the superconservatives who do their own thing and think that going to school with whites is a part of obtaining a *higher* education. There are the street niggers who are boisterous, smoke dope and drink wine. People don't realize that the way the street brothers make money is for reasons of survival and not a means of acting cool. Then there are the intellectual niggers who go to class, sit and rap about intellectual things. And fourth there are the bourgeoisie niggers who can't understand why street niggers are like they are, but who want to imitate them because they think that "acting" like street niggers is cool.

Rod, the above student, is a newcomer to a white university. In his all-black high school, he was in the upper 15 percent of his graduating class and was in a college preparatory track. He initially came to the Counseling Center because he thought that he was "going crazy." He had had several "white-outs" in the past, but they were increasing in frequency. These episodes always occurred at night, when he was relaxing in his room with his mind on nothing in particular. His girl friend has been present and has witnessed the episodes. Although he talks during these periods of "altered consciousness," he has no memory of it. He feels that he was thrown into a "sea of white people," when he came to the University, and that it "blew" his mind. He will not be returning next quarter.

Case No. 2

Vernon attended an all-black junior high school and a high school that was 75 percent black and 25 percent oriental. Except on two or three occasions, he had no encounters with whites during his high school years. When he came to the University he was conscious of his blackness, but in a proud sense. He made a point during the first year to emphasize his blackness—as something different from whiteness—and attempted to make white students aware of that difference. His associations during the first year were predominantly white and this presented no difficulties. During his second year, he noticed that his writings were "increasingly white oriented rather than black oriented." When he came to the Student Health Center, his main complaint was that people were influencing him in nonverbal ways, making him write things that he did not really mean, and act in ways that were contrary to his wishes. Vernon completed two quarters without major difficulties, but chose to leave the University to live closer to home. During our talks, he longed for a closer communication with his father who worked at two jobs and who was never home. His father never answered letters, and when he talked with his son over the phone, he allegedly emphasized the "negative rather than the positive." But Vernon admired his father because he got along with people, and Vernon believed that his *father* could have prevented his difficulties in the past, as well as future difficulties, if he would talk with him more.

Case No. 3

Ellen is a 25-year-old unmarried black student who has two children aged four and five. She attended parochial schools until she entered the University. Her mother never really wanted her, she said, but favored her older sister. When she was 20 years old, Ellen was raped and became pregnant. Because she felt that the baby needed a father, she proceeded to find someone whom she thought might become a father. She became pregnant again but marriage never materialized. While at the University, Ellen became pregnant again but obtained a therapeutic abortion. She had been in counseling with a white therapist before coming to the University, and felt that he had helped her. "I was told that I had to stop hating myself," she said. Now she feels comfortable in telling people that she has a "shrink." "People didn't believe me when I said I had to leave the party to keep an appointment with my shrink," she reported as she entered the office.

Case No. 4

Debra is a 28-year-old black senior student who has a six-year-old daughter. Prior to coming to the University, she had two years of marital counseling and became "insanely dependent" on her "white, republican therapist," as she put it, and he on her. "He did not know how to deal with it," so she says. Once she went to a white community hospital and a white doctor asked her if she knew what a "Pap smear" was. She became indignant and did not go back. She wants a "black" pediatrician for her overweight daughter and wants a "black" child psychiatrist, if the cause of the obesity is thought to be psychogenic. On weekends, Debra sells Black Panther papers and attends political education meetings in another city. She leaves her daughter at home alone, since the campus neighborhood is safe. The daughter comes to her mother for help and attention, but the mother is tired and not interested.

Case No. 5

Sylvia is a 26-year-old black student who is the second of three children. She has an older sister whom the mother prefers. She feels that the students at the University are "guinea pigs" and that people are "conducting experiments" on them. She also has had a lot of disappointments in the past which are "buried too deep to talk about now." She is a loner, and it bothers her to hear others tearing each other apart on their "ego trips," because she wishes she had an "ego to be torn apart." After high school she worked in the east and her parents and relatives called her "a gypsy." She feels that she has nothing at the University that is her own, not even her apartment which she must share and where she cannot legally keep her cat.

During her first two years at a junior college, Sylvia had a white counselor who was interested in her and whom she felt comfortable with. She is generally suspicious and distrustful of people. As she began to feel comfortable with me, she said, "I'll have to give a party and invite you." After she was accepted for a foreign exchange program, she wondered if I would give her the required physical examination, because "others" were giving her "a rough time," telling her that she did not have the "right" papers. "All you have to do is take my blood pressure and pulse and things like that," she assured me.

Case No. 6

Janet asked over the phone, "Are you a doctor who helps people when they feel depressed?" This black student complained of being "depressed, insecure and paranoid" and somehow these feelings were related to her *family*, she believed. She had no complaints about the University. I listened and asked questions and she hung up. When I next saw her roommate, she asked, "What did you do to Janet? She looked so elated after talking with you." Later, a worker from the Counseling Center told me that Janet had been impressed by me because I *listened* to her. Although I was not consciously trying to be therapeutic, the message was that many of these students were constantly hearing advice as to what they should do, how they should think, and what they should be. They wanted to be accepted for who they were and where they were, in their inner and outer worlds. This was in June, and Janet wondered if I would be around during the summer because she would be willing to commute to see me, even if it were on the weekend.

There were many other students whom I came to know—students who had problems with each other, problems such as they had had before they came to the University and such as their mothers and fathers had. They argued about interracial sex and lack of group cohesiveness and about political apathy. One black student explained it this way:

"The black male–white female thing is a major source of polarization between black brothers and sisters on every campus. The brothers have more ego insecurities than sisters when it comes to blackness, and they exclude the sisters from their politics. They spend a lot of time ego tripping and trying to impress each other."

History, perhaps, has made it so, and the black Yale psychiatrist and Associate Dean, James Comer, believes that some of this disunity and intergroup conflict might stem from the days of slavery.[4] And the black Boston University psychoanalyst, Charles Pinderhughes has stated:

"It is commonly recognized that persons who have felt inferior, criticized and discriminated against in one situation unconsciously carried these feelings with them into new situations where they behaved as if they were still subjected to the same treatment." [16]

And the black Harvard University psychiatrist, Chester Pierce has cautioned that the most that young people might expect in the next decade is to become members of a subminority group—the *least* disadvantaged.[15]

DISCUSSION

As black students have increasingly demanded the recruitment of more black faculty, they also have insisted on more black counselors and therapists. Where these demands have been met, there are suggestions that minority students have been less reluctant to come to traditional facilities for help.[2] There are also suggestions that when the "black-skinned" helper is traditional and middle class in his total approach, he might be "white-balled," so to speak, by black students.[14] The black Harvard psychiatrist and Associate Dean, Alvin Poussaint, predicted over two years ago that there would probably never be enough trained black professionals to be available to the burgeoning number of black students on white campuses.[17] If there is hope, it will be in the ability of the educational institutions to become flexible and sensitive in their overall approach to students.

A number of investigators have concerned themselves with the unique challenges posed by culturally different patients.[20] The conclusions are conflicting. Some have reported that the helping person need not be a minority group member. Others feel that white helpers often have difficulty relating therapeutically to minority group persons because of counter-transference issues. Even black therapists are not immune from having nontherapeutic attitudes.[2]

My own observations have been that no single current position is adequate as a valid generalization, and that the above observations represent parts of the total picture. When I first arrived at the University Health Center, I was told by black administrators that black students needed health care, but had strong reservations about coming to the "lily white" health center. Even though I was available on a limited basis and students were referred to me by other blacks on campus, they did not come to the Health Center until a Counseling Center with minority staff was established, and then only after constant prodding by the counselors. Psychiatry has its unique burden of stereotypes, and a black skin is often not enough to assure a young black client who has read *One Flew Over the Cuckoo's Nest*, and who feels he is "about to go crazy," that he will not receive electroshock treatments in a mental institution.[11] Once the students did make contact with me, they continued to come and referred each other. By the end of the spring quarter, just prior to the summer recess, a steady stream of black students were seeing me at the Student Health Center and wondering if I would be around to see them during the summer.

SOME CONCLUDING HYPOTHESES

Although no statistically valid conclusions can be made from the students I saw, certain patterns were suggested. Black female students were significantly less reluctant than black male students to utilize the Counseling and Health Centers. Female students who were older and who had had previous counseling or therapy were most apt to seek out help, even though their previous therapists had usually been white.

Black males, who did not utilize helping facilities, were more suspicious and distrustful, and relied more on "home and street remedies." One male student made the following response:

"I went to the Health Center once and they got all flustered—like I was some kind of an animal or something. It was too much of a strain on me *mentally* to go there for something *physical*. The rooms aren't sound proof. You hear others and you feel that they hear you."

Another black male said the following:

"You got to realize, Doc, that we may be at a modern university, but we are still mentally in the Dark Ages. We can't close *all* of the gaps in one single leap. Most of us were brought up on home remedies. People only went to the hospital when it was time to die. Put a *sister* at the receptionist window, and you'll have brothers coming several times a day, even when they are in perfect health."

As for myself, the participant and observer, I became aware of a number of myths I had about myself and other black people. I learned to understand and appreciate the coping styles of blacks who were different from me.[14] Some of my pet biases were shaken. Those whom I listened to and questioned extensively have thanked me for what they say I have done for *them*. But I would imagine that

I have good cause to thank them for what they have *allowed* me to do and learn about them and about things and conditions and situations that the public media never talk about when they report on the liberalization of admission policies to white universities.

A FINALE

Recently when I had a free moment, I dropped by a student's room, as I often do, and asked, "What's happening, brother? What's going on, my man?"

"Marvin Gaye's latest album is what's happening around here.[8] You go in anyone's room, any brother's or sister's room, that is, and you will hear Marvin Gaye. People around here wake up and go to bed by Marvin Gaye. That's what's happening, *now*, my man, around here."

> *Father, father, everybody thinks we're wrong*
> *But who are they to judge us*
> *Simply because our hair is long*
> *You know we've got to find a way*
> *To bring some lovin' here today*
> *For only love can conquer hate*
>
> *Father, father, talk to me so you can see*
> *What's going on, talk to me.**

Special acknowledgment is given to Dr. Gerald Sinykin, Director of University Health Services, and to Dr. Louis Gottschalk, Chairman of Department of Psychiatry and Human Behavior, whose joint sensitivity and guidance made this study possible.

REFERENCES

1. BERNARD, V. W. Some principles of dynamic psychiatry in relation to poverty, *Amer. J. Psychiat.* 122:254–266, 1965; HOLLINGSHEAD, A. B. AND REDLICH, F. C. *Social Class and Mental Illness*, New York, John Wiley and Sons, 1958; MINUCHIN, S., MONTALOVO, B., GUERNEY, B. G., ROSMAN, AND B. L. SCHUMER, F. *Families of the Slums*, New York, Basic Books, 1967.
2. CALNEK, M. Racial factors in the countertransference: the black therapist and the black client, *Amer. J. Orthopsychiat.* 40:39–46, 1970.
3. COLES, R. Observation or participation: the problem of psychiatric research on social issues. *J. Nerv. Ment. Dis.* 141:274–284, 1965; HIRSCH, S. J., AND KENISTON, K. Psychosocial issues in talented college dropouts, *Psychiatry*

33:1–20, 1970; COMER, J. P. Research and the black backlash, *Amer. J. Orthopsychiat.* 40:9–11, 1970.

4. COMER, J. P. Social power of the Negro, *Sci. Amer.* 216(4):21–27, 1967; COMER, J. P. Individual development and black rebellion: some parallels, *Midway* 9:33–48, 1968.
5. EDWARDS, H. *Black Students,* New York, Free Press, 1970.
6. ERIKSON, E. H. *Identity: Youth and Crisis,* New York, Norton, 1968.
7. FARNSWORTH, D. L. *Mental Health in College and University,* Cambridge, Harvard University Press, 1957; BLAINE, G. B., JR. AND McARTHUR, C. G. *Emotional Problems of the Student,* New York, Appleton-Century-Crofts, 1961; FARNSWORTH, D. L. *Psychiatry, Education, and the Young Adult,* Springfield, C. C. Thomas, 1966; BOYCE, R. M., AND THURLOW, H. J. Characteristics of university students with emotional problems, *Canad. Psychiat. Ass. J.* 14:490–491, 1961.
8. GAYE, M., CLEVELAND, A., AND BENSON, R. What's going on, Jobete Music Company, Inc. 1971.
9. HAMMOND, C. D. Paranoia and prejudice-recognition and management of the student from a deprived background, *Int. Psychiat. Clin.* 7:35–48, 1970.
10. JONES, W. P. Education of the black man in the United States, *Sch. Soc.* 98:467–470, 1970.
11. KESEY, K. *One Flew Over the Cuckoo's Nest,* New York, Signet Books, 1962.
12. LEAVITT, A., CAREY, J., AND SWARTZ, J. Developing a mental health program at an urban community college, *JACHA* 19:289–292, 1971.
13. MACKEY, E., JR. The psychosocial plight of the black psychiatrist in the black colony, *J. Nat. Med. Ass.* 63:455–459, 1971; MACKEY, E., JR. The black scholar and the black student: negotiating an alliance, *J. Nat. Med. Ass.* 64:23–31, 1972.
14. McCORD, W., HOWARD, J., FRIEDBERG, B., AND HARWOOD, E. *Life Styles in the Black Ghetto,* New York, Norton, 1969; RAINWATER, L. (Ed.) Black Experiences: Soul, Trans-action Books-6, 1970.
15. PIERCE, C. M. Problems of the negro adolescent in the next decade, in *Minority Group Adolescents in the United States,* edited by Brody, E. Baltimore, Williams and Wilkins, 1968, pp. 17–47.
16. PINDERHUGHES, C. A. Effects of ethnic group concentration upon educational process, personality formation, and mental health, *J. Nat. Med. Ass.* 56:411, 1964.
17. POUSSAINT, A. Personal communication, Symposium on the Black Psychiatrist in the Black Community, Los Angeles, June, 1969.
18. ROBINSON, A. L., FOSTER, C. C., AND OGILVIE, D. H. (Eds.) *Black Studies in the University,* New York, Bantam Books, 1969.
19. SOBEL, R. Special problems of late adolescence and the college years, in *Modern Psychoanalysis: New Directions and Perspectives,* ed. by Marmor, J. New York, Basic Books, 1968, pp. 476–492; COLES, R. Students who say no: black, radicals, hippies, *Internat. Psychiat. Clin.* 7:3–14, 1970; HAMMOND, C. D. *op cit.*
20. THOMAS, A. Pseudo-transference reactions due to cultural stereotyping, *Amer. J. Orthopsychiat.* 32:894–900, 1962; GRIER, W. H., AND COBBS, P. M.

Black Rage, New York, Basic Books, 1968; BANKS, G. P. The effects of race on one-to-one helping interviews, *Soc. Serv. Rev.* 45:137–146, 1971; SILVER-MAN, P. R. The influence of racial differences on the negro patient dropping out of psychiatric treatment, *Psychiat. Opin.* 8:29–36, 1971.

21. THOMAS, C. W. Black-white campus issues and the function of counseling centers, in *Psychology and the Problems of Society*, ed. by KORTEN, F. F., COOK, S. W., AND LACEY, J. I. Washington, American Psychological Association, 1970, pp. 420–426.

22. TORREY, E. F., VAN RHEENAN, F. J. AND KATCHADOURIAN, H. A. Problems of foreign students: an overview, *JACHA* 19:83–86, 1970.

Developmental Life Transitions: Intimacy, Marriage, and Parenthood

The increasing openness and flexibility of our society is changing the nature of the relationships involved in marriage and parenthood, and making available other options like living together and divorce. This increases the importance of effective coping and adaptation, as fewer rules are given and people are required to make more fundamental decisions about the structure of their relationships. Establishing an intimate relationship (with or without marriage) and becoming a parent are major transitions presenting individuals with developmental tasks that are stressful but that also have significant rewards. Unlike earlier generations when marriage was quickly and almost inevitably followed by parenthood, modern birth-control methods and new role opportunities for women have made marriage and parenthood separate issues.

The normal developmental processes of increasing emotional ties to peers and of decreasing emotional dependence on the family eventually lead to a significant affectionate or love relationship between two unrelated individuals. Through marriage, or the increasingly common practice of living together without marriage, couples encounter the problems (and rewards) of adapting their individual differences in life style to each other. Erik Erickson[2] groups these various tasks and situations under the concept of "intimacy." Intimacy involves not only sexual relations and more long-term couple planning, such as for parenthood or owning a home, but also day-to-day issues such as mealtime practices and housekeeping standards.

In the first article Norman Lobsenz dicusses couples who live together without being married, their problems and the differences between their relationships and marriage. Lobsenz interviewed couples in their twenties and thirties, who saw their arrangement as a permanent alternative to marriage. The majority of college students who have tried living-together arrangements have been in briefer relationships; they still consider marriage a possible future alternative. For these people, living together is a new life-cycle stage coming between adolescence and marriage. It is not "trial marriage" in the sense that the couple expect to marry each other, but it serves as preparation for marriage nonetheless by giving experience in close couple living and in developing communication and coping skills that will be helpful regardless of the eventual marriage partner. The limited nature of the commitment among the more temporary arrangements is evident in that the majority of these couples maintain at least the fiction (and often the reality) of separate residences, and in their unwillingness to inform their parents of their new status.[1,5,6]

For the most part, Lobsenz's couples were involved in more stable arrangements in which significant adaptation was required. They encounter some of the same issues married couples do, despite the supposed freedom of their relationship from problems like sexual jealousy or falling into marital, gender-defined roles. For some people a living-together arrangement is a practical way of coping with a fear of marriage and the long-term commitment to a relationship that marriage implies. The absence of this commitment is the real issue with which all unmarried couples must come to grips. The ease with which the arrangement can be terminated often discourages serious efforts to work out painful disagreements. In addition, if the decision to remain together must in theory be renewed every day there is no emotional security for either individual. The widespread agreement by participants that living-together arrangements are worthwhile suggests that they do offer an opportunity for couples to confront and work out an intimate relationship, even when they are not prepared to make the long term commitment that marriage involves.

In the second article Harold Raush, Wells Goodrich, and John Campbell discuss in more detail the process of establishing a marital relationship in an open-structure society. In such a society, where the task is "not . . . adapting to what *is*, but . . . working out what *is to be*," couples must find their own methods of resolving conflict and making decisions, of dealing with the issue of autonomy versus mutuality, and of determining acceptable levels of empathy and support. For many couples the traditional honeymoon provides a special setting, withdrawn and isolated from parents and friends, in which to begin the development of intimacy.[7] Daily routines must be adjusted as for living-together couples, but married couples must also establish satisfactory relationships with each other's families and decide whether and when they will become parents.

During the early months and even years of marital closeness when couples are actively exploring each other's capacities and limitations—the "psychic honeymoon"—the essential structure and character of the marriage begins to take shape. For example, some balance is struck between individual autonomy and compromise for the sake of mutual agreement. The quality of communication skills the partners can draw on or develop determines to a large extent their ability to cope with the demands for change that marriage creates.

Much of the discussion over the last fifteen years about the impact of becoming parents for the first time has been in the context of family crisis theories. It has long been known that some people, because of emotional instability, immaturity, or unfavorable social or economic circumstances, have great difficulty functioning effectively as parents. The "crisis" studies have revealed that even those young adults who have no special handicaps, and seem well-suited to parenthood, find the transition a stressful one. LeMasters[4] reported that 83 percent of the new parents he interviewed considered their experience to have constituted either an "extensive" or "severe" crisis.

As with other developmental transitions, like going away to college or getting married, there are rewards as well as stresses. In a study of new fathers Greenberg and Morris[3] describe a phenomenon they call "engrossment," in which fathers become totally and pleasurably absorbed and preoccupied with their newborn infants in a way they had not anticipated. However, one of the features of this transition that sets it apart from the others is its irreversibility.[8] Also, the parents' ability to cope is dependent not just on themselves but on the unpredictable inherent nature of the child. As Russell[9] points out, the parents of a calm, even-tempered child are likely to have a much easier transition than those of an active, "fast-excitable" type of infant.

The third article, by Everett Dyer, identifies some of the factors affecting the degree of difficulty perceived by new parents and perhaps the effectiveness of their coping behavior. Dyer found 53 percent of his thirty-two couples experiencing "extensive" or "severe" and 38 percent "moderate" crises. The responsibilities and restrictions of the new parental role were frequently mentioned as problems. For example, new mothers were troubled by feelings of being "tied down" and by exhaustion due to loss of sleep. Fathers worried about the increase in expenses, just as the wife's income had stopped, and felt somewhat neglected by their wives. A lower level of crisis was associated with self-ratings of good marital relationships, with having taken preparation-for-marriage courses in school, with having been married three or more years, and with planned rather than random parenthood (see also Russell[9]).

Many of these factors are related to the issue of preparation. As we saw earlier in Signell's article on kindergarten entry (Part II) and Silber and others on college entry (Part IV), some problems can be anticipated and prepared for. Also, the skills learned in previous experiences can be applied to the new situations. Thus, for example, couples with a good marriage may have already developed useful coping techniques that they can apply as new parents. Similarly, preparation-for-marriage courses help couples in anticipating the problems they may encounter. Even for those who did have serious problems, though, a large majority made a reasonably good adaptation, reorganizing and adjusting to incorporate their new roles as parents into their lives.

REFERENCES

1. ARAFAT, I., AND YORBURG, B. On living together without marriage. *The Journal of Sex Research*, 1973, *9*, 97–106.
2. ERIKSON, E. H. *Childhood and society* (2nd Ed.). New York: W. W. Norton, 1963.
3. GREENBERG, M., AND MORRIS, N. Engrossment: The newborn's impact upon the father. *American Journal of Orthopsychiatry*, 1974, *44*, 520–31.
4. LEMASTERS, E. E. Parenthood as crisis, In H. L. Parad, ed., *Crisis Intervention: Selected readings.* New York: Family Service Association of America, 1965.
5. MACKLIN, E. D. Cohabitation in college: Going very steady. *Psychology Today*, November 1974, pp. 53–59.
6. PETERMAN, D. J.; RIDLEY, C. A.; AND ANDERSON, S. M. A comparison of cohabiting and noncohabiting college students. *Journal of Marriage and the Family*, 1974, *36*, 344–54.
7. RAPAPORT, R., AND RAPAPORT, R. N. New light on the honeymoon. *Human Relations*, 1964, *17*, 33–56.
8. ROSSI, A. S. Transition to parenthood. *Journal of Marriage and the Family*, 1968, *30*, 26–39.
9. RUSSELL, C. S. Transition to parenthood: Problems and gratifications. *Journal of Marriage and the Family*, 1974, *36*, 294–301.

Article 12 Living Together

A Newfangled Tango or an Old Fashioned Waltz?

NORMAN M. LOBSENZ

A funny thing happened on the way to unmarried happiness: Countless couples discovered that the very problems they hoped to avoid were waiting for them as they crossed the threshold.

At a recent meeting of sociologists, a group of experts on what was referred to as "cohabiting couples" spent much of their time trying vainly to define that term. Some of the experts contended that in order to qualify, a couple must have shared a bedroom a specific number of nights for a minimum of weeks. Others said that a couple could be considered living together "if they thought they were." This nitpicking extends to other facets of the subject as well. Researchers bedeviled by the need to have a shorthand way of labeling cohabitors suggested such acronyms as CUs ("consensual unions"), LTUs ("living-together unmarrieds") and even UNMALIAS ("unmarried liaisons").

All this adds up to another example of science (in this case, sociology) scurrying to catch up with life. For the reality, of course, is that increasing numbers of young men and women are choosing to live together rather than to marry. The latest U.S. Census figures list a quarter of a million such couples—a total that presumably is just the official tip of a huge statistical iceberg.

In view of the growing number of these couples, and of the many others who might be considering joining their group, the editors of *Redbook* felt that it would be useful to try to learn what living together is all about. Do the ideals work out in practice? Does the relationship live up to expectations or is it booby-trapped with unpleasant surprises? Does living together require the same depth of emotional commitment that marriage presumably does? And if the commitment of couples who live together differs, how does it do so?

In an attempt to find some meaningful answers to these questions, I've spent

---

several months talking with scores of couples in their 20s and 30s who are living together—not those who wander through a series of brief affairs, but men and women who see their arrangement as a serious alternative to formal wedlock.

The responses were, as you might expect, a mixed bag. Hard and fast conclusions are difficult to come by, much less support. Moreover, there is no such thing as a "typical" living-together couple, and thus no way of judging to what extent the experiences and reactions of those I spoke with are representative. Nevertheless, the answers do help to provide a realistic view of a phenomenon that is very much with us.

In a college community in North Carolina I met a couple, both in their early 20s, who had decided to call off their scheduled wedding because the ritual began to seem hypocritical. Their explanation sums up the reason many couples give for their decision not to marry: "We want to share our lives, but we think we can do that more honestly by living together. You don't need a license to love, and what is in our hearts is more important than the words in a ceremony."

Other couples object to the legalistic aspects of marriage. They don't see why they should need a "piece of paper from the state" in order to join their lives. Their attitude is not so much a rebellion against social controls as it is a matter of practicality. Statistics indicate that one out of three marriages ends in divorce, so, these men and women say, why bother to get a marriage license when it's likely that you'll have to get a divorce decree later?

If it were possible, I wondered, would any of these couples be interested in legalizing their living-together status in some form other than marriage? They considered this possibility, and admitted that they were aware of the advantages: fewer complications with leases and insurance, perhaps some tax savings, less friction with parents. One woman observed: "If I could show my mother something legal that says it's okay for me to have sex, it would make things a lot easier for her." Still, nobody seemed enthusiastic; the consensus is that any legal device is unnecessary.

To a third group of couples the key motivation for not marrying is their wish to maintain their emotional freedom and integrity. "We don't want to get hung up on possessiveness and jealousy," one woman said. "Love should be a gift, not an obligation."

In all this there is a good deal of truth and common sense. But when I talked at greater length with some of these couples, I learned that often they had decided to live together and not marry for one reason—and continued to avoid marriage for quite another.

(In fact, many men and women did not consciously *decide* to live together at all; rather, they gravitated into the relationship. "Jerry would stay overnight once or twice a week," a 26-year-old secretary in the Midwest told me. "Soon he was staying over four or five nights. After a while it seemed foolish for him to go back to his own place at all, or to pay rent on an apartment he hardly used.")

Take Fred and Nancy, for example. They are a Boston couple who have been living together for three years. During that time Fred, a bearded advertising copy writer, became increasingly demanding, and Nancy, a slender, blond teacher, finally rebelled. Because neither wanted to leave the other, they went to a family service agency for counseling. (In some cities, by the way, couples who live together account for as much as 20 per cent of the caseload of marriage counselors.) After some weeks Nancy and Fred were able to admit to themselves their real motive for living together.

Nancy grew up in a cold, unaffectionate family, with a bossy father and a submissive mother. "My idea of a wife was a child-ridden woman who mostly did laundry and cooked," she says. "I wasn't going to be squashed into being that kind of person." And Fred, an only child whose parents had always given him whatever he wanted, subconsciously thought he would lose his power in a marriage. So when he felt anxious he made demands on Nancy in order to reassert his control—and Nancy fought back to maintain her independence. For them, living together was essentially a way of dealing with their hidden fears about marriage.

There is no question that couples who see living together as a chance to combine emotional closeness with emotional freedom work hard to achieve that goal. Linked by love and mutual concern, they are able to give each other areas of both spiritual and physical privacy. For some, however, there is a major conflict between theory and practice, especially when it comes to sex. Most of the couples I spoke with said they believed in the idea of individual sexual freedom but were practicing monogamy. "The theory is that we're both free to sleep with other people we find interesting or attractive," one woman said. "But he'd be jealous as hell if I did, and vice versa."

Another woman—30 years old and previously married and divorced—who has been living with her partner for more than a year, said she does feel free to date other men, but doesn't want to because the possibility of sex with them makes her uneasy. And a man of 31, an attorney, said he and his partner had ruled out sexual freedom, but not because they felt jealous or possessive. "We're working hard at living together," he declared, "and if either of us gets involved with someone else, we'll deprive our relationship of emotional energy."

Most couples who live together are aware of the irony implicit in remaining "faithful" when physical faithfulness is one of the conventional moralities. Some believe that jealousy is just too rooted in human nature to be overcome easily. After all, they say, we are products of our past, and we carry the cultural baggage of that past around even though we may not want to be burdened by it.

A few individuals who have sexual relations with people other than their usual partner are surprised to find that they feel guilty about it. "I didn't think you *could* feel guilty about something you had permission to do," one woman observes. "When you know intellectually that it's okay, that you've both agreed to it," a man says, "it's even more upsetting to discover that lump of guilt."

Another development that surprises a good many of the couples who live together is that as time passes it becomes more and more difficult for them to avoid falling into marital "roles" and patterns. Take household chores, for instance. No matter how firmly living-together couples set out to share them, to avoid traditional sex roles, there is a disillusioning tendency for the woman to inherit most of them. Even if she also holds a job, it may not be long before she is doing the cooking, cleaning and laundry, while he performs the traditional American male task—taking out the garbage.

When living-together couples socialize, their friendships tend to be couple-oriented—the same two-by-two social life married persons have. It seems, however, that there's also far more tolerance for each partner's separate friends, and for friendships that cross sexual lines, than exists in most marriages.

Part of the reason, perhaps, that some couples are drawn into traditional patterns is that they tend to assume—albeit with varying degrees of reluctance—the camouflages of marriage in order to avoid legal and social problems or family pressures. Gradually this pretense becomes the shape of their reality. This is especially true in small towns that do not offer the anonymity and privacy of large cities. To avert difficulties with landlords, utility companies, employers or neighbors, couples may pretend to be man and wife.

Apart from sex and sex roles, another area where theory and practice often conflict is money. The couples I spoke with were divided almost equally in the way they handled their finances. Some deliberately do not pool their income or keep track of who pays for what; others combine their assets and share expenses.

"We keep our money separate and it really bugs me," a 25-year-old woman in New England said. "I believed that living together meant sharing everything. But when I suggest to Bob that we pool our earnings—and I earn more than he does—he gets furious. I suppose it hurts his pride. He can't see it as my wanting to *give*."

Pooling incomes and sharing expenses often works well unless the balance is suddenly upset with one person's earning more than the other or becoming extravagant. A woman graduate student had been getting an allowance from her parents, who didn't know she was living with a man. When they found out, they promptly cut off her allowance.

"I figure if Peter and I economized, we'd make out all right," she said, "and I could finish my year's schooling. But when I broke the news to Peter, he suggested we stop sharing and each pay our own way from then on. I was appalled. He pointed out that he didn't *have* to share his income with me. After all, he said, it wasn't as if we were *married*.

"We're still together," the woman continued, "but obviously it's not the same as it was before. I'm just scrambling to hold on and get my degree, and then I'll have to do some hard thinking about our future."

Although lack of money is, of course, one reason many of these couples live with few luxuries or comforts—a painting, a good piece of furniture, certain electric appliances—there seems to be another reason for their hesitancy to make that kind of investment. It is the secret worry that they may break up. Betsy, a serious young woman in Washington, D.C., mentioned that she and her partner Donald, during their three years together, had surges of enthusiasm for buying some nice things for their apartment and then surges of thinking, let's not. "I wonder about the emotional significance of our never having bought anything of real worth," she said. "I think it is a sign that deep down we were not entirely sure of going on together."

I pointed out that married couples planned for the future even though they realized the possibility of divorce.

"That's different," said Betsy.

"Why?"

She searched her mind, finally shrugged. "I don't know. . . . Maybe they have more of a sense of commitment."

Commitment. There finally is the key word. The echoes of its meaning not only underlie the paradoxes that complicate the process of living together; it is, in a way, the *ultimate* paradox. These couples pledge their future to the belief that they are giving witness to greater mutual commitment than couples who marry. They phrase it in various ways, but the message remains the same: "I'm there every morning because I want to be, not because I have to be"; "we work harder on our love because we aren't forced to."

Moreover, some couples say, living together makes it easier to be honest about exploring differences. "In marriage," one man said, "a serious fight can be terribly frightening, with its implications of a breakup. But Janet and I are separate to begin with, so we can fight without that fear."

A young woman in a New York suburb told how she and her partner fought so bitterly that they decided on several occasions to end their relationship. "But each time, the argument somehow opened a door to a new closeness. One night we had a fight in a restaurant, and I said I was leaving him. 'I don't want you to,' he told me, 'but if you feel you must, all right.' Then he asked very gently if I wanted him to spend that night in a motel while I packed my things. I looked at him and said, 'I want to go home with you.' If we'd been married, I don't think that would have happened. That kind of fight, my words, his resentments . . . they all would have had a deadly finality. And maybe I would not have been brave enough to say them at all."

Yet this continuing daily freedom to stay or go often has an effect surprisingly different from what living-together couples expect. It can, for example, sow doubts. If one person is always free to leave—"to pack my toothbrush in the morning and say good-by," as one woman put it—the other lives under a constant threat, never entirely sure where he or she stands emotionally. Moreover, if the door is always open, the disenchanted partners don't have to try so hard to improve things. And even if they do try, often it's hard to know where to begin. As one marriage counselor put it, "When these men and women find that they are not quite so 'free' as they imagined, when they encounter sexual difficulties or when they are disillusioned because their expectations fall short of reality, they do not know how to either strengthen or end their relationship."

Finally, the couple's right to choose each other anew every day can become an enormous emotional burden. For one thing, there is the implicit need to *make* that daily decision, even if it is made only subconsciously. For another, should the time come when one partner does not want to end the relationship, the breakup often seems to involve more guilt and pain than many divorces do.

In Washington, D.C., Betsy is about to break up with Donald after three years. "I think it would be easier if we could get a real divorce," she said. "It would help me explain what happened in terms that other people—our families especially—could accept. There would be social supports. This way is messier."

It seems that for most couples who live together, the basic conflict has to do with commitment—what it means to each partner and how each reacts to it psychologically. The majority draw a line between emotional and legal commitment. Marriage, they feel, is a form of social pressure to stay together that they are free to ignore (except, most say, if they are having a child). If we are secure in our love, they feel, we don't *need* the technical commitment of marriage.

If the marriage ceremony is as meaningless as some couples say, why do they think it represents such a commitment? And if marriage is a pointless form of that commitment because one can easily divorce, why isn't living together just as pointless, since either person can leave at any time? And don't most married couples stay together because they *want* to, rather than because they signed a paper or took a vow? These are hard questions that must be answered by anyone who is making a choice of life-styles.

For no relationship can survive for long unless the partners show some concrete evidence of commitment to it. And if it continues, it can grow and deepen only if each person makes an increasingly greater commitment. The Catch-22 for couples who live together is that the ultimate commitment in this progression, is, of course, marriage.

Dr. James R. Ramey, director of the Center for the Study of Innovative Life-styles, in New City, New York, says that commitment occurs when two people value the *bond* between them more than anything else in their relationship. A number of men and women who choose to live together rather than to marry indicate this attitude is missing for them. They talk in terms of "freedom for self-growth"or "wanting to have my needs met." They are, in a word, *self*-involved.

One facet of this self-involvement is an often unacknowledged wish to avoid certainties in emotional relationships. Getting close to a person isn't as frightening as staying close. "But when you feel deeply," says a wife who is also a marriage counselor, "you want to tie your feelings down. Inside my wedding band is engraved 'Always,' not 'Probably always.' "

And the late Dr. Nathan W. Ackerman, a pioneer in family therapy, said, "Marriage is yes or no, not maybe. It is the symbol of readiness to cope with commitment and responsibility."

One young woman, assessing the long-range potentials for successful living together, summed it up this way: "If you have a good relationship, it's probably as rare as a good marriage—and it probably will lead to one." Otherwise, she implied, it probably will end, as most not-so-good relationships do. Most of the couples I talked with seemed to be coming, with great reluctance, to much the same conclusion. They were gentle, loving, serious young men and women—not rebels, not exploiters of other people's egos, not social visionaries.

Like any two people who live together, married or not, their abilities and limitations are being tested. They are disturbed by what is happening to their hopes and dreams for a new kind of loving relationship, but they are trying hard to maintain them.

For some of the couples, it seems likely that if they do not move toward marriage, they eventually will move away from each other. Other couples say they are going to keep striving to make this new structure work. The most heartening fact is that both groups of couples are convinced that living together is a good thing to have done—they have learned, they have grown, they have changed. And I wonder: Can one ask much more from any relationship?

Article 13 Adaptation to the First Years of Marriage

HAROLD L. RAUSH

WELLS GOODRICH

JOHN D. CAMPBELL

This paper is an outgrowth of a program of research on the early stages of family development, in which a pilot study has been undertaken of young middle-class couples in the early months of their marriages. We conceptualize this period as a stage in the life cycle, which has, like other stages, its characteristic functions, problems, and tasks. We have tried to evaluate the effectiveness of adaptation of the couples in meeting these tasks, both in terms of the present and in terms of laying the groundwork for the future developmental stages of the family. The two couples described in this paper are coping with the problems of initial marital adaptation within what we call an "open" structure—that is, one in which a great many of the solutions are not predetermined by the society, and it is left open to the couple to decide, for example, exactly where the sex-role boundaries shall lie, and what their relationships with their own families shall be. We hypothesize that the open structure places a heavier burden upon the effectiveness of interpersonal communication between the marital partners than would be true in a traditional and precedent-bound structure, and for this reason our evaluation of adaptation depends strongly upon a study of the quality of their communication.

THE CONCEPTS OF DEVELOPMENTAL STAGE AND ADAPTATION

Although most of our study involves "harder" data, the crude fact appeared, from the beginning of our work, that some couples seemed to us to be coping more

HAROLD L. RAUSH, PH.D., is Professor of Psychology, University of Massachusetts, Amherst.

WELLS GOODRICH, M.D., is Professor of Psychiatry and Pediatrics, University of Rochester Medical School, Rochester, New York.

JOHN D. CAMPBELL, PH.D., is with the Laboratory of Socio-Environmental Studies, National Institute of Mental Health, Bethesda, Maryland.

Source: Psychiatry, Vol. 26 (November 1963), 368–80. Copyright by The William Alanson White Psychiatric Foundation, Inc. Reprinted by special permission of The William Alanson White Psychiatric Foundation, Inc.

effectively with the issues of early marriage, and others seemed to be coping less effectively. Notions such as effectiveness or adaptation are value-oriented, yet difficult to get away from. They may form an unassimilated cluster of assumptions influencing both research and theory. It thus seems necessary to recognize our basic concern with effective coping and adaptation and to attempt to give these concepts greater precision.

An appropriate context for appraising coping and adaptation is provided by the notion of develomental stages in the marital career. A developmental stage is neither purely biological nor purely social. A stage is rather a biosocial phenomenon, defined by the meeting of biological and social forces. Biological development places demands upon and challenges the social structure. Social structures, in turn, limit, define, and challenge the directions of biological development. Studies of adaptation, then, become concerned with the modes by which a synthesis is attempted in the dialectic between a given biological phase and a given social milieu.

The sequence of interactions set in motion around a transition to a stage can be abstracted to a unit which we have called a developmental transaction. For example, the neonatal development of a pattern of feeding is a developmental transaction, and among the tributaries to this transaction are the effectiveness of the infant's feeding system,[1] the mother's attitudes toward the many components of infant feeding and the intensity of these attitudes, and other social forces impinging on the feeding situation—such as the presence or absence of other children, the supportiveness or disruptiveness of other kin, and any of a variety of socioeconomic considerations.

From a functional view—of forces to be integrated, of tasks to be met, of problems to be solved—the effectiveness of the specific transaction can be evaluated in its own terms; in this case, effectiveness can be measured in terms of the extent to which the achieved synthesis results in an integrated pattern of feeding by which the infant derives nourishment, and the mother some measure of satisfaction, both without undue strain. The system is effective to the extent that its function is accomplished within the energy available to the system and without unnecessary strain on other ongoing systems. It is ineffective to the extent that it is not viable—at the extreme, when the infant fails to obtain the nourishment necessary for life.

Reuben Hill and others have shown the heuristic value of considering developmental events in relation to the requirements of a stage within the family life cycle.[2] Each such stage confronts the family members and the family system with culturally characteristic tasks and situations which define the adaptations that they are challenged to work out. An individual's or a family's mode of response to a given stage may be either more or less effective from the point of view of adaptation. Whether defined as overt behavior patterns, conscious attitudes, or unconscious intrapsychic changes, these modes of response may represent effective or ineffective coping with the biosocial demands of the stage. The tasks and situations to be coped with at different stages are not identical, nor are the criteria for effectiveness of coping necessarily similar at different stages. Each stage has its central developmental issue; that is, it has its demand, or set of demands, which acts in a specific direction on an individual or on a system of relationships between individuals, and is relevant to the process of further development. As each new stage is ushered in by a transition event, new demands are placed on the family members and new developmental issues assume critical importance, becoming the foci for growth during subsequent developmental transactions.[3] As Erik Erikson has stated, the newly married couple is faced with a variety of new tasks and situations to be worked out which may be subsumed under the concept of *intimacy* as the central developmental issue. The subsequent stage of child-rearing presents the issue of *generativity*, which subsumes a qualitatively different set of new tasks and situations to which the couple must adapt.[4]

One aspect of adaptation has to do with the fulfillment of the biological and social functions of a particular stage in an appropriate manner, within a limited time span. But the transaction at a given stage must be evaluated also in terms of whether it fosters or hinders the transition to the next stage, and in terms of its contribution to subsequent stages. Erikson, in particular, has delineated the strands by which the transactions of each stage are interwoven to make the mesh of later stages.[5] His concepts of trust, autonomy, initiative, and identity, while they have particular relevance to particular life phases, take their major meaning through consideration of, and in context with, the entire life cycle. A judgment of adaptation, then, must be geared not only to the functions of a particular stage, but to the functions of the future. This is particularly difficult in times of rapid cultural change, when, by the time adulthood is reached, developments fostered in childhood may have become obsolete or dysfunctional, and new functions for which the organism is unprepared or misprepared may be especially called for.[6]

FUNCTIONS IN EARLY MARRIAGE

In longitudinal research on early family development, we have been investigating two developmental stages, the initial phase of marriage and initiation into parenthood, and the transition between them.[7] We are interested in describing various modes of integrating the demands and challenges of these stages and in learning something about the effectiveness of these modes.

The functional relationships between biological and social forces are in some ways clearer, and effectiveness is more easily conceptualized, for the earlier stages of life, as in the infant-mother feeding example we have cited. The narrow biological and environmental ranges of infancy expand and differentiate, in the later stages of life, into great diversity of biological and social functions and diversity and subtlety of modes for integrating these functions. Yet most social scientists agree that the family is an institution designed to solve a number of common human needs.[8] Procreation and child care are most often referred to, regulation of sexual union is often mentioned, and there are frequent references to economic, educational, and religious functions.

The problem of specifying functions and evaluating the effectiveness with which these functions are fulfilled becomes more manageable if one restricts oneself to a rather homogeneous segment of a single society, as we have done in our selection for a pilot study of recently married middle-class couples who fall within a narrow age range.[9] We hope that the same criteria are at least partially relevant for most of these young people and that the differences are in the modes and the effectiveness with which they deal with similar problems.

For these couples marriage, as a transition event, signifies the formal initiation of a series of problems to be coped with. Some of these involve working out and modifying behaviors and attitudes with respect to the following: (1) The sexual relationship. (2) The establishment and maintenance of a household. (3) Relationships with each other's families. (4) Relationships with friends and the establishment of new friendships. (5) Educational, occupational, or career plans for both husband and wife. (6) Plans for future parenthood. (7) Mealtime rituals and relationships concerning food. (8) The handling of money. (9) Styles of avocational, political, or religious activity and attitude. (10) Situations of physical intimacy, including nudity, dressing habits, sleeping and waking habits, and so on.

But if we are to describe and evaluate the effectiveness of particular couples in coping with these *specific* areas of interaction, it becomes necessary to examine other, more *general* functions. These general functions involve questions of adaptation and personal satisfactions and fulfillment, including, for example, the fit of individual identity with available actual social roles, the clash of inner conscience or value system with cultural norms, and the relation of available societal challenges and rewards to personal needs.

In more operational terms, general functions involve such processes as negotiating a balance between individual autonomy and mutuality, mediating one another's needs, and integrating modes of communication. Of particular relevance in early marriage is the development of patterns for resolving interpersonal conflict and making decisions. Each of these functions is in turn related to such variables as empathy, supportiveness, understanding, and effectiveness of communication. And all of these aspects as they are coordinated with and reflected in the *specific* functions listed above contribute to the coalescence which Erikson describes broadly as intimacy.[10]

STRUCTURES IN EARLY MARRIAGE

In research and in theoretical discussions, the concepts referred to as *general* functions become particularly relevant in modern societies in which traditional sex-role boundaries have become somewhat diffuse, and marriage has shifted from what we shall call a closed or predefined structure to what we shall call an open or emergent structure. The concepts of closed and open structures may be applied to several levels of analysis of personal and social phenomena. Thus, one may speak of closed and open structures within persons, or dyads, or societies. Similarly, the concepts can be applied to interrelations among levels. Closed and open structures are, however, used here to refer primarily to the husband-wife relationship. As shall be seen, the terms closed and open might be thought of as distinguishing between self-regulating and self-organizing systems, respectively.

We may speak of effectiveness in the sense of coping with what is predefined. The predefinition may be an externally imposed one of social or economic limitations, it may be a matter of clear-cut traditions and values, or it may be by personal identifications, character development, and unconscious pressures.[11] In some aspects of life, the developmental choices and the barriers and the roles are more clearly delineated than in others. Different social groups differ in the aspects delineated, in the extent and rigidity of delineations, and, of course, in available choices and barriers.[12] For example, in Catholic marriages the moral barrier against divorce reduces the availability of escape from the marriage as a reaction to conflicts. When the solutions to stage-relevant problems are more defined, effectiveness of coping demands consideration of different variables than when solutions are more undefined.

Some examples may make this clearer. In a society in which social relationships with kinship groups are clearly prescribed—as is true of many societies—the relationships of a specific couple with their parents are to a large part determined. When the economy does not provide the newlywed couple with the opportunity to set up a separate household, they must adapt to the roles available in the family of orientation. Similarly, where food is highly limited in quantities and varieties—as is true throughout most of the world—there is little room for choice or for conflicts between marital partners as to what meals are to be like; the solutions to the problems of diet are already defined by the economy.

There may still be much to be coped with. Establishing and maintaining the prescribed forms depends on socialization experiences and their adequacy for the inculcation of the appropriate roles. Coping not only involves gratifications and opportunities but may impose strains, defenses against these strains, and inhibitions on the development of personal capacities. An evaluation of the effectiveness of coping must consider, then, the psychic expense of various modes of dealing with prescribed paths and barriers.

Thus if a social system is closed at a given point, effectiveness of coping involves the nature of the adaptation to what *is*. For the investigator, this involves an evaluation of the strains and inhibitions imposed by various modes of adaptation (including, of course, efforts to change the system). Since the interpersonal field in the closed system is relatively fixed and stable, the demand is upon the person to alter his unique needs and capacities so as to adapt to the system. Ineffective coping will tend to be associated with intrapsychic conflict and the use of ego defenses such as repression and denial; overt signs of such ineffective coping will include manifest anxiety and forms of symptomatic behavior characteristic of intrapsychic conflict. And although such symptomatic behavior may be enacted on the interpersonal scene, the primary struggle is *intrapersonal*. Conversely, effective coping may be recognized by the efficiency and smoothness with which the specific functions of marriage are accomplished. But with successful as with unsuccessful coping, the primary arena in a closed system is likely to be the intrapersonal.

In a system which is open at a given point, effectiveness of coping does not involve adapting to what *is*, but requires working out what *is to be*. If the solution to a problem is not prescribed, adaptive ego functions and interpersonal factors can take a larger role. Thus, if there are few rigid traditions about child-rearing roles, the area can become an arena for interplay between individual desires. There is more room for discussion, more interplay in decision, and a greater possibility of clash between individuals. Consideration of effectiveness of coping must involve, then, not only the adequacy of a solution to a problem—in terms of integrating the biological and social forces of the particular stage, and in terms of preparing for future stages—but it must also involve considering those gratifications, strains, and inhibitions which are part of the *process* of *working out* a solution between two unique personalities.[13]

We sometimes see the difference between open and closed solutions in different sections of our interviews with recently married couples. When we ask the couples about how they have happened to work out their particular arrangements about cooking and serving food, they often look blank and cannot answer. It is traditional that women cook and serve food, and even though this may involve personal strains or limitations, there is little interpersonal *working out* of this particular aspect. When we talk with them, however, about relationships with their families or plans about parenthood, they are often much more explicit about decision-making, conflict or mutuality, understanding or lack of understanding of one another. Since in these areas there is more room for individual choice, the arena for coping is more likely to be *interpersonal*.

Family life in the United States has been and apparently still is undergoing rapid cultural shifts.[14] The solutions to specific functional problems, such as those of the sexual relationship, the establishment of the household, and the determination of occupational and friendship patterns, are no longer so traditionally defined. We do not usually see the close-knit social network that Elizabeth Bott describes for British factory-worker families.[15] Even among those of our couples who have close relationships with their immediate families, the attitude is usually that the newlyweds are to work out their own solutions to their problems.

The openness of most of the systems for our middle-class, urban couples requires the consideration of variables related to the *general* functions described above. Where decisions are all to be made without the guide of clear precedent, adequacy of the decision-making process becomes important. Where a couple is to work out its own solution to a stage-relevant problem, concepts such as intimacy, empathy, supportiveness, mutual understanding of activities, and communication become critical for adaptation to initial marriage. It is perhaps the importance of these process variables in an open system of interpersonal relations that leads to what has been called the "psychic honeymoon"—the sometimes extended postmarital period when through close intimacy and joint activity the couple can explore each other's capacities and limitations. One hypothesis for research is that when the paths and barriers and roles in the solutions of problems are well defined, these process variables receive less emphasis and the phenomenon of the psychic honeymoon is less extended. Be that as it may, in studying newlywed coping in middle-class, urban couples, evaluation of effectiveness of coping must include such aspects.

COMMUNICATION IN AN OPEN MARITAL STRUCTURE: TWO COUPLES

With the above considerations in mind, we shall attempt to describe the patterns of communication of two couples. With the assumption that both marriages involve open marital structures, one of these patterns seems to us to illustrate very effective coping, the other pattern ineffective coping. Our aim in this examination is toward a phenomenological description of some modes of coping and toward a clarification of concepts, particularly as these concern affective communication between husband and wife.

If one looks solely at the specific functions, it is apparent that Mr. and Mrs. Gerhart (the names, of course, are changed) have many things in common. They have much the same values and interests. They are pleased with their housekeeping arrangements in their apartment, and they are agreed about the kind of house they want in the future. As with almost all of our couples, both partners work. Both agree that Mrs. Gerhart should stop working after the first child is born. Meanwhile, Mr. Gerhart does a considerable share of the work around the house. Arrangements about food and eating, handling of the budget, and relationships with friends have been worked out in a mutually satisfactory fashion. Mrs. Gerhart wants to have a child in a year, and although Mr. Gerhart is less certain, he goes along with her wishes. They both are ostensibly satisfied with the way things are working out in the sexual area, but the discrepancy between their reports in the individual interviews leads one to suspect that it does not go as well as they say. Mr. Gerhart would prefer that Mrs. Gerhart's mother be less critical of her; he claims, however, not to be upset by the situation.

The Vreelands also have many interests and values in common. They too both work, and, in addition, both are continuing their education. Mr. Vreeland helps less around the house than does Mr. Gerhart, but Mrs. Vreeland is well satisfied with the arrangements. The planning and handling of the food situation is very satisfactory to both Mr. and Mrs. Vreeland, and while their budget offers some problems, they are facing these realistically and jointly and without disagreement. They agree that they want to have a child in the very near future, and they seem more informed about infant care than are the Gerharts. Unlike the Gerharts, in the individual interviews both Mr. and Mrs. Vreeland report some problems in relation to sex; they do not see these problems as at all insoluble; their sexual relationship is, they say, improving, and they are very optimistic about the future. There are no problems in their relationships with their immediate families.

As far as these bare outlines go, both couples seem, then, to be coping adequately with the requirements of this stage of early marriage. They have worked out different patterns, but these seem to fit with the values, interests, and capacities of each couple. It is only when we turn from the content to note the processes within the relationship that we see a major difference in effectiveness of coping. When new situations arise for which stereotyped answers will not do, when the couple is faced with choices which can be resolved only by relying on individual identity and style rather than upon cultural traditions, when there is interpersonal conflict, our expectation is that the Vreelands will be highly effective in coping and that the Gerharts will be highly ineffective.

Interview Responses
It is here we turn to differences in patterns of communication in these two pairs. Apparent in the very first interview—in the home with both partners— is the high level of interaction between the Vreelands. They are relaxed and spontaneous with each other. Their position in relation to the interviewer is a comfortable balance between individual autonomy and mutuality as a couple. That is, they are capable of expressing their individual views, yet they are aligned with one another. The words *we* and *us* are often used. In contrast, the Gerharts interact much less, and seem much stiffer. Their expressions convey neither a comfortable autonomy nor a comfortable jointness. There is a quality of inhibition and defensiveness, not only in relation to the interviewer but in relation to one another. The words *I* and *me* are prominent.[16] That these are not solely stylistic differences or solely reflections of comfort in the interviewing situation is shown in parallel examples which occur very early in the first interview.

Mrs. Gerhart is asked about major changes in her family when she was a child. After some initial description, she begins very hesitantly to talk about her mother's tuberculosis which occurred when she herself was about nine, and which was followed by a period of some three years in which the mother spent varying lengths of time shifting between hospital and home. Mrs. Gerhart's anxiety makes it difficult for her to describe the situation very clearly, and the interviewer asks several questions to clarify some of the changes that occurred at the time. He asks: "[This happened] from the time you were nine to the time you were about twelve, huh?" Mrs. Gerhart says: "No. Nine until twelve, about twelve—" Mr. Gerhart has been silent during this discussion, which is obviously a difficult one for his wife. At this point, however, he interrupts to say: "You even have me confused, if you want to know the truth." She defends herself by saying that she cannot help being mixed up.

Again, sometime later in the interview Mrs. Gerhart is talking about a conflict with her family when she began going seriously with Mr. Gerhart before their marriage. She refers back to her concern about her mother's illness and to difficulties the mother had with the maternal grandmother. Mrs. Gerhart's talk again gets a bit disjointed. Apologetically, she says to the interviewer: "... if I'm rambling on, say so." Mr. Gerhart says at this point: "[It] sounds like a drama you hear on the radio—in the afternoon," and he laughs.

In the first home interview with the Vreelands, Mrs. Vreeland is discussing her school years. She talks about a change in schools and about a decline in her interest in school and in her grades when she was about sixteen. When asked by the interviewer why the change occurred, Mrs. Vreeland has some difficulty. She talks about being unhappy in the big school she had been going to, about feeling that she had to make a change, and then she adds: "... I worked very hard and, uh, I—I just know I wasn't getting the recognition at home that I needed, I wasn't getting the support, so I just gave up. I know that's the reason. So—uh—I just made the change." The interviewer begins to ask another question, and at this point Mr. Vreeland interrupts. Softly he says: "There's something she hasn't mentioned yet—it probably doesn't stand out—as a change—and that's that her father drank." Mrs. Vreeland then talks about the difficulty this created during her school years, and about the financial insecurity in the home. Mr. Vreeland adds very, very softly, "And he gambled." And Mrs. Vreeland begins to describe how the problem of her parents related to the decline in her school grades. At a later point in the interview, Mrs. Vreeland speaks of bringing home a magazine article which described some of the problems that children of alcoholics might have. She and her husband read the article together and with it as a stimulus they discussed some of her earlier difficulties. She talks about crying during this discussion, and her husband comments: "She doesn't cry about just anything." In describing their marriage, the wife speaks of their encouragement of one another and of their each bringing the other "the right kind of comfort."

We have a cue, then, from the fact that when there is a block in communication during the interview, Mr. Gerhart bolsters the block through distancing himself and, in one instance, through sarcasm. Mr. Vreeland, in contrast, acts so as to help his wife with the flow of communication and to assist her to clarify her memories while talking with the interviewer about anxiety-laden matters.

Responses in Experimental Situations

Very similar differences in husband-wife communication patterns can be obtained under conditions of experimental interpersonal conflict.[17] In one of these investigative procedures for studying modes of resolving conflicts and making decisions, the couple is requested to improvise in playing out the resolution to a series of four standard interpersonal conflicts. The couple's actions, gestures, and shifts in position are observed and recorded, and verbal exchanges are tape-recorded. This procedure is one of several in which experimentally evoked communication patterns are obtained for comparison with the interview sequences described above.

Efficiency Versus Inefficiency. When we examine the behaviors of the two couples in these experimental situations, there is a clear contrast between the efficiency of the Vreelands and the inefficiency of the Gerharts in resolving conflicts and making decisions. The wastefulness and energy consumption of the communication process in the Gerharts is illustrated in their failure to resolve two of the scenes, and in their strained and contrived resolutions of the other two. In contrast, not only do the Vreelands resolve the issues, but also they use half the time used by the Gerharts, and this despite the fact that from all other indications the Vreelands are the much more verbal couple of the two. One cannot argue, then, that the Gerharts are simply less dependent on communication. If their generally lower level of communication were a function of their operating within a closed marital structure in which roles were clearly defined, one could expect their decision-making to be highly efficient. Their relative ineffectiveness in dealing with interpersonal issues lends plausibility to the assumption that the Gerharts, like the Vreelands, are coping within an open marital structure. The inadequacies in the Gerharts' communication with one another interfere with this coping.

Differentiation Versus Spread of Conflict. For the Vreelands a problem is a problem; for the Gerharts a problem acts as a fuse to inflame a wide range of sensitivities. The Vreelands, once they specify the issues, deal with them directly. There are only a few digressions, and these are never very far from the issue. In contrast, the Gerharts do not keep to an issue. An issue is not discussed or argued on its own merits, but expands so that it incorporates much more than what is ostensibly involved. For example, in one of the scenes, a conflict about Mrs. Gerhart's overeating turns into an issue of trust versus suspicion and of who is to do what for whom. The issue shifts so that it is no longer a specific disagreement or even an intense argument over a specific issue; rather the issue is transformed so that it represents threats and counterthreats to the balance of power and affection. The conflict becomes one involving basic bonds of the marriage, and the whole relationship.[18]

Mutual Recognition and Awareness Versus Manipulation and Defense. With the Vreelands communication serves to increase mutual awareness; with the Gerharts communication serves functions of manipulation and defense. The Gerharts talk a great deal. But although they talk *to* one another they do not talk *with* one another. In one of the experimental conflict situations, Mr. Gerhart attempts to convince his wife that *she* doesn't want to do the dishes and that *she* doesn't want to cook. These are attempts at manipulation which are ineffective since they do not at all reflect his wife's needs. She, on the other hand, attempts to trap his commitment by trying to convince him that she has a secret purpose which will benefit him if he yields to her. It is a complex game in which the husband pretends that he is interested solely in his wife's welfare and she engages in a play of which the theme seems to be: Give in to me for *your* sake; if you submit I will nurture you and give you a pretense of power. He pretends to be strong, she pretends to be weak; but she wields the power and he receives the nurturance, and neither is able to recognize the nature of this relationship. With the Gerharts many statements are designed to manipulate the other person. And the other person's statements are listened to only as threats to be parried or for gaps in the other's defensive line.

For the Vreelands, talk and listening do not have this defensive function. In one of the experimental conflict situations, Mrs. Vreeland helps her husband discuss his irritation with his job. She recognizes and helps him to communicate his feelings and then helps him to differentiate his feelings about his job from his feelings about home. In another such situation, Mr. Vreeland supports Mrs. Vreeland's expression of her subjective attitudes despite her early denial of the objective importance of her feeling. That is, when she says that what is bothering her is "probably not very important," he says: "That doesn't make any difference. Evidently you think it's important because of the way you've been acting lately, and if it's that important I want to know about it." With the Gerharts there is a lack of awareness or rejection of the subjective reality of the other's experience; each can only assert and reassert his own experience. In contrast, each of the Vreelands reinforces the subjective reality of the other's experience, and each works to specify and clarify what the issues are for the other. This mutual recognition makes for a marked efficiency when the Vreelands cope with a situation which is new, stressful, and interpersonally conflictual.

THE FUNCTIONS OF COMMUNICATION

Adaptation to the Present

The varied functions of communication are a fruitful area for an empirical ego psychology. Comparing the speech of hyperaggressive with that of normal children, Raush and Sweet suggest that for the hyperaggressive children words have an actional function. That is, words are produced and heard as signals directly touching off an affective or motor response. In speculating on a developmental-psychodynamic approach to speech function, they propose that in the less disturbed neurotic child, speech is not used actionally, but rather is used excessively for anxiety reduction—the defensive function overshadowing its function in need-gratification and problem solving.[19] This defensive function is prominent in the communications of the Gerharts. In contrast, speech function for the Vreelands has more often a transactional quality. That is, language is used for communicating specific content and for problem-solving.[20] In addition, there is at times in the speech of the Vreelands another function—which probably seldom, if ever, appears in the speech of children—that of conveying an aspect of communion related to the "I-thou" principle that Buber speaks of. This function can be apperceived in the "closeness" of the Vreeland's talk at times with each other, which differs from the appropriately transactional quality of their talk with the interviewer.

When, as in an open unstructured marriage, people must make many decisions not only about specific tasks, but also about what their goals are to be and

⊙

how they are to work together, the affective components of communication will become especially salient. How they feel about one another will form a major part of the relationship structure and will be a major determinant in the syntheses they reach. Since the affective bonds between the communicants have high salience, an effective synthesis requires that both speaking and listening must be infused—at least at times—with the special quality which Edith Weigert has described as sympathy.[21] With respect to communication, sympathy involves an empathic awareness of and respect for what the other person is feeling. In a psychoanalytic sense, it would involve momentary introjection of the affective components of the other's communications. The empathic recognition of the other can ease and expand the communication process at those times when the process might be threatened. It enables, then, at critical points, one person to foster learning rather than inhibition in another.

The Vreelands give an apt description of the value of this process in an openly structured marriage. Asked how his marriage is different from what he thought it would be, Mr. Vreeland says:

> . . . I hadn't really contemplated although I had been told. . . . I didn't really see that the adjustments in life would be quite as great as what I am beginning to realize they are now. And although I in no way think that it is something that I can't cope with and that it's going to hinder me in any great respect, I didn't really contemplate that you would have to change your life quite as much as what I am finding out now you really have to. . . . The big things we more or less agreed on before we were married, but it is the little everyday living things which, of course, we have no way of knowing before . . . that are popping up now from time to time and we are having to talk over. One way or another our ideas differ and we just have to agree. Some of the things are so petty you never would have thought of them— some of the day-to-day living such as making beds, eating meals, doing dishes, and this sort of thing. I don't know, a lot of it is my own fault, back on the farm where I grew up it was the woman's job and the woman's always. . . .

A bit later he says:

> . . . most of the things I've been of a firm belief . . . we should discuss, and I stress this, and when I know something is bothering her I try to get her to let me help about it. Likewise, when something bothers me I let her know about it and try to discuss things, and I think this is one of the things, I am of the opinion, will help us more than anything else that we do.

In her interview, Mrs. Vreeland talks about the change in responsibilities that come with marriage and the adjustment required to these responsibilities. Asked about any problems which they have difficulty in discussing, she says:

> . . . I guess each couple has discussions in which we, they, come to one another with little petty complaints or little grievances. . . . This was difficult

at first, maybe the first couple of times, to get one of these little problems up on the surface and discuss them, but this doesn't seem to be too much of a problem any more because we can both sense from the other when something is bothering us and we usually get it right out in the open.

The interviewer, who has already learned about Mrs. Vreeland's early problems with her family, asks whether she has any difficulty in discussing this with her husband. She says:

No, I have discussed my relatives with my husband pretty extensively because I feel that by understanding them and my reaction toward them now, in our present life, he can understand me more. I find it very easy to discuss my relatives with him because he sometimes gives me an assuring word when it is necessary, and if I am not quite right in my opinion he senses it and doesn't say anything. This makes me sort of think about what I am saying and understand that I am not totally right.

The Past and the Future

What facilitates the intimacy of the Vreelands as reflected in the effectiveness of their communications with each other? And what interferes with intimacy in the Gerharts so that they seem to move in tangential circles of isolation? For both couples there is evidence of parental psychopathology and childhood stress. Yet with each of the Vreelands one is impressed by a striving for mastery of the past and by an at least partial success in this striving. For each of the Gerharts, the past seems vague and unintegrated and characterized by diffusion rather than mastery, and the nature of defenses against anxiety seems more primitive. The Vreelands seem to have achieved a sense of identity and the Gerharts have not. Erikson states: "... it is only after a reasonable sense of identity has been established that real intimacy ... is possible." [22]

As marriages go on in time, as the partners get to know each other well, as they evolve more differentiated role-functions in the marriage, one would expect their involvement in the communication *process* to decrease. Communication will continue to be necessary as new areas of life come into the foreground. Growth problems of children, job difficulties, "middle-age," departure of children, old age—changing circumstances bring new crises and each requires continued and changing communication. But problems of *how* to communicate and how to relate to each other's modes of communication should presumably become less central and less a matter of broad exploration. The stage of early marriage enables interpersonal communicative modes to be established, and an open marital structure requires this. Both couples discussed above are newly married. We have suggested that the Gerharts have not as yet worked out effective modes of communicating with one another. This lack of resolution at this stage will, we expect, create difficulties for them in the future. Nevertheless, one cannot predict with assurance that the Vreelands will have fewer difficulties in their married life than will the Gerharts. The Gerharts have worked out a *modus vivendi*—albeit a somewhat restrictive one—in their marriage.[23] They may meet fewer—and, indeed, they are likely to avoid—novel situations, stressful situations, and interpersonal conflicts. But since it is unlikely that one can wholly avoid such problems, we anticipate that the ineffectiveness in the Gerhart's coping will show at certain developmental transitions. One of these critical transitions will occur, we would guess, when they become parents.

In the course of discussing communication, we have referred, at least connotatively, to a variety of other concepts relevant to adaptation to the stage of early marriage. Among these have been empathy, mutuality, supportiveness, autonomy, trust, inhibition, and defense. As dicussed earlier, each of these is relevant to intimacy. The Vreelands' capacity for intimacy should stand them in good stead for meeting the problems of the next stage, involving (in Eriksonian terms) generativity versus self-absorption. The Gerharts must still struggle with their psychological isolation, and in their interview discussion of parenthood we are able to see the contrast between their self-absorption and the Vreelands' budding generativity. The Vreelands will, of course, have troubles. There are in their individual characters some points of vulnerability. But the Vreelands are able to help each other learn. Their marriage has a capacity for growth, in contrast to the static quality of the Gerharts' marriage.

GENERAL CONSIDERATIONS

The general functions of early marriage, such as developing modes for resolving conflicts and making decisions, become particularly salient in cultures which foster an open marital structure. In contrast to a closed marital structure, in which coping involves an adaptation to what *is* and the primary conflicts and resolutions are *intrapersonal*, the open marital structure requires working out what *is to be*, and the primary conflicts and resolutions are *interpersonal* processes required for defining the development of the dyadic system. It is from this viewpoint that we have tried to evaluate initial marital adaptation within an open structure, to illustrate relatively effective and ineffective modes of communication, and to present some tentative criteria for such an evaluation.

There remains a general problem of the evaluation of open versus closed structures. This is only partly a matter of values. For given a world where the pace of social and technological changes is ever-increasing, new modes of adaptation are increasingly demanded. Systems must be sufficiently open to respond to these demands. The open marital structure is, then—whether one likes it or not—a social response to the requirements of modern societies. The positive side of this development has recently been emphasized by the Russells.[24] They note the evolutionary, social, and personal advantages to systems which are "progressive," "intelligent," and "exploratory," in contrast to systems which are "instinctive," "automatic," and "specialized." The open system allows greater freedom for personal and social development, and it has a greater capacity for creating as well as adapting to social change.

Yet the open structure has its problems. The Dutch psychiatrist van den Berg discusses the consequences of the progressive dissolution of stability over the course of history.[25] He notes the increasing multivalency of institutions, and the loss of universality and the gain in personalization of meanings, so that even very basic words such as *mother, father, child, husband, wife, work,* and *faith* lack communality of interpretation.[26] He notes, furthermore, the plurality in our lives, and the diversity presented to the growing child—whose classmates may all dress differently and come from different types of homes, with different values, habits, and rituals.[27] This lessened stability and increased diversity, van den Berg suggests, are related to the change in symptomatology of those who currently seek psychotherapy, as opposed to those who sought it in former times. The problems of identity diffusion which are so pervasive these days seem to be related to the transition from closed to open structures.[28]

The closed system may place a considerable burden on the defensive functions of the ego. In addition it may limit the development of individual capacities and adaptive work. Thus, the stable authoritarian family structure may be expected to foster a more rigid superego and to provide fewer opportunities for individual choice and individual development. On the other hand, the burdens of an open structure on the adaptive functions of the ego and on the interpersonal process can be considerable. The less stable, more permissive family structure may provide such a range of choices as to lead to periods of confusion and to the development of unclear ego boundaries.

Furthermore, in an open system the extent of individual autonomy is a matter to be worked out. For example, where specific aspects of child-rearing are more clearly defined as primarily a woman's function, she is free to develop her own modes of action within the defined cultural prescriptions. As the functional arrangements of responsibilities for child-rearing shift to a more open structure, the wife is freer to enlist the aid of her husband, but she must also be prepared to brook interference with her own favored modes of responding to children. Equality of status may thus, in a sense, frustrate needs for autonomy. And, as we have noted, the open as compared to the closed structure requires greater sensitivity to the potentials of the other, and more adequate communication. Attempts to escape from the burdens of the open structure may lead, as Fromm and Riesman suggest, to a search for peer-determined standards of conformity or to a search for the charismatic leader to whom these burdens can be abdicated.[29]

In the long run, the open structure seems to have the potential for fuller utilization of experience and for fuller differentiation among sources of information. In a rapidly changing world it will therefore allow greater possibility of adapting to change. And in terms of the maximization of human potentialities, we would judge the open system in marriage to be more effective. But under what circumstances it can be maintained is another question.

REFERENCES AND NOTES

1. RICHARD Q. BELL, "Relations Between Behavior Manifestations in the Human Neonate," *Child Development* (1960) 31:463–477.

2. REUBEN HILL AND D. A. HANSEN, "The Identification of a Conceptual Framework Utilized in Family Study," *Marriage and Family Living* (1960) 22:299–311.
3. J. D. CAMPBELL, D. W. GOODRICH, AND H. L. RAUSH, "Toward a Conceptual Scheme for the Study of Adaptation in Marriage," presented at the Seventh International Seminar of Family Research, Washington, D.C., September 9, 1962.
4. ERIK H. ERIKSON, "Identity and the Life Cycle," *Psychological Issues* (1959) 1:1–171; and *Childhood and Society*; New York, Norton, 1950.
5. See footnote 4.
6. The fact of rapid cultural change also increases the fallibility of investigators' judgment about functions likely to be useful to the developing personality.
7. See D. W. GOODRICH, "Possibilities for Preventive Intervention during Initial Personality Formation," pp. 249–264, in *Prevention of Mental Disorders in Children*, edited by Gerald Caplan; New York, Basic Books, 1961.
8. See, for example: GEORGE PETER MURDOCK, *Social Structure*; New York, Macmillan, 1949. TALCOTT PARSONS, "The Stability of the American Family System," pp. 93–97; CLYDE KLUCKHOHN, "Variations in the Human Family," pp. 45–51; AND FLORENCE R. KLUCKHOHN, "Variations in the Basic Values of Family Systems," pp. 304–315; in *A Modern Introduction to the Family*, edited by N. W. Bell and E. F. Vogel; Glencoe, Ill., Free Press, 1960.
9. These are 50 volunteer couples, all embarking on first marriages. They are initially seen jointly and individually on four evenings after three months of marriage and then are studied up to several months following the birth of the first child. Couples are in their early twenties, and we select only those in which the wives have not become pregnant by the third month of marriage. Various requirements favor a predominantly middle-class sample. The design of the study follows the general principles described in Richard Q. Bell, "Retrospective and Prospective Views of Early Personality," *Merrill Palmer Quarterly* (1959–60) 6:131–144.
10. See footnote 4.
11. The value orientations which Kluckhohn and Strodtbeck have used to characterize the attitudes held by a culture toward man, nature, time, and work clearly illustrate such environmental predefinitions. FLORENCE R. KLUCKHOHN AND FRED L. STRODTBECK, *Variations in Value Orientations*; Evanston, Ill., Row, Peterson, 1961.
12. KURT LEWIN, *Field Theory in Social Science*; New York, Harper, 1951. D. R. MILLER, "Personality and Social Interaction," pp. 271–300, in *Studying Personality Cross-Culturally*, edited by Bert Kaplan; Evanston, Ill., Row, Peterson, 1961.
13. For a beautiful discussion of the relationships between intrapersonal and interpersonal adaptive processes, see Erik Erikson, *Young Man Luther*; New York, Norton, 1958.
14. See Urie Bronfenbrenner, "The Changing American Child: A Speculative Analysis," *J. Social Issues* (1961) 17:6–18.
15. ELIZABETH BOTT, *Family and Social Network*; London, Tavistock Publications, 1957.

16. A count of *we–us* versus *I–me* expressions confirms that the Vreelands differ significantly from the Gerharts in the suggested directions.
17. D. W. Goodrich and D. S. Boomer, "Experimental Modes of Conflict Resolution," *Family Process* (1963) 2:15–24.
18. This progressive spread of conflict is seen in much intensified form in the communications of disturbed children; see Harold L. Raush and Blanche Sweet, "The Preadolescent Ego: Some Observations of Normal Children," *Psychiatry* (1961) 24:122–132. This is not to suggest that the Gerharts are like such children, but rather that one of the phenomena associated with unconscious conflict is lack of differentiation and a tendency for conflict to spread.
19. See footnote 18.
20. See Heinz Werner, *Comparative Psychology of Mental Development* (Rev. Ed.); New York, Internat. Univ. Press, 1957.
21. Edith Weigert, "The Nature of Sympathy in the Art of Psychotherapy," *Psychiatry* (1961) 24:187–196.
22. See "Identity and the Life Cycle," in footnote 4; p. 59. The problem of establishing an identity is not the same as the problem of taking or modifying a role. The Gerharts try valiantly at times to take each other's role. But if one is in serious doubt about who he is, a pressure for role-taking is likely to threaten the integrity of character. The attempt to shift one's role, even momentarily, will be anxiety-producing—for example, in the obsessive. If the struggle for identity is given up, as with the psychopath, shifts in roles may be rapid and empty, and role variation will fail to contribute to character integration. Only the person who is somewhat certain about who he is can experiment or play with various roles so as to expand his self-definition. In childhood, role-taking can be played out within a microcosm, and if this playing-out is successful it can perhaps contribute to later more complete formation of identity. But in adulthood the formation of identity would seem to be a precondition for adequate role-taking.
23. Recent studies by D. R. Miller indicate that if one wishes to predict or understand stability in a marriage, the dynamic relations among a number of components must be considered. Miller implies that stability, trait compatibility, or trait complementariness are not in themselves sufficient criteria for marital effectiveness, in the sense that the term has been used here. See footnote 12.
24. Claire Russell and W. M. S. Russell, *Human Behavior*; Boston, Little, Brown, 1961.
25. J. H. van den Berg, *The Changing Nature of Man*; New York, Norton, 1961.
26. See footnote 25; p. 40.
27. See footnote 25; pp. 71*ff*.
28. We would suggest, too, that the current preoccupation, popular as well as in research, with such concepts as empathy and communication derives from those same considerations.
29. Erich Fromm, *Escape from Freedom*; New York, Rinehart, 1941. David Riesman, *The Lonely Crowd*; New Haven, Yale Univ. Press, 1950.

Article 14 Parenthood as Crisis: A Re-Study

EVERETT D. DYER

Over the past thirty or more years family sociologists have devoted considerable attention to the study of various kinds of family crises. Many investigators have concerned themselves with crises of extra-family origin, such as war, depression, or unemployment. (e.g., Angell, Cavan and Ranck, Koos, Komarovsky.)[1] Others have studied crises whose sources were essentially intrafamily in origin. Crises of this kind could be divided into two subtypes:[2] (1) those due to "dismemberment," or loss of some family member (e.g., by divorce, desertion, or death); and (2) those due to "accession," or the addition of an unprepared-for member (e.g., re-marriage of widow or widower, return of deserter, or birth of an unprepared-for child). Most of the studies here have dealt mainly with the effects of dismemberment rather than accession. (e.g., Eliot, Waller, Goode, Hill.)[3]

One of the very few studies devoted entirely to accession was recently made by E. E. LeMasters, who investigated the various effects of the addition of the first child to the family.[4]

PURPOSE

Purpose. The research reported on here was prompted by the LeMasters study, and the dearth of studies of intra-family crises of the accession type. While not a true replication of the LeMasters study, the present one concerns itself with the same basic problems, i.e., what effects does the arrival of the first child have upon the family roles and relationships? Is it true, as LeMasters suggests,[5] that the arrival of the first child constitutes a "crisis" or "critical event"? Does the addition of this new member to the family system constitute a structural change which will force a drastic reorganization of statuses, roles, and relationships in

EVERETT D. DYER is Professor of Sociology, University of Houston, Texas.

Source: *Marriage and Family Living*, Vol. 25 (May 1963), 196–201. Copyright 1963 by National Council on Family Relations. Reprinted by permission.

order to re-establish an equilibrium in the family system? How does the crisis (if there be such) manifest itself? How may it be related to various social and demographic variables (such as age, number of years married, marriage-preparation courses in school, planned parenthood, etc.)? How does the family cope with the crisis, and go about establishing a new equilibrium?

Patterned along the lines suggested by Hill,[6] and followed by LeMasters in general, the present study seeks to investigate (1) the state or level of the family organization up to the time of the crisis, (2) the impact of the crisis upon the family, and (3) recovery and the subsequent level of family reorganization.[7]

METHOD

Sample

To seek answers to some of these questions, a group of young first-time parents were given questionnaires. Husbands and wives each had separate questionnaires, administered separately.

In an effort to obtain a homogeneous sample (and one quite similar to that obtained by LeMasters in his study),[8] each couple had to meet the following qualifications: (1) unbroken marriage; (2) urban or suburban residence (all were living in Houston); (3) ages, 35 or under; (4) college education for husband and/or wife; (5) husband's occupation middle class; (6) wife not employed after birth of child; (7) must have had their first child within two years of the time studied.

Applying the above criteria, a final sample of 32 couples was obtained by asking people in the community to supply names of young first-time parents. Each couple was then asked to fill out the questionnaires. Of those contacted, 74 percent met the qualifications and agreed to participate in the study (i.e., the final 32 couples). It is recognized that this sample has its limitations both as to representativeness of the urban middle-class and as to size. The data were obtained from the couples in 1959.

Definition of Crisis

"Crisis" may be defined in various ways. For the purposes at hand, Reuben Hill's definition should suffice. In his study of *Families Under Stress*, Hill defines crisis as "any sharp or decisive change for which old patterns are inadequate . . . . A crisis is a situation in which the usual behavior patterns are found to be unrewarding and new ones are called for immediately." [9]

Method of Measuring Crisis

A Likert-type scale was devised for the purpose of measuring the extent to which the arrival of the first child represented a crisis to each couple.[10] Items for the scale were drawn from areas of marriage and family life upon which the advent of the first child was felt most likely to have disruptive effects, according to previous studies and professional opinion.[11] Item analysis was employed in selecting those items which would yield an internally consistent scale.[12] The crisis score for each couple was based upon responses to the items in this scale. (i.e., the crisis score for the couple was the average of the summed item scored for both husband and wife.)

The scores were then used to indicate the position of the family on a 5-point continuum similar to that employed by LeMasters:[13] (1) no crisis, (2) slight crisis, (3) moderate crisis, (4) extensive crisis, and (5) severe crisis.[14]

The reliability of the scale was tested by the split-half method.[15] Correlating the odd-even responses yielded a coefficient of .84, which, when corrected by the Spearman-Brown formula, gave a reliability coefficient of .94. Insofar as reliability can be based upon this single method, the scale would appear to be reliable.

An effort was made to assess the validity of the scale by the "jury opinion" method.[16] Confirmation of the logical validity of the measure was sought from a jury of 6 young married couples each having one or more small children.[17] All agreed that the scale should yield a valid measure of the extent of crisis experienced by new first-time parents. It is recognized that this is a relatively limited kind of evidence for validity and that the measure has need to be validated against external criteria.

FINDINGS

The State of the Organization of the Family up to the Time of the Birth of the First Child

Studies of family crisis have shown that the impact of the crisis will depend not only upon the nature of the crisis event but also upon the state of the organization of the family at the time the crisis occurs, and the resources the family has to draw upon to help meet the crisis event.[18] Accordingly, a series of questions were asked to determine the strength of the marriage up to the time of the child's arrival. On the basis of these questions each couple was rated on a four-point scale ranging from "excellent" to "poor." [19] The 32 couples were distributed as follows: Excellent–40.5 percent; Good–50 percent; Fair–9.5 percent; Poor–none. This suggests that the state of the marriage and the family organization was average or above for the large majority of these couples, up to the advent of their first child.

A comparison of these family organization scores and the crisis scores (to be discussed below) shows a direct correlation between the two. This suggests that those couples whose marriage is stronger and who have more resources to draw on tend to experience less crisis when their first child is born.

Impact of the First Child on the Family

To what extent did the arrival of the first child represent a crisis to these couples? In what ways did the crisis manifest itself?

1. Distribution of Families According to Crisis Scores. As indicated above, each family was given a crisis rating or score. The 32 families were distributed as

follows: (1) No crisis—none; (2) Slight crisis—9 percent; (3) Moderate crisis—38 percent; (4) Extensive crisis—28 percent; (5) Severe crisis—25 percent. By comparison, LeMasters found only 17 percent of his families in the first categories, and the remaining 83 percent in the Extensive and Severe crisis categories.[20]

It is recognized that caution must be exercised in making the above comparisons. Although the effort was made to draw a sample from a population defined so that it would be similar to that of LeMasters, the samples were somewhat different with respect to both education and the number of years possible between the birth of the child and the date of the interview.[21] Other unknown sample differences which might bias the findings could have been present. Also, the "crisis" distribution in the two samples was likely affected by differences in defining and classifying crisis, and by the above-described differences in measuring crisis.

For the present sample, as for that of LeMasters, the evidence tends to support the hypothesis that adding the first child to the urban, middle-class married couple does constitute a crisis event to a considerable degree.

2. How Did the Crisis Manifest Itself?

A. Experiences, problems, and reactions reported *by the new mothers* in adjusting to the first child, starting with items most frequently mentioned: (1) Tiredness and exhaustion (87 percent); (2) Loss of sleep, especially during the first 6–8 weeks (87 percent); (3) Feelings of neglecting husband, to some degree (67 percent); (4) Feelings of inadequacy and uncertainty of being able to fill the mother role (58 percent); (5) Inability to keep up with the housework (35 percent); (6) Difficulty in adjusting to being tied down at home; curtailing outside activities and interests (35 percent). Here are some typical expressions by the mothers: "There is not enough time to be housekeeper, wife, and mother"; "I had very little prior realization of the vast amount of time and attention the baby requires"; "We were unable to foresee the number and drasticness of the changes in our lives the child would bring"; "I'm not able to go out with my husband anymore."

B. Experiences, problems, and reactions reported *by the new fathers* in adjusting to the first child: The fathers repeated many of the above-mentioned reactions of their wives, but felt more strongly about certain things and less so about others. They also added a few items of their own. Following are the items, starting with those most frequently mentioned: (1) Loss of sleep, up to 6 weeks (50 percent); (2) Adjusting to new responsibilities and routines (50 percent); (3) Upset schedules and daily routines (37 percent), one father ruefully exclaiming, "I expected not too much change in daily routine—Ha!"; (4) Ignorance of the great amount of time and work the baby would require; (5) Financial worries and adjustments for the majority of the families, involving adjustment to one income with the added expenses of the child, from two incomes before the child came. Sixty-two percent of the wives had been employed before having the child.

Some of the fathers expressed their feelings in these words: "You must get used to subjugating your feelings and desires to those of the child"; "Getting used to being tied down was our big problem"; "Wife has less time for me."

C. Most Severe Problems: Each husband and wife was asked to indicate which of the problems encountered was considered the most severe. Eighty-seven percent of the wives admitted to one or more severe problems. Among those mentioned: (1) Adjusting to being tied down or being restricted to the home was most frequently indicated; (2) "Getting accustomed to being up at all hours"; (3) "Inability to keep up with the housework . . . everything piled above my head!"; (4) "The feeling of anti-climax, or let down, after the birth of the child. It is a black feeling while it lasts."

Eighty percent of the husbands admitted to one or more severe problems. Among those mentioned: (1) "Adjusting to one income after my wife quit her job and the baby came"; (2) "Adjusting to the new demands of parenthood"; (3) "Getting used to the new routines"; (4) "Sharing with grandparents and other relatives."

Relationship Between Crisis and Other Social and Demographic Variables

It was hypothesized that some middle-class couples would be better equipped and prepared than others for the advent of the first child, and thus experience a lesser degree of crisis. The rationale was that those couples would probably be better able to meet the demands of parenthood who were better educated, who had studied about and planned for parenthood, who were not too young and had not started their families too quickly after marriage, where the husband and wife were quite close in both age and education, and where the wife had not become too attached to a work role outside the home. It was also felt that those couples whose marital adjustment was very good would likely experience less crisis on becoming new parents. Accordingly, correlations were made between the crisis scores and variables selected on the basis of the above-mentioned considerations. Chi-square tests were used to determine the presence or absence of a significant relationship. (It should be noted that the relative homogeneity of the sample tended to rule out much range on certain variables, such as education and age.)

Significant relationships were found between "crisis" and the following variables: a. Marital adjustment rating of the couple after the birth of the child,[22] those rating their marriage as excellent having experienced significantly less crisis; b. Preparation for marriage courses in high school or college, those taking the courses having experienced less crisis; c. Number of years married, those married three years or more having experienced less crisis; d. Education of husband but not wife, couples where the husband was not a college graduate having experienced greater crisis; e. "Planned parenthood," crisis being less among those who had "planned" their parenthood and followed their plan, and greater among those who had no plan or had failed to follow their plan; f. Age of the child,

Table 1. Crisis by Marital Adjustment, Preparation for Marriage Courses, Years Married, Husband's Education, Planned Parenthood, and Age of First Child, in 32 Families*

	Slight and Moderate	*Extensive and Severe*
Marital Adjustment		
Excellent	12	3
Good, and Fair	3	14
Preparation for Marriage Courses		
None	4	12
Husband and/or Wife	11	5
Years Married		
Under 3 Years	0	10
3 Years and over	15	7
Husband's Education		
High School and Some College	2	11
College Graduate	13	6
Planned Parenthood		
Followed Plan	14	5
Failed to Follow Plan, and No Plan	1	12
Age of First Child		
Under 6 Months	1	10
6 Months and Over	14	7

* Differences in each of the above analyses are statistically significant at the .05 level.

couples whose child was under six months were still experiencing more crisis, problems, etc., than those whose child was six months or over.

No significant relationships were found between "crisis" and the following variables: a. Employment of the wife before the child arrived; b. Ages of husband or wife; c. Number of years between marriage and birth of the child; d. Husband and wife differences in preparation for parenthood; e. Educational differences between husband and wife; f. Age differences between husband and wife; g. Education of the wife. It may be that higher education for the wife has some dysfunctional aspects relative to her adjustment to her mother role. Her needs and expectations may become so oriented to extra-family values and other roles that it is harder for her to stay home as a new mother. LeMasters found that the mothers with professional work experience suffered extensive or severe crisis in every case.[23]

What Was the Subsequent Level of Family Reorganization?

Each couple was given a "recovery and reorganization score" based upon (1) the duration of the crisis problems specified, and (2) the couple's success in solving their problems up to the time of the study.[24] The sample was distributed as follows: (1) Excellent recovery and reorganization: 19 percent; (2) Good to fair recovery and reorganization: 65.5 percent; (3) Poor recovery and reorganization: 15.5 percent.

A large majority (80 percent) of both husbands and wives admitted that things were not as they expected them to be after the child was born. Forty percent of the couples indicated they were still experiencing problems at the time the study was made, which was on the average 12 months after the birth of the child. Only 24 percent of the mothers and 38 percent of the fathers felt they had largely overcome the "crisis" at the time of the study. Those couples manifesting the greatest crisis were experiencing the most difficulty in recovering, quite understandably.

Discussion and Interpretation

The findings will be discussed as they relate to various questions and hypotheses that have been raised about parenthood as crisis. Some comparisons will be made with LeMasters' findings and interpretations.

1. It has been suggested that many American parents—especially middle-class parents—experience some incompatibility between their parental roles and certain other roles.[25] It would seem that, if true, such incompatability would be acutely felt with the arrival of the first child. LeMasters found this to be true in his study.[26] Although the scope of the present study was not sufficient for a thorough testing of this hypothesis, the findings do appear to support it to some degree. Both husbands and wives expressed feelings of loss at being tied down with the baby and thus being less free to do other customary things together. However, although 62 percent of the wives had been employed before their child was born, only 12 percent expressed feelings of loss since quitting, and not one wife said she wanted to return to her job!

2. Is there a significant lack of preparation or "training" for parenthood among middle-class couples? And, if so, is this lack related to the crisis?

LeMasters' findings on these questions were in the affirmative. His interpretations: "One can see that these couples were not trained for parenthood, that practically nothing in school, or out of school, got them ready to be fathers and mothers—*husbands* and *wives*, yes, but not *parents*." [27]

In the present study 38 percent of the mothers and 65 percent of the fathers admitted to no formal or informal preparation for parenthood. Only 35 percent of the wives and 12 percent of the husbands had taken any courses in high school or college involving preparation for parenthood (generally as part of a preparation for marriage course); 25 percent of the couples had had a Red Cross course, and 50 percent of the wives and 12 percent of the husbands mentioned such informal preparation as care of younger brothers and sisters, Girl Scout activities, and "some reading."

While a majority said they had thought they were adequately prepared before their child arrived, still a large majority (80 percent) admitted that things were not what they had expected after the child was born. In offering advice to other parents-to-be a frequently expressed item was a heart-felt "Be prepared!"

As noted above, a correlation was found between "crisis" and formal courses taken in high school or college, those having had such courses experiencing less crisis. This lends support to LeMasters' contention that more formal education in preparation for parenthood is needed by middle-class parents-to-be.[28]

3. Is the change from the pre-parenthood husband-wife pair relationship to the husband-wife-child triad relationship a difficult adjustment for the middle-class husband and wife to make?

The husband-wife courtship and pre-parenthood marriage relationship has become routinized and satisfying to the couple over the years, and then the intrusion of a third member (nonsocialized and all-demanding) calls inevitably for substantial readjustments in the husband-wife interaction patterns. Will the child's claim of priority on the mother make the husband feel he is the third party in the trio, a semi-isolate perhaps, as LeMasters suggests?[29] Or, in some instances, does the wife come to feel that her husband is more interested in the child than in her now?

The present findings supported this hypothesis in some degree. While 37 percent of the husbands felt their wives never neglected them for the baby, 50 percent felt she sometimes did, and another 12 percent said she often did. Recall that among the more frequently mentioned "crisis" problems of the wives were such items as "husbands grew tired of being second," and "less time to give to my husband,"etc. Only 12 percent of the wives felt their husbands sometimes neglected them for the child, however.

4. LeMasters suggests "that parenthood (not marriage) marks the final transition to maturity and adult responsibility in our culture. Thus the arrival of the first child forces the young married couples to take the last painful step into the adult world,"[30] jarring them out of the honeymoon stage of marriage, as it were.

In the present study 50 percent of the husbands and 87 percent of the wives felt that parenthood had indeed been a maturing experience.

Indication of the maturing function of parenthood, and increased awareness of adult responsibilities, may be seen in the comments and advice they offer to other young couples expecting their first child, e.g.: (1) "Realize your life will be different because of addition of the baby, but it will be a better and more complete life"; (2) "If the husband and wife want the child and will share the responsibilities, I think they will be happier, and the child will deepen their love for each other. It is a new but very rewarding experience." There were many other similar expressions. All of the husbands and all but two of the wives said they now felt much better prepared for any subsequent children they might have.

CONCLUSIONS

1. The findings tend to support the main hypothesis that the addition of the first child would constitute a crisis event for these middle-class couples to a considerable degree, forcing each couple to reorganize many of their roles and relationships. A majority of couples experienced extensive or severe crisis.

2. The degree to which the advent of the first child represents a crisis event appears to be related to: (a) the state of the marriage and family organization at the birth of the child; (b) the couple's preparation for marriage and parenthood; (c) the couple's marital adjustment after the birth of the child; and (d) certain social background and situational variables such as the number of years married, "planned parenthood," and the age of the child.

3. The large majority of the couples appear to have made a quite satisfactory recovery from the crisis, although this often followed a difficult period of several months. Eighty-one per cent made "fair to good" or "excellent" on their recovery scores. As would be expected, those couples experiencing the severest crisis were having the hardest time recovering.

4. More study is needed of the characteristics and backgrounds of the families experiencing the greater and the lesser crises. In addition to social and demographic background and situational variables, certain personality variables may be important, not only with respect to the parents, but also the child. Many of us who have been blessed with a "fast-excitable" type of infant who never seems to need much sleep and who demands constant attention, would be willing to wager that our parenthood crisis traumas would have been greatly diminished had we been blessed with a child of the "easy going-passive" type—the kind with which our friends generally seem to be blessed.

5. The findings in the present study lead the writer to concur with LeMasters that by emphasizing preparation for parenthood, family life educators can probably make an important contribution to young married couples who are contemplating parenthood.

REFERENCES AND NOTES

1. ROBERT C. ANGELL, *The Family Encounters the Depression*, New York: Charles Scribner's Sons, 1936; RUTH CAVAN AND KATHERINE RANCK, *The Family and the Depression*, Chicago: University of Chicago Press, 1938; E. L. Koos, *Families in Trouble*, New York: King's Crown Press, 1946; MIRRA KOMAROVSKY, *The Unemployed Man and His Family*, New York: Dryden Press, 1940.

2. REUBEN HILL, *Families Under Stress*, New York: Harper and Brothers, 1949, pp. 9–10.
3. THOMAS D. ELIOT, "Bereavement: Inevitable But Not Insurmountable," in *Family, Marriage, and Parenthood*, edited by Howard Becker and Reuben Hill, Boston: D. C. Heath and Company, Second Edition, 1955; WILLARD WALLER, *The Old Love and the New*, New York: Liveright, 1930; WILLIAM J. GOODE, *After Divorce*, Glencoe: The Free Press, 1956. Hill, *op. cit.* Hill's study is one of the few concerned with both dismemberment and accession.
4. E. E. LeMASTERS, "Parenthood as Crisis," *Marriage and Family Living*, 19 (November, 1957), pp. 352–55.
5. *Ibid.*, p. 352.
6. Hill, *op. cit.*, Ch. 2.
7. LeMasters, *op. cit.*, p. 352. Also see Ernest R. Mowrer, "Social Crisis and Social Disorganization," *American Sociological Review*, 15 (February, 1950), p. 61, on the cyclical nature of crises.
8. LeMasters, *op. cit.*, p. 352.
9. Hill, *op. cit.*, p. 51. The present study is concerned primarily with the exploration of the crisis experience itself, and is not designed to explore the new behavior patterns that are worked out by the family following recovery from the crisis period.
10. While LeMasters arrived at a "crisis rating" for each couple jointly with the couple during the interview (*op. cit.*, p. 353), the writer sought an objective way of identifying and measuring crisis.
11. Areas of family life from which the items were drawn: (1) Husband-wife division of labor; (2) Husband-wife division of authority; (3) Husband-wife companionship patterns; (4) Family income and finances; (5) Home-making and housework; (6) Social life and recreational patterns; (7) Husband and wife mobility and freedom of action; (8) Child care and rearing (i.e., anxieties, difficulties, burdens, etc.); (9) Health of husband, wife, and child; and (10) Extra-family interests and activities. See LeMasters, *op. cit.*, pp. 353–54; ROBERT O. BLOOD, *Marriage*, Glencoe: The Free Press, 1962, pp. 415–25; ROBERT O. BLOOD AND DONALD M. WOLFE, *Husbands and Wives*, Glencoe: The Free Press, 1960, pp. 138–45; RUTH CAVAN, *The American Family*, New York: Crowell Co., 1953, pp. 504–17; EVELYN M. DUVALL, *Family Development*, New York: Lippincott Co., 1957, pp. 187–227; WILLARD WALLER, *The Family*, New York: The Dryden Press, 1938, pp. 375–93; HARVEY J. LOCKE, *Predicting Adjustment in Marriage*, New York: Henry Holt and Company, 1951, pp. 158–70.
12. The discriminative power of each item was calculated, and a final scale of 16 items, each having a discriminative power of 0.50 or higher, was obtained. Of 26 original items, 10 failed to show a discriminative power of 0.50 or higher and were dropped. Each item represented a 5-point continuum with values assigned from 0 to 4, with the largest value indicating the greatest degree of crisis. See William J. Goode and Paul K. Hatt, *Methods in Social Research*, New York: McGraw-Hill, 1952, pp. 274–81.
13. LeMasters, *op. cit.*, p. 353.
14. The score ranges were: (1) No crisis: 0; (2) Slight crisis: 1–16; (3) Moderate crisis: 17–32; (4) Extreme crisis: 33–48; (5) Severe crisis: 49–64.

15. Goode and Hatt, *op. cit.*, p. 236.
16. *Ibid.*, pp. 237–38. This method seeks a confirmation of the logical or face validity of the scale through the judgment of "experts" in the field where the scale applies.
17. The sole qualification of these couples as experts lies in their experiences as parents of young children.
18. Hill, *op. cit.*, Ch. 2.
19. The couple's family organization score was based upon: (1) a self-rating of the couple's marital adjustment up to the birth of the child; (2) the degree of confidence expressed by the husband and wife in their ability to perform their respective roles as husband and wife; and (3) the couple's evaluation of their economic and financial adequacy. It should be noted that the method used in assessing the state of the family organization has yet to be tested as to reliability and validity.
20. LeMasters, *op. cit.*, p. 353.
21. In the LeMasters study, all of the husbands had graduated from college, and each couple must have had its first child within 5 years of the date interviewed; while in the present study 59 per cent of the husbands had graduated from college, and each couple must have had its first child within two years of the time interviewed.
22. The marital adjustment rating for each couple was based upon a self-rating scale similar to that used by Judson Landis, "The Length of Time Required to Achieve Adjustment in Marriage," *American Sociological Review*, 11 (December, 1946), pp. 666–76. See p. 674. Each husband and wife was asked to rate his marriage on a 5-point scale: very happy, happy, average, unhappy, or very unhappy. The separate ratings of the husband and wife were then combined and averaged to give the marital adjustment rating for the couple.
23. LeMasters, *op. cit.*, p. 354.
24. A four-point rating scale was used: (1) Excellent recovery and reorganization; (2) Good recovery and reorganization; (3) Fair recovery and reorganization; (4) Poor recovery and reorganization. Neither the reliability nor the validity of this measure has been established; accordingly, not too much weight can be given to the findings based upon its use here.
25. ARNOLD W. GREEN, "The Middle-class Male Child and Neurosis," *American Sociological Review*, 11 (February 1946), pp. 31–41.
26. LeMasters, *op. cit.*, p. 354.
27. *Ibid.*
28. *Ibid.*
29. *Ibid.*, p. 355.
30. *Ibid.*

Developmental Life Transitions: Relocation and Migration

VI

Residential change over both short and long distances is becoming an increasingly common occurrence in modern society. Between 1955 and 1960 about 50 percent of American families moved at least once. People often experience significant stress in leaving their homes, close friends, and familiar places, and in encountering new physical conditions, customs, and interpersonal relationships. This is true even when the move is made to improve living standards and is therefore seen as beneficial. Forced moves such as relocation due to urban renewal create special problems not associated with voluntary moves. Migration, a term implying greater distances and usually the crossing of cultural boundaries, has its own unique features. All these moves, though in varying degrees, require individuals to deal with a loss of the old and the known, to adjust to the strange and the unfamiliar, and to cope with the stress generated by both tasks.[4]

In the first article Marc Fried discusses some of the deep feelings people have about their homes and their attempts to deal with forced relocation. The primary problems of disrupted social ties and loss of familiar physical spaces are also found with voluntary moves, but the special feelings Fried described as "grief"—longing for the lost home, helplessness, depression, anger, and a tendency to idealize the lost place—usually are not. Forty-six percent of the women and 38 percent of the men interviewed experienced fairly severe grief reactions. For this particular group, their deep commitment to the area, the West End of Boston, was a major determinant of their adjustment to relocation. The greater their involvement in the vast social network and their perception of the neighborhood as an extension of their own homes, the more likely they were to experience a severe reaction.

Cagle and Deutscher[3] found in their study of a predominantly black slum clearance project that commitment to the neighborhood was not significant; for these people the quality of the new housing situation determined their success in adjusting to relocation. Fried also noted the importance of the postrelocation experience but pointed out that selection of a new home could be affected by feelings about the original home and by personality factors like the ability to anticipate one's needs and opportunities. The primary task for West Enders, and probably for most people who must shift their homes, is to maintain some sense of continuity: physical (stay in general vicinity), social (move among relatives or friends), or individual (accentuate remaining roles, such as spouse, parent, worker).

Loss of continuity is clearly a problem for school children who change residence in the midst of a school term. Murray Levine examines the problem of the child's adjustment to changing home and school in the second article. Levine suggests that the rate of moving is highest among those who already have social and educational problems, for example, the unemployed, urban ghetto residents, and those experiencing family problems like divorce or widowhood. The specific problems a school child must deal with are gaining acceptance by a new group of peers, learning the rules and customs of a new environment, and relearning material affected by curriculum variations (for example, different methods of teaching math). Most schools do little to orient the new child in these spheres. Teachers are further handicapped by the failure of school records to give adequate information about incoming transfer students. Levine suggests the need for orientation and preventive mental health programs to assist children through the transition and foster healthy adjustment.

Migration across cultural and international boundaries has been and continues to be a prominent feature of the North American scene. People have left their homes in Europe, Asia, Africa, and Central and South America for political, economic, and religious reasons. The nature of the reason for emigrating has a significant impact on the adjustment experience, as do the nature of the cutural differences encountered[1,5] and the kind of reception met in the new environment.[2] People who left their homes for political reasons (for example, East Europeans) or for religious reasons (for example, European Jews) faced their new environments with a different set of resources, internal and external, from those who were motivated primarily by difficult economic conditions (for example, Mexicans).

In the third article Saul Levine discusses a group of American expatriates— young draft evaders and military service deserters—and their adaptation to self-exile in Canada. Based on interviews with sixty healthy and twenty-four disturbed expatriates Levine identifies some aids to a good adjustment: having lived away from home before, having been to college, having spent some time planning the move to Canada, having moral support from wives or parents, and having some ideological framework for the decision to move. Factors contributing to a poor adjustment were family rejection, impulse as the basis of flight, lack of preparation that is often associated with having unrealistic expectations, poor educational and occupational background, and a history of psychological problems.

These factors help determine the rate of progress in adaptation, a process Levine divides into four stages. "Disorganization," the first stage, is a time of confusion, ambivalence, loneliness, and general distress. Advance planning (for example, lining up a job or a place to live) and meeting friends who can help provide a sense of belonging are the most useful aids to coping at this point. In the "acting out" stage the individual emphasizes noninvolvement, building superficial and exploitive interpersonal relationships and sometimes displaying aggressive antisocial behavior (such as stealing, pushing drugs, violence). The third stage is a period of "searching," taking stock, maturing, and looking for closer interpersonal ties. In the fourth phase, "adaptation and integration," the individual has become totally involved in his new life and sees himself as a Canadian rather than an American-in-exile. External circumstances like the changing political climate in Canada and in America interact with the individual factors outlined earlier to determine the pace and intensity of the coping stages.

The basic problems faced by people who move are the same—the loss of a "home," the disruption of a social network, and the need to find a new place for oneself amid unfamiliar places and people. Whether the move is long or short, voluntary or forced, these dual stresses are present. Individual personality characteristics and external circumstances determine the appropriate strategy for coping with this disruption of spatial and social continuity.

REFERENCES

1. BAILYN, L., AND KELMAN, H. C. The effects of a year's experience in America on the self-image of Scandinavians: A preliminary analysis of reactions to a new environment. *Journal of Social Issues*, 1962, 18, 30–40.
2. BRODY, E. B., ed. *Behavior in new environments: Adaptation of migrant populations.* Beverly Hills, Calif.: Sage Publications, 1969.
3. CAGLE, L. T., AND DEUTSCHER, I. Housing aspirations and housing achievement: The relocation of poor families. *Social Problems*, 1970, 18, 243–56.
4. KANTOR, M. B., ed. *Mobility and mental health.* Springfield, Ill.: Charlces C. Thomas, 1965.
5. SOMMERS, V. S. Resolution of an identity conflict in a Japanese-American patient. *American Journal of Psychotherapy*, 1969, 23, 119–34.

Article 15 Grieving for a Lost Home

MARC FRIED

THE NATURE OF THE LOSS IN RELOCATION: CASE ANALYSES

The dependence of the sense of continuity on external resources in the working class, particularly on the availability and local presence of familiar places which have the character of "home," and of familiar people whose patterns of behavior and response are relatively predictable, does not account for all of the reaction of grief to dislocation. In addition to these factors, which may be accentuated by depressive predispositions, it is quite evident that the realities of *post*-relocation experience are bound to affect the perpetuation, quality, and depth of grief. And, in fact, our data show that there is a strong association between positive or negative experiences in the post-relocation situation and the proportions who show severe grief. But this issue is complicated by two factors: (1) the extent to which potentially meaningful post-relocation circumstances can be a satisfying experience is *affected* by the degree and tenaciousness of previous commitments to the West End, and (2) the post-relocation "reality" is, in part, *selected* by the people who move and thus is a function of many personality factors, including the ability to anticipate needs, demands, and environmental opportunities.

In trying to understand the effects of pre-relocation orientations and post-relocation experiences of grief, we must bear in mind that the grief reactions we have described and analyzed are based on responses given approximately two years after relocation. Most people manage to achieve some adaptation to their experiences of loss and grief, and learn to deal with new situations and new experiences on their own terms. A wide variety of adaptive methods can be employed to salvage fragments of the sense of continuity, or to try to re-establish it

---

MARC FRIED, PH.D., is Research Professor of Human Sciences and Director, Laboratory of Psychosocial Studies, Boston College, Chestnut Hill, Mass.

Source: From chapter 12 (pp. 159–71) of *The Urban Condition*, ed. L. Duhl. New York: Basic Books, 1963.

on new grounds. Nonetheless, it is the tenaciousness of the imagery and affect of grief, despite these efforts at dealing with the altered reality, which is so strikingly similar to mourning for a lost person.

In coping with the sense of loss, some families tried to remain physically close to the area they knew, even though most of their close interpersonal relationships remain disrupted; and by this method, they appear often to have modified their feelings of grief. Other families try to move among relatives and maintain a sense of continuity through some degree of constancy in the external bases for their group identity. Yet others respond to the loss of place and people by accentuating the importance of those role relationships which remain. Thus, a number of women report increased closeness to their husbands, which they often explicitly relate to the decrease in the availability of other social relationships for both partners and which, in turn, modifies the severity of grief. In order to clarify some of the complexities of pre-relocation orientations and of post-relocation adjustments most concretely, a review of several cases may prove to be instructive.

It is evident that a very strong pre-relocation orientation to the West End is relatively infrequently associated with a complete absence of grief; and that, likewise, a negative pre-relocation orientation to the area is infrequently associated with a strong grief response. The two types which are numerically dominant are, in terms of rational expectations, consistent: those with strong positive feelings about the West End and severe grief; and those with negative feelings about the West End and minimal or moderate grief. The two "deviant" types, by the same token, are both numerically smaller and inconsistent: those with strong positive pre-relocation orientations and little grief; and those with negative pre-relocation orientations and severe grief. A closer examination of those "deviant" cases with strong pre-relocation commitment to the West End and minimal post-relocation grief often reveals either important reservations in their prior involvement with the West End or, more frequently, the denial or rejection of feelings of grief rather than their total absence. And the association of minimal pre-relocation commitment to the West End with a severe grief response often proves on closer examination to be a function of a deep involvement in the West End which is modified by markedly ambivalent statements; or, more generally, the grief reaction itself is quite modest and tenuous or is even a pseudo-grief which masks the primacy of dissatisfaction with the current area.

GRIEF PATTERNS: CASE EXAMPLES

In turning to case analysis, we shall concentrate on the specific factors which operate in families of all four types, those representing the two dominant and those representing the two deviant patterns.

1. The Figella family exemplifies the association of strong positive pre-relocation attachments to the West End and a severe grief reaction. This is the most frequent of all the patterns and, although the Figella family is only

one "type" among those who show this pattern, they are prototypical of a familiar West End constellation.

Both Mr. and Mrs. Figella are second-generation Americans who were born and brought up in the West End. In her pre-relocation interview, Mrs. Figella described her feelings about living in the West End unambiguously: "It's a wonderful place, the people are friendly." She "loves everything about it" and anticipates missing her relatives above all. She is satisfied with her dwelling: "It's comfortable, clean and warm." And the marriage appears to be deeply satisfying for both husband and wife. They share many household activities and have a warm family life with their three children.

Both Mr. and Mrs. Figella feel that their lives have changed a great deal since relocation. They are clearly referring, however, to the pattern and conditions of their relationships with other people. Their home life has changed little except that Mr. Figella is home more. He continues to work at the same job as a manual laborer with a modest but sufficient income. While they have many economic insecurities, the relocation has not produced any serious financial difficulty for them.

In relocating, the Figella family bought a house. Both husband and wife are quite satisfied with the physical arrangements but, all in all, they are dissatisfied with the move. When asked what she dislikes about her present dwelling, Mrs. Figella replied simply and pathetically: "It's in Arlington and I want to be in the West End." Both Mr. and Mrs. Figella are outgoing, friendly people with a very wide circle of social contacts. Although they still see their relatives often, they both feel isolated from them and they regret the loss of their friends. As Mr. Figella puts it: "I come home from work and that's it. I just plant myself in the house."

The Figella family is, in many respects, typical of a well-adjusted working-class family. They have relatively few ambitions for themselves or for their children. They continue in close contact with many people; but they no longer have the same extensiveness of mutual cooperation in household activities, they cannot "drop in" as casually as before, they do not have the sense of being surrounded by a familiar area and familiar people. Thus, while their objective situation is not dramatically altered, the changes do involve important elements of stability and continuity in their lives. They manifest the importance of externally available resources for an integral sense of spatial and group identity. However, they have always maintained a very close marital relationship, and their family provides a substantial basis for a sense of continuity. They can evidently cope with difficulties on the strength of their many internal and external resources. Nonetheless, they have suffered from the move, and find it extremely difficult to reorganize their lives completely in adapting to a new geographical situation and new patterns of social affiliation. Their grief for a lost home seems to be one form of maintaining continuity on the basis of memories. While it prevents a more wholehearted adjustment to their altered lives, such adjustments would imply forsaking the remaining fragments of a continuity which was central to their conceptions of themselves and of the world.

2. There are many similarities between the Figella family and the Giuliano family. But Mrs. Giuliano shows relatively little pre-relocation commitment to the West End and little post-relocation grief. Mr. Giuliano was somewhat more deeply involved in the West End and, although satisfied with the change, feels that relocation was "like having the rug pulled out from under you." Mr. and Mrs. Giuliano are also second-generation Americans, of similar background to the Figellas'. But Mrs. Giuliano only moved to the West End at her marriage. Mrs. Giuliano had many objections to the area: "For me it is too congested. I never did care for it . . . too many barrooms, on every corner, too many families in one building. . . . The sidewalks are too narrow and the kids can't play outside." But she does expect to miss the stores and many favorite places. Her housing ambitions go beyond West End standards and she wants more space inside and outside. She had no blood relatives in the West End but was close to her husband's family and had friends nearby.

Mr. Giuliano was born in the West End and he had many relatives in the area. He has a relatively high status manual job but only a modest income. His wife does not complain about this although she is only moderately satisfied with the marriage. In part she objected to the fact that they went out so little and that he spent too much time on the corner with his friends. His social networks in the West End were more extensive and involved than were Mrs. Giuliano's. And he missed the West End more than she did after the relocation. But even Mr. Giuliano says that, all in all, he is satisfied with the change.

Mrs. Guiliano feels the change is "wonderful." She missed her friends but got over it. And a few of Mr. Guiliano's hanging group live close by so they can continue to hang together. Both are satisfied with the house they bought although Mrs. Giuliano's ambitions have now gone beyond this. The post-relocation situation has led to an improved marital relationship: Mr. Guiliano is home more and they go out more together.

Mr. and Mrs. Guiliano exemplify a pattern which seems most likely to be associated with a beneficial experience from relocation. Unlike Mr. and Mrs. Figella, who completely accept their working-class status and are embedded in the social and cultural patterns of the working class, Mr. and Mrs. Giuliano show many evidences of social mobility. Mr. Giuliano's present job is, properly speaking, outside the working-class category because of its relatively high status and he himself does not "work with his hands." And Mrs. Giuliano's housing ambitions, preferences in social relationships, orientation to the class structure, and attitudes toward a variety of matters from shopping to child rearing are indications of a readiness to achieve middle-class status. Mr. Giuliano is prepared for and Mrs. Giuliano clearly desires "discontinuity" with some of the central bases for their former identity. Their present situation is, in fact, a transitional one which allows them to reintegrate their lives at a new and higher status level without too precipitate a change. And their marital relationship seems sufficiently meaningful to provide a significant core of continuity in the process of change in their patterns of social and cultural experience. The lack of grief in this case is quite understandable and appropriate to their patterns of social orientation and expectation.

3. Yet another pattern is introduced by the Borowski family, who had an intense pre-location commitment to the West End and relatively little post-relocation grief. The Borowski's are both second-generation and have four children.

Mrs. Borowski was brought up in the West End but her husband has lived there only since the marriage (fifteen years before). Her feelings about living in the West End were clear: "I love it—it's the only home I've even known." She had reservations about the dirt in the area but loved the people, the places, and the convenience and maintained an extremely wide circle of friends. They had some relatives nearby but were primarily oriented towards friends, both within and outside the West End. Mr. Borowski, a highly skilled manual worker with a moderately high income, was as deeply attached to the West End as his wife.

Mr. Borowski missed the West End very much but was quite satisfied with their new situation and could anticipate feeling thoroughly at home in the new neighborhood. Mrs. Borowski proclaims that "home is where you hang your hat; it's up to you to make the adjustments." But she also says, "If I knew the people were coming back to the West End, I would pick up this little house and put it back on my corner." She claims she was not sad after relocation but, when asked how she felt when the building she lived in was torn down, a strangely morbid association is aroused: "It's just like a plant . . . when you tear up its roots, it dies! I didn't die but I felt kind of bad. It was home. . . . Don't look back, try to go ahead."

Despite evidences of underlying grief, both Mr. and Mrs. Borowski have already adjusted to the change with remarkable alacrity. They bought a one-family house and have many friends in the new area. They do not feel as close to their new neighbors as they did to their West End friends, and they still maintain extensive contact with the latter. They are comfortable and happy in their new surroundings and maintain the close, warm, and mutually appreciative marital relationship they formerly had.

Mr. and Mrs. Borowski, and particularly Mrs. Borowski, reveal a sense of loss which is largely submerged beneath active efforts to deal with the present. It was possible for them to do this both because of personality factors (that is, the ability to deny the intense affective meaning of the change and to detach themselves from highly "cathected" objects with relative ease) and because of prior social patterns and orientations. Not only is Mr. Borowski, by occupation, among the highest group of working-class status, but this family has been "transitional" for some time. Remaining in the West End was clearly a matter of preference for them. They could have moved out quite easily on the basis of income; and many of their friends were scattered throughout metropolitan Boston.

But while they are less self-consciously mobile than the Giuliano's, they had already shifted to many patterns more typical of the middle class before leaving the West End. These ranged from their joint weekly shopping expeditions to their recreational patterns, which included such sports as boating and such regular plans as yearly vacations. They experienced a disruption in continuity by virtue of their former spatial and group identity. But the bases for maintaining this identity had undergone many changes over the years; and they had already established a feeling for places and people, for a potential redefinition of "home" which was less contingent on the immediate and local availability of familiar spaces and familiar friends. Despite their preparedness for the move by virtue of cultural orientation, social experience, and personal disposition, the change was a considerable wrench for them. But, to the extent that they can be categorized as "over-adjusters," the residue of their lives in the West End is primarily a matter of painful memories which are only occasionally reawakened.

4. The alternate deviant pattern, minimal pre-relocation commitment associated with severe post-relocation grief, is manifested by Mr. and Mrs. Pagliuca. As in the previous case, this classification applies more fully to Mrs. Pagliuca, since Mr. Pagliuca appears to have had stronger ties to the West End. Mr. Pagliuca is a second-generation American but Mrs. Pagliuca is first-generation from an urban European background. For both of them, however, there is some evidence that the sadness and regret about the loss of the West End should perhaps be designated as pseudo-grief.

Mrs. Pagliuca had a difficult time in the West End. But she also had a difficult time before that. She moved into the West End when she got married. And she complains bitterly about her marriage, her husband's relatives, West Enders in general. She says of the West End: "I don't like it. The people . . . the buildings are full of rats. There are no places to play for the children." She liked the apartment but complained about the lady downstairs, the dirt, the repairs required, and the coldness during the winter. She also complains a great deal about lack of money. Her husband's wages are not too low but he seems to have periods of unemployment and often drinks his money away.

Mr. Pagliuca was attached to some of his friends and the bars in the West End. But he didn't like his housing situation there. And his reaction tends to be one of bitterness ("a rotten deal") rather than of sadness. Both Mr. and Mrs. Pagliuca are quite satisfied with their post-relocation apartment but are thoroughly dissatisfied with the area. They have had considerable difficulty with neighbors: ". . . I don't like this; people are mean here; my children get blamed for anything and everything; and there's no transportation near here." She now idealizes the West End and claims that she misses everything about it.

Mr. Pagliuca is an unskilled manual laborer. Financial problems create a constant focus for difficulty and arguments. But both Mr. and Mrs. Pagliuca appear more satisfied with one another than before relocation. They have four children, some of whom are in legal difficulty. There is also some evidence of past cruelty toward the children, at least on Mrs. Pagliuca's part.

It is evident from this summary that the Pagliuca family is deviant in a social as well as in a statistical sense. They show few signs of adjusting to the move or, for that matter, of any basic potential for successful adjustment to further moves (which they are now planning). It may be that families with such initial difficulties, with such a tenuous basis for maintaining a sense of continuity under any circumstances, suffer most acutely from disruption of these minimal ties. The Pagliuca family has few inner resources and, having lost the minimal external resources signified by a gross sense of belonging, of being tolerated if not accepted, they appear to be hopelessly at sea. Although we refer to their grief as "pseudo-grief" on the basis of the shift from pre-relocation to post-relocation statements, there is a sense in which it is quite real. Within the post-relocation interviews their responses are quite consistent; and a review of all the data suggests that, although their ties were quite modest, their current difficulties have revealed the importance of these meager involvements and the problems of re-establishing anew an equivalent basis for identity formation. Thus, even for Mr. and Mrs. Pagliuca, we can speak of the disruption in the sense of continuity, although this continuity was based on a very fragile experience of minimal comfort, with familiar places and relatively tolerant people. Their grief reaction, pseudo or real, may further influence (and be influenced by) dissatisfactions with any new residential situation. The fact that it is based on an idealized past accentuates rather than mimimizes its effect on current expectations and behavior.

CONCLUSIONS

Grieving for a lost home is evidently a widespread and serious social phenomenon following in the wake of urban dislocation. It is likely to increase social and psychological "pathology" in a limited number of instances; and it is also likely to create new opportunities for some, and to increase the rate of social mobility for others. For the greatest number, dislocation is unlikely to have either effect but does lead to intense personal suffering despite moderately successful adaptation to the total situation of relocation. Under these circumstances, it becomes most critical that we face the realities of the effects of relocation on working-class residents of slums and, on the basis of knowledge and understanding, that we learn to deal more effectively with the problems engendered.

In evaluating these data on the effect of pre-relocation experiences on post-relocation reactions of grief, we have arrived at a number of conclusions:

1. The affective reaction to the loss of the West End can be quite precisely described as a grief response showing most of the characteristics of grief and mourning for a lost person.
2. One of the important components of the grief reaction is the fragmentation of the sense of spatial identity. This is manifest, not only in the pre-location experience of the spatial area as an expanded "home," but in the

varying degrees of grief following relocation, arising from variations in the
pre-relocation orientation to and use of local spatial regions.

3. Another component, of equal importance, is the dependence of the sense
   of group identity on stable, social networks. Dislocation necessarily led
   to the fragmentation of this group identity which was based, to such a
   large extent, on the external availability and overt contact with familiar
   groups of people.

4. Associated with these "cognitive" components, described as the sense
   of spatial identity and the sense of group identity, are strong affective
   qualities. We have not tried to delineate them but they appear to fall into
   the realm of a feeling of security in and commitment to the external
   spatial and group patterns which are the tangible, visible aspects of these
   identity components. However, a predisposition to depressive reactions
   also markedly affects the depth of grief reaction.

5. Theoretically, we can speak of spatial and group identity as critical foci of
   the sense of continuity. This sense of continuity is not *necessarily* contingent on the external stability of place, people, and security or support. But
   for the working class these concrete, external resources and the experience
   of stability, availability, and familiarity which they provide are essential
   for a meaningful sense of continuity. Thus, dislocation and the loss of the
   residential area represent a fragmentation of some of the essential components of the sense of continuity in the working class.

It is in the light of these observations and conclusions that we must consider problems of social planning which are associated with the changes induced
by physical planning for relocation. Urban planning cannot be limited to "bricks
and mortar." While these data tell us little about the importance of housing or
the aspects of housing which are important, they indicate that considerations
of a non-housing nature are critical. There is evidence, for example, that the
frequency of the grief response is not affected by such housing factors as increase
or decrease in apartment size or home ownership. But physical factors may be
of great importance when related to the subjective significance of different spatial
and physical arrangements, or to their capacity for gratifying different sociocultural groups. For the present, we can only stress the importance of local areas
as *spatial and social* arrangements which are central to the lives of working-class
people. And, in view of the enormous importance of such local areas, we are
led to consider the convergence of familiar people and familiar places as a focal
consideration in formulating planning decisions.

We can learn to deal with these problems only through research, through
exploratory and imaginative service programs, and through a more careful consideration of the place of residential stability in salvaging the precarious thread
of continuity. The outcomes of crises are always manifold and, just as there is an
increase in strain and difficulty, so also there is an increase in opportunities for
adapting at a more satisfying level of functioning. The judicious use of minimal

resources of counseling and assistance may permit many working-class people to reorganize and integrate a meaningful sense of spatial and group identity under the challenge of social change. Only a relatively small group of those whose functioning has always been marginal and who cannot cope with the added strain of adjusting to wholly new problems are likely to require major forms of intervention.

In general, our results would imply the necessity for providing increased opportunities for maintaining a sense of continuity for those people, mainly from the working class, whose residential areas are being renewed. This may involve several factors: (1) diminishing the amount of drastic redevelopment and the consequent mass demolition of property and mass dislocation from homes; (2) providing more frequently for people to move within their former residential areas during and after the renewal; and (3) when dislocation and relocation are unavoidable, planning the relocation possibilities in order to provide new areas which can be assimilated to old objectives. A closer examination of slum areas may even provide some concrete information regarding specific physical variables, the physical and spatial arrangements typical of slum areas and slum housing, which offer considerable gratification to the residents. These may often be translated into effective modern architectural and areal design. And, in conjunction with planning decisions which take more careful account of the human consequences of urban physical change, it is possible to utilize social, psychological, and psychiatric services. The use of highly skilled resources, including opportunities for the education of professional and even lay personnel in largely unfamiliar problems and methods, can minimize some of the more destructive and widespread effects of relocation; and, for some families, can offer constructive experiences in dealing with new adaptational possibilities. The problem is large. But only by assuring the integrity of some of the external bases for the sense of continuity in the working class, and by maximizing the opportunities for meaningful adaptation, can we accomplish planned urban change without serious hazard to human welfare.

REFERENCES

1. ABRAHAM, K., "Notes on the Psycho-analytical Investigation and Treatment of Manic-Depressive Insanity and Allied Conditions" (1911), and "A Short Study of the Development of the Libido, Viewed in the Light of Mental Disorders" (1924), in *Selected Papers of Karl Abraham*, Vol. I, New York: Basic Books, 1953; BIBRING, E., "The Mechanisms of Depression," in *Affective Disorders*, P. Greenacre, ed., New York: International Univ. Press, 1953; BOWLBY, J., "Processes of Mourning," *Int. J. Psychoanal.*, 42:317–340, 1961; FREUD, S., "Mourning and Melancholia" (1917), in *Collected Papers*, Vol. III, New York: Basic Books, 1959; HOGGART, R., *The Uses of Literacy: Changing Patterns in English Mass Culture*, New York: Oxford Univ. Press, 1957; KLEIN, M., "Mourning and Its Relations to Manic-Depressive States," *Int. J. Psychoanal.*, 21:125–153, 1940; LINDEMANN, E., "Symptomatology and

Management of Acute Grief," *Am. J. Psychiat.*, 101:141–148, 1944; MARRIS, P., *Widows and Their Families*, London: Routledge and Kegan Paul, 1958; ROCHLIN, G., "The Dread of Abandonment," in *The Psychoanalytic Study of the Child*, Vol. XVI, New York: International Univ. Press, 1961; VOLKART, E. H., with S. T. MICHAEL, "Bereavement and Mental Health," in *Explorations in Social Psychiatry*, A. H. LEIGHTON, J. A. CLAUSEN, AND R. N. WILSON, eds., New York: Basic Books, 1957.

2. FRIED, M., AND GLEICHER, P., "Some Sources of Residential Satisfaction in an Urban Slum,".*J. Amer. Inst. Planners*, 27:305–315, 1961.

3. GANS, H., *The Urban Villagers*, New York: The Free Press of Glencoe, 1963; GANS, H., "The Human Implications of Current Redevelopment and Relocation Planning," *J. Amer. Inst. Planners*, 25:15–25, 1959; HOGGART, R., *op. cit.*; HOLE, V., "Social Effects of Planned Rehousing," *Town Planning Rev.*, 30:161–173, 1959; MARRIS, P., *Family and Social Change in an African City*, Evanston, Ill.: Northwestern Univ. Press, 1962; MOGEY, J. M., *Family and Neighbourhood*, New York: Oxford Univ. Press, 1956; SEELEY, J., "The Slum: Its Nature, Use, and Users," *J. Amer. Inst. Planners*, 25:7–14, 1959; VEREKER, C., AND MAYS, J. B., *Urban Redevelopment and Social Change*, New York: Lounz, 1960; YOUNG, M., AND WILLMOTT, P., *Family and Kinship in East London*, Glencoe, Ill.: The Free Press, 1957.

BIBLIOGRAPHY

ERIKSON, E., "Ego Development and Historical Change," in *The Psychoanalytic Study of the Child*, Vol. II, New York: International Univ. Press, 1946.

ERIKSON, E., "The Problem of Ego Identity," *J. Amer. Psychoanal. Assoc.*, 4:56–121, 1956.

FIREY, W., *Land Use in Central Boston*, Cambridge, Mass.: Harvard Univ. Press, 1947.

Article 16 Residential Change and School Adjustment

MURRAY LEVINE

A stranger was before him—a boy a shade larger than himself. A newcomer
of any age or either sex was an impressive curiosity in the poor shabby little
village of St. Petersburg.

Neither boy spoke. If one moved, the other moved—but only sidewise, in
a circle; they kept face to face and eye to eye all the time. Finally Tom said,
"I can lick you!"

"I'd like to see you try it."

"Well, I can do it."

"No, you can't either." . . .

Tom chased the traitor home, and thus found out where he lived. He then
held a position at the gate for sometime, daring the enemy to come outside,
but the enemy only made faces at him through the window and declined. At
last the enemy's mother appeared, and called Tom a bad, vicious, vulgar
child, and ordered him away. So he went away, but he said he " 'lowed" to
"lay" for that boy.

Tom Sawyer
Mark Twain

This paper will discuss residential change, a prominent facet of American life
having important implications for community mental health practice and for
preventative services. Moving represents a stress experience requiring one to
give up old relationships and adapt to new conditions. The point is clear when
one considers an immigrant from a foreign country, when one thinks of a rural,
southern Negro moving into a northern urban environment; or when one
thinks of the wife of the junior executive, born and bred in the big city, who
is suddenly transported to a relatively small town in the deep South. However,
long distance moves constitute a relatively small fraction of all residential

MURRAY LEVINE, PH.D., is Professor of Psychology, State University of New York, Buffalo.
Source: Community Mental Health Journal, Vol. 2, No. 1 (Spring 1966), 61–69. Reprinted
by permission.

changes. About 70 per cent of all moves are made within the same city, or within the same county.[24] Such moves have received considerably less attention because it is believed that a local move does not result in very much of a change in the way of life. While such a proposition may be true for adults, every move involving children requires a change of school and a change of friends.

It is the thesis of this paper that *any* move represents both a problem in adaptation for children, and an opportunity for the development of preventative mental health programs. There is also an important implication for community mental health practice in the sense that residential changes reach such proportions in some areas that the feasibility of neighborhood-based clinical services, or services fixing on a neighborhood-based elementary school must be examined closely. It is the purpose of this paper to bring some of the facts and the problems associated with residential moving to the attention of those concerned with community mental health.

EXTENT OF RESIDENTIAL AND SOCIAL CHANGES

The extent and nature of residential change is well documented in U.S. Census Bureau figures. The 1960 Census showed that approximately 25 per cent of all persons five years old and over had moved at least once in the 15 months preceding the census. Somewhat more than 12,000,000 school age children changed residence, and thereby changed schools at least once in the 15 months preceding the census. Of these, 8.2 million were of elementary school age. Over a five-year period, from 1955 to 1960, approximately 50 per cent of all school age children moved at least once.

The Census does not provide information about the number of school changes experienced by individual children. Rakieten[15] studied children in 17 of 31 New Haven, Connecticut schools, in 1959–1960. She used grades three to six and reported that 35 per cent of 10-year-olds had been in *two* or more previous schools, 24 per cent of 11-year-olds, 35 per cent of 12-year-olds, and 48 per cent of 13-year-olds. Levine, Wesolowski, and Corbett[11] studied children in one elementary school in the inner city of New Haven. They reported that 35 per cent of the children currently enrolled in the sixth grade had been in two or more previous schools, while less than one in four now in the sixth grade had begun in that same school. Twenty-two per cent of kindergarten children, 36 per cent of first graders and 63 per cent of second grade children had already been enrolled in at least *one* previous school. Greene and Daughtry[5] studied high school juniors in Savannah, Georgia. They reported that 74 per cent of the junior class experienced at least one change of school due to a family move, while 45 per cent had experienced two or more such changes. Greene and Daughtry[5] provided data for a white population retained in school. The majority of children then experience multiple changes of schools (beyond changing from elementary to junior high, and from junior high to high school) during the course of their school careers. None of these studies dealt with the families of migrant workers who represent special problems.

The People Who Move

Who moves, and what's the nature of the moves that are made? In addition to Census data, the most extensive surveys of geographic mobility are conducted by the U.S. Department of Commerce, because of their concern with characteristics of the labor force. Such surveys report movement in and out of *labor market areas*. A labor market area may be either a county, or a space whose limits are defined by the fact that people who live within a designated geographic area find employment there. Movement between labor markets tends to be high for young people and for college-educated people.[9] Similar findings hold for Negro in-migrants to northern cities. For the period 1955–1960, Negro in-migrants to northern cities had a level of education equal to or exceeding the resident white population. Taeuber and Taeuber[21] concluded that "as the Negro population has changed from a disadvantaged rural population to a metropolitan one of increasing socioeconomic levels, its patterns of migration have changed to become very much like those of the white population."

While such data would suggest moves associated with economic advancement, long distance moves constitute a small fraction of all moves. The vast bulk of moving is local in nature. Moves across state lines account for about one move in eight, while local moves account for close to 70 per cent of all residential changes. If it were primarily the well-educated, young person who was moving, problems associated with moving would not figure so prominently in the thinking of urban education officials.[2] Overall, the rate of moving seems to be highest in that segment of the population where a high proportion of social problems and educational difficulties are found. Sexton[19] showed that pupil turnover figures in elementary schools in a large midwestern city varied from 49 per cent in schools serving low income populations to 16.7 per cent in schools serving upper income areas.

According to 1960 Census figures for the City of New Haven, 55 per cent of those earning under $3,000 a year and 29 per cent of those earning more than $15,000 a year moved at least once in a five-year period. Moves were more frequent among nonwhites than among whites (47 per cent and 29.5 per cent respectively). The differential in the population of children is quite striking. In 1960, 50 per cent of white males, and 58 per cent of white females were living in the same house as five years ago. Only 34 per cent of nonwhite males, and 23.4 per cent of nonwhite females were living in the same house as five years ago. Most moves were local in nature. Among the white population, 64 per cent moved within the city, while 95 per cent of nonwhites moved within city limits. Similar figures hold for other cities.

Levine, Wesolowski, and Corbett[11] found a turnover rate of about 50 per cent in an inner city elementary school serving a disadvantaged neighborhood.

In the course of a school year, the elementary school lost about 25 per cent of its pupils, and these were replaced by another 25 per cent. Reflecting census figures, the vast bulk of transferring children entered from another New Haven school. Some of these figures may reflect urban renewal programs, but it should be noted that Abbott and Breckinridge,[7] studying school attendance in Chicago in the early 1900's, commented that it is well known that people who live in poorer neighborhoods move a great deal. While urban renewal may account for some proportion of the current figures, it is a reasonable hypothesis that frequent moves, particularly among poorer people, are somehow associated with the conditions of urban life. In any event, each move, local or long distance, for any reason whatsoever, involves a child in a change of schools. Children from poor, nonwhite families are apparently exposed to such moves more often than children from wealthier, white families.

Why Moves Are Made

Why do people move? According to Lansing and Barth,[9] the threat of unemployment, or a search for steadier employment was the reason for moving in about one out of five cases. Moves to obtain a better job were made by about 17 per cent. Family reasons (desire to be closer to relatives, health considerations, death in a family, or the desire to get away from someone, as in a divorce) were mentioned by 26 per cent of those moving. For some substantial number of people, a move occurred in relation to economic insecurity, or because of feelings about a family situation.

A most detailed study of moves and movers has been conducted by Rossi. Rossi selected four census tracts in Philadelphia for study. The census tracts consisted of two relatively low and two relatively high status areas. Two of the tracts were areas with high mobility, and two were areas with low mobility. Rossi specifically excluded tracts with high proportions of nonwhites, or foreign born from consideration. He interviewed probability samples in each of these four tracts, and repeated his interviews eight months later. Rossi found that most of his respondents could predict their moves rather well. The major reason for moving had to do with complaints about the adequacy of physical space and the neighborhood. In general, younger heads of households and larger households tend to be more mobile. Rossi concluded that mobility, in the sample he studied, was primarily related to the housing needs of the family in relation to the life cycle of the family unit. Rossi was dealing with a more stable population than is generally found in the inner cities. Remember Rossi specifically excluded areas with greater than 20 per cent Negro or foreign born from consideration. Nonetheless, for many children, moves may have been accompanied by recent changes in family composition, including the arrival of new siblings. While benefits accrue in terms of better housing, for many children the moves may take place in relation to a situation of exacerbated sibling rivalry.

There is no information comparable to Rossi's for inner city populations.

Work in an inner city school with children who come to a psychologist's attention provides a highly biased sample. However, an informal assessment of that experience suggests that a rather large number of moves are associated with divorce and separation, or conversely with a mother establishing a new household with a new "husband." Levine, Wesolowski, and Corbett[11] found a statistically significant relationship between the number of previous schools attended and the intactness of the family at the time of original enrollment. Children from broken homes had moved more frequently. Sometimes moves take place in relation to economic distress, and attempts to dodge bill collectors, or other officials, including zealous psychologists who pursue parents to help neglected children. In one family the move was precipitated by the mother's arrest. In still others the moves follow upon months of struggle with inadequate space, heat, and plumbing. There is no information at all about how parents prepare children for moves, but in the inner city population, we have been impressed with the precipitate nature of many moves. It is our impression that children are frequently given little or no reason for a move, including any explanation of what happened to the father in the event of a separation.

A study of reasons for moving suggested that a good number of moves in all socioeconomic groups, but particularly among poorer people, take place under less than ideal conditions. Economic insecurity, family difficulties, and enlarging family size account for a rather large percentage of all moves. For a sizeable number of children, a move may well be taken as a signal of distress in the family.

Moving and Adjustment

What are the effects of moving? At one level, moving and migration have been related to rates of hospitalization for mental illness. Results however are inconsistent. Malzberg and Lee[12] reported a significantly high rate of hospitalization for migrants in New York State; but Jaco[7] found little or no relation to migration in Texas. Hollingshead and Redlich[6] found no relationship of foreign birth to the diagnosis of schizophrenia. However, Roberts and Myers[17] reported that the foreign born are represented out of proportion to their numbers in the total psychiatric population. Most of these studies define mobility in terms of long distance moves, and they do not usually deal with the frequency of moves within a local area. They also deal with gross hospitalization rates, and do not indicate anything at all about the problems these moves pose for children. The available evidence would support a view of a move as a stressful situation with some likelihood of being associated with mental disturbance in adults in the United States.[8]

There is far less evidence concerning the effects of moves on the mental health of children. Tietze, Lemkau, and Cooper[22] studied the prevalence of mental disturbances of all kinds in children in Baltimore. Mobility was defined in terms of the number of years the family lived in the same house, and the number of years the family was located in Baltimore. They found a significant negative relationship between the prevalence of each of the kinds of disturbances they studied and *duration of residence in the same house. Duration within the City of Baltimore* was not significantly related to the prevalence of disturbance. No matter what the breakdown, rates were highest for the shortest residents, and lowest for the longest residents. However, the Negro population conformed less closely to the general pattern. How restrictions in the possibilities for moves by Negroes contributed to their results is not clear. The study was completed in Baltimore in the pre-war years, before the onset of the South-to-North pattern of migration, and before the beginning of major urban renewal programs. Data from the pre-World War II period may not be applicable to today's problems. Moreover, disturbance was defined in terms of a child who came to the attention of a social agency or guidance clinic. One cannot say anything about the effect of moving in relation to school adjustment and academic performance from their results.

Kantor[8] interviewed more than 400 families before and after a short-distance move. She found that before the move, families who move are more likely to have children who manifest behavior problems than families who do not change residence. However, following the move, there is no significant change in the number of symptoms mothers report. In other work, Kantor reports that teachers rate mobile children as less well adjusted than non-mobile children. Kantor suggests that some families with problem children may move because the child is not getting along in the neighborhood or school.

Gordon and Gordon[4] studied the prevalence of treated disturbances of children from case files in clinics and in the offices of private practitioners, in relation to the migration rate in four suburban New Jersey counties. They reported an increasing rate of emotional disturbance with increasing mobility in the county. As mobility increases, the ratio of boys to girls also increases; but paradoxically, Negro girls outnumber boys six to one for treated emotional disturbance in these mobile suburban counties. The finding is curious in view of the generally greater incidence of all forms of disturbances in childhood among boys. Gordon and Gordon argued that the higher rates among suburban Negro girls are due to the high rates of working mothers among Negroes living in the suburbs. Why the lesser availability of the mothers should be a greater problem for girls than for boys is unexplained by them. They also gave several case studies of children whose problems seemed to be related to moving, and they recommended that the community and helping agencies offer practical assistance in these instances.

Pedersen and Sullivan[14] have studied the relationship between mobility, parental attitudes, and emotional disturbances in children whose parents are in military service. Military families tend to move a great deal; but according to Pedersen and Sullivan, it is not the move per se which is related to emotional disturbance. Parents who seemed more closely identified with military life, and who accepted the military way of life, moved a great deal; but they did not have children who were disturbed. Parents of children who were disturbed, in the sense that they were seen at a psychiatric clinic, tended to be less closely identified with the military way of life. Pedersen and Sullivan concluded that something in the parental attitudes about moving may mediate the emotional effects for the children. Comparable data do not exist for children of nonmilitary families. Freedman[3] argued for a concept of mental mobility, by which he meant the extent to which a person is adapted to and accepts a high rate of change and new experience as a normal part of life to be assimilated to his more usual way of life. For such a person, mobility may be routine and not at all disorganizing. However, if so, from the point of preventative mental health, and from the point of view of the theory of ego adaptation, it would be very useful to understand just exactly how such mentally mobile people adapt to change.

There is very little data, on a relatively microscopic level, of the emotional significance of moves for children, or of the problems moves pose for children. Stubblefield[20] reported several case studies of emotional disturbances in children precipitated or exacerbated by family moves. He pointed out at least four conditions which seem related to disturbances in his patients: (a) Children are ignored or placed out of the home while parents are settling in. (b) The children are ignored or actively rejected by the peer group when moving in. (c) The children may experience grieving reactions when separated from old playmates. (The new peer group may be difficult to join, leading to some overvaluation of the lost friends.) (d) In some instances, children also seem to suffer "shock" when not prepared for the move, or when they are told at the very last minute.

Stubblefield,[20] like Gordon and Gordon, also recommended that practical assistance in settling in, and friendly acceptance by teachers, church, family doctor, and neighbors can be important in alleviating the stressful effects of moves. Stubblefield indicated that the problem of isolation is marked, and feels there is an important need for the individual to have a cognitive map of his new physical and social environment in order to help orient himself again. As we shall see, the school, a major social institution affected by and affecting children who move, has no systematic program for orienting new children.

ENTERING A NEW SCHOOL

What happens when a child enters a new school? According to Rakieten's[15] survey almost all children feel apprehensive about the prospect of entering a new school. Most are not so much worried about doing the school work, as they are about how peers will accept them, and how they will find out what is expected of them in school. In sociometric study, at least one-third of those who were highly mobile (had been in at least two other schools), were isolates. Kantor[8] cites other studies in support of Rakieten.

Rakieten's children reported getting a rather abrupt introduction to the new school in at least 50 per cent of the cases. Particularly in lower class schools, the child was likely to be sent to the new classroom unaccompanied by anyone. Only

20 per cent of the children recalled getting any information at all from either principal or teacher on rules and routines of the new school. Sixteen per cent indicated they found out the rules only when they had made a mistake and had been called down for it by the teacher. Few recalled getting any help in making friends or in getting to know their classmates. According to the children's recollections, teachers would frequently assign children to seats on the basis of availability, but without any real introduction, or any encouragement to the other children to be friendly and helpful to the newcomer.

Rossi[18] said that a high mobility school had found it helpful to institute an orientation program for new students, but he did not describe the program at all. A pamphlet developed by the Association for Childhood Education[16] offered hints to parents and teachers about how one might welcome and orient children to the new school; but it is not clear that any schools have made systematic attempts to introduce children to the new school. The ACE pamphlet points out there is a dearth of information on the adjustment problems of children entering new schools, and there is no basis in fact for developing meaningful programs.

In Rakieten's study the children themselves indicated they would prefer an orientation period in which they could just watch the classroom to familiarize themselves with what was going on. The self-observations of these children are reminiscent of Murphy's[13] reports of nursery school children entering new situations. Apparently many of them went through an orientation period in which they just watched the new situation. The orientation period has not been evaluated systematically as a technique for introducing children into schools, and in fact on first blush, it seems to be just the opposite of what any action-oriented intervention program might try. What else should be done, or should not be done is far from clear.

Academic Performance

Moving away from the problem of adjustment, is there also a relationship between academic performance and moving? There are enough variations in curriculum and materials, even within the same school system, to suggest a child might have to unlearn some things while he is learning new procedures for the same task. Some examples from our experience will make the point clearly. In some schools, cursive writing is introduced earlier than in other schools. A child coming from a school which continues printing might be faced with the problem of learning cursive writing at the same time he is exposed to new spelling words. In another instance, in a traditional school, a teacher insisted a child form his letters by making the strokes in a given order. The child seemed to have learned to make the letter perfectly adequately his own way, but the requirements of curriculum and method demanded he unlearn and relearn. In another instance a child had learned to divide by the newer method of approximating the first digit in long division. Introduced into a class still using the old method, he had to reorient himself to that technique. Reading series and reading techniques also differ from system to system, and with experimental programs being introduced into schools at a rapid rate, the reading program may differ considerably within a grade in the same school system from building to building. One building had teachers available to provide tutoring. Other buildings in the same school system did not have such a program, and children moving away from the one school who had such individual attention would now miss it.

We have also been impressed that school records contain a bare minimum of information about the work children have been exposed to, and a minimum of information about any special problems a child presents. The records may also be slow in arriving. Parents will not always have the opportunity to meet the teacher and explain the special situation until the problem has manifested itself. The minimum of information does not offer teachers too much help in programming for given children. Moreover, differences in curriculum and in approach can be received unfavorably by parents who can transmit their attitudes and feelings to the children. We have had occasion to speak to parents who interpreted a difference in the placement of subjects in the curriculum as evidence of the inferiority of the present school when compared to the previous one.

While it seems reasonable to expect a relationship between the frequency of moves and academic performance, some studies do not obtain the expected result.[8] Greene and Daughtry[5] studied the relationship between an index of mobility based upon a weighting for frequency and distance and school grades among 434 high school juniors in Savannah, Georgia. Savannah has considerable mobility stemming from the flow of military personnel, industrial workers, executives, and rural-to-urban migrants. They present no data on the actual composition of the school, but since they do not indicate otherwise, it is safe to assume they studied an all white high school. Essentially, they found few relationships between their index of mobility and 102 other measures. However, those correlations which were significant showed that students who had moved more, or who had moved longer distances, had more favorable grades.

Kantor[8] cites data showing that the area from which a family migrated, the social status of the family, and the child's intelligence level are important variables. Transient children of professional parents may gain from mobility, while children of unskilled laborers may do considerably less well after moving.

Levine, Wesolowski, and Corbett[11] studied the population of an inner city elementary school. In their study, there was a significant negative relationship between the number of previous schools attended and school grades in citizenship and in academic subjects. The relationship seemed strongest in girls. The inner city elementary school was comprised of about two-thirds Negroes, and about a third of the group, or more, came from broken homes. Those coming from broken homes tended to move more than those coming from stable homes. The index, number of previous schools, did not weight the distance of moves, although there was no relationship between place of birth (New Haven, southern states, other states) and grades. While Greene and Daughtry found moves were related, if anything, to better school grades among white high school juniors, Levine and associates found moves were correlated with poorer grades among poor, nonwhite, younger children. It seems clear that moving per se is not the most important issue; but factors associated with the move may be vital in determining the effect on the child in school.

Implications

In so far as children entering a new school are concerned, it seems perfectly feasible to develop programs to help induct and orient the new child. One such program has been introduced on a pilot basis in two elementary schools. Upper grade children have been recruited and trained to serve as guides to children who are entering the school. The older child provides a tour of the building, instructs the new child in the general rules of the school, and discusses facilities in the school and neighborhood. An assessment of the effects of such an orientation program on children's feelings about entering the new school is in progress.

It should also be possible to counsel parents in the preparation of children for moving since it is clear from Rossi's[18] data that most families can predict their moves. However, such a counseling program depends upon adequate publicity and parental cooperation, and upon knowledge of how to achieve mental mobility. Such cooperation may be difficult to obtain from Negro parents in the inner city when approached by white professionals. However, the possibility for such a program is there.

Finally, the implications of the high turnover in inner city schools for neighborhood or school-based mental health programs are clear. The Psycho-Educational Clinic, centered as it is on consultation within the elementary school building, has lost a number of children due to moves after considerable time and effort was expended with the children and their teachers. In some instances, therapists who started with children in one school found it necessary to expand their sphere of operation to other public schools in order to follow their children. A problem-preventative tutoring program, involving student teachers and first-grade children lost fully one-half of the students due to moves in the course of a school year. Locating the clinical service in the school, strategic to the manifestation of the problem, was not entirely sufficient because of the problems created by geographic mobility in the inner city school. The problem is clear, but the solution is far from obvious.

REFERENCES

1. ABBOTT, E., & BRECKINRIDGE, S. P. *Truancy and non-attendance in the Chicago schools.* Chicago: Univer. of Chicago Press, 1917.
2. CHASE, W. W. *Problems in planning urban school facilities.* Washington, D.C.: U.S. Dept. HEW, OE 21023, Bulletin, 1964, No. 23.
3. FREEDMAN, R. Cityward migration, urban ecology, and social theory. In

E. W. Burgess & D. J. Bogue (Eds.) *Contributions to urban sociology.* Chicago: Univer. of Chicago Press, 1964.

4. Gordon, H. E. & Gordon, K. K. Emotional disorders of children in a rapidly growing suburb. *Int. J. Soc. Psychiat.*, 1958, 4, 85–97.

5. Greene, J. E., Sr. & Daughtry, S. L. Factors associated with school mobility. *J. educ. Sociol.*, 1961–62, 35, 36–40.

6. Hollingshead, A. B. & Redlich, F. C. Schizophrenia and social structure. *Amer. J. Psychiat.*, 1954, *110*, 695–701.

7. Jaco, E. G. *The social epidemiology of mental disorders.* New York: Russell Sage Foundation, 1960.

8. Kantor, M. B. (Ed.) *Mobility and mental health.* Springfield, Ill.: Charles C. Thomas, 1965.

9. Lansing, J. B. & Barth, N. *The geographic mobility of labor.* Washington, D.C.: Economic Redevelopment Research, U.S. Department of Commerce, Area Development Administration, 1964.

10. Lansing, J. B. & Ladd, W. *The propensity to move.* Washington, D.C.: U.S. Dep't of Commerce, Area Development Administration, 1964.

11. Levine, M., Wesolowski, J. C., & Corbett, F. J. Pupil turnover and academic performance in an inner city elementary school. *Psychology in the schools.* 1966 (in press).

12. Malzberg, B. & Lee, E. S. *Migration and mental disease.* New York: Social Science Research Council, 1956.

13. Murphy, L. *The widening world of childhood.* New York: Basic Books, 1962.

14. Pedersen, F. A. & Sullivan, E. J. Relationships among geographic mobility, parental attitudes and emotional disturbances in children. *Amer. J. Orthopsychiat.*, 1964, 34, 575–580.

15. Rakieten, H. The reactions of mobile elementary school children to various elementary school induction and orientation procedures. Unpublished doctoral dissertation, Teachers College, Columbia Univer., 1961.

16. Rasmussen, H. (Ed.) *When children move from school to school.* Washington, D.C.: Association for Childhood Education, International, 1960.

17. Roberts, B. H. & Myers, J. F. Religion, national origin, immigration and mental illness. *Amer. J. Psychiat.*, 1954, *110*, 759–764.

18. Rossi, P. H. *Why families move.* Glencoe, Illinois: Free Press, 1955.

19. Sexton, P. *Education and income.* New York: Viking Press, 1961.

20. Stubblefield, R. L. Children's emotional problems aggravated by family moves. *Amer. J. Orthopsychiat.*, 1955, 25, 120–126.

21. Taeuber, K. K. & Taeuber, A. F. The changing character of Negro migration. *Amer. J. Sociol.*, 1965, 70, 429–441.

22. Tietze, C., Lemkau, P., & Cooper, M. Personality disorder and spatial mobility. *Amer. J. Sociol.*, 1942, 48, 29–40.

23. U.S. Bureau of the Census, U.S. Census of Population, 1960. *Summary: general social and economic characteristics.* Washington, D.C.: U.S. Govt. Printing Office, 1963.

24. U.S. Bureau of the Census, U.S. Census of Population, 1960. *Mobility for metropolitan areas.* Washington, D.C.: U.S. Gov't. Printing Office, 1963.

Article 17 Draft Dodgers

Coping with Stress, Adapting to Exile

SAUL V. LEVINE

The young American who is drafted has a choice: he can become a soldier or he can go against the system by going underground, going to jail, or leaving his country. Recent articles and books have stressed the terrible dilemmas that these young men face.[4,7,9] This paper will concentrate on those who have chosen to go to Canada.

Harboring draft dodgers, deserters, and dissenters is not a new phenomenon in the modern history of the relationship between Canada and the United States. In the early part of World War II many like-minded Canadians fled to the United States and were not extradited.[5,12] Since the Vietnam conflict expanded to a full-scale war, and social unrest has increased in the United States, Canada has served as a destination and home base for an estimated 60,000 to 100,000 young Americans in varying states of relationship to the United States Armed Forces. This has done more to reverse the "brain drain" than any other factor, although emigration of other Americans (women, older adults with children) to Canada has also increased markedly in recent years. The reception of Canadians to these immigrants has varied. A new Immigration Act was passed in 1967, which utilized a "point system" in establishing criteria for admission to Canada. Parameters such as education, vocational skills, employment offers, and marital status were weighted, and the total determined one's eligibility for immigration, obviously favoring the middle-class applicant. In spite of this, immigration officers at the border gave many of the American applicants for landed immigrant status considerable difficulties if they were the "wrong type" (draft dodgers, leftists, hippies). The situation persisted in spite of a public clamor until May 1969,

SAUL V. LEVINE, M.D., C.M., F.R.C.P.(C), is Associate Professor of Psychiatry and Psychology, University of Toronto.

Source: American Journal of Orthopsychiatry, Vol. 42 (1972), No. 3 (April), 431–40. Copyright © 1972 the American Orthopsychiatric Association, Inc. Reproduced by permission.

when the government enunciated a policy that directed border officials to desist in their harassment of these individuals.

There are thus many American expatriates in this country. Most are landed immigrants, and men outnumber women in the ratio of about five to one. It has been estimated that, in addition, another 25,000 or more young Americans are here without official landed immigrant status.

A disproportionate number of the draft dodgers and deserters live in Toronto (perhaps half of the total in Canada), partly due to its size (two million people), heterogeneity, liberalism, accessibility, language (English), and position as business and cultural center of English Canada. The draft dodgers in Toronto live in a variety of settings: urban or rural communes, shared rooms, homes, Free University, boarding houses, apartments. From time to time various sources in the large antidraft network referred such individuals in crisis for purposes of psychiatric assessment. In the course of over two years, 24 draft evaders and deserters were assessed in various stages of emotional disturbance, and they were all followed over the ensuing months. The similarity in many of their stories led me to interview a large "normal" population in the search for common modes of coping with the stress of leaving home suddenly and moving to a new country. Approximately 60 American military expatriates were interviewed for this purpose.

DRAFT DODGER SCENE

It has been difficult to estimate just how many draft dodgers, deserters, and dissenters have entered Canada in the last few years; estimates vary from 60,000 to 100,000 American expatriates in this country as landed immigrants or as visitors. Most of them are young (17 to 28), obviously in the draftable years. Most of them are also white Anglo-Saxon Protestant and from middle-class backgrounds. They have also been a fairly well educated group, with the majority having completed high school and gone on to college. In these respects they have been a fairly uniform group, at least until quite recently when increasing numbers of deserters and decreasing numbers of draft dodgers have made their way here. By and large, the deserters have been younger, less educated, from lower socioeconomic groups and have come to Canada with much less preparation and support than their draft evader counterparts. We have been struck by the relative absence of young black expatriates in the exodus to Canada; de facto segregation seems to permeate even the young, idealistic groups. Women are obviously not draft dodgers and deserters in the usual sense, but many have come up for ideological or personal reasons, and their presence has often been of great support to their boyfriends or husbands.

There are a number of factors that seem to determine how the individual draft dodger will cope with the stress of leaving his homeland and adapt to his new country. In general those who have lived away from home previously, have gone to college, have planned the move to Canada beforehand, have some moral backing from their families, have contacts in this country, and who think in ideological terms (social, political, religious, or philosophical) have had an easier time in their first few months here. Those who have been relatively unprepared, have had unrealistic expectations (Utopia, total freedom, warm welcome), have not lived away from home or gone to college, have had previous psychological difficulties, or who have been rejected by their immediate families or those closest to them, have had greater difficulties in coping with the inevitable stresses.

The nature of the extrinsic stress that these young people undergo is fairly obvious. Certainly if they receive no financial or, even more important, moral support from their families, the move up here can be an alienating and frightening experience. In some cases the desire of the individual to come to Canada has precipitated total rejection on the part of his parents or important reference groups. Some of those who have come to Canada (especially deserters) feel "on the lam," hunted or persecuted by American or Canadian authorities. While often this is exaggerated, there is evidence that this type of harassment has occurred.[15] In addition to "official" surveillance, the new arrival often faces open hostility on the part of some Canadians, or more subtle disdain by many others.

If the individual has come with little or no preparation, he is often at a loss as to where to begin. The availability of contacts in Canada—official organizations, receptive expatriates, or clergy—makes the transition to this new country much easier. If he has been "placed" in a commune, for example, with no prior information or planning, and has not lived away from home, this new life style can be quite unnerving: if some planning has occurred the communal living arrangement offers much in the way of friendship, nurturance and support. One problem faced by many of the recent newcomers is lack of job opportunities and money. Subsistence income or less contributes considerably to maladaptive, self destructive, or even antisocial behavior. Some of the new American arrivals have complained that American-owned firms in Canada will not hire draft evaders or deserters (this accounts for more than 50% of the Canadian economy). Aside from this, jobs in general are scarce, unemployment rates are high, and there is a general slowdown in the economy. The highest unemployment rates are in the 18-to-25 age group, which obviously hits the draft evaders hardest.

The dodgers vary in their reasons for coming to Canada, and their motivation for the move in fact plays an important role in their ability to cope. Some would not fight in what they call an immoral war. Some have come to escape what they see as an oppressive socio-political system, and see themselves as political refugees even though the draft might not have been a personal threat. Others have left an intolerable personal or family situation, and occasionally this has been done irresponsibly and destructively. Some were not "making it" in school, peer groups, or society and felt that Canada would be a fresh start. Others were seeking adventure, or were hoping to find their nirvana here. In many cases there was a combination of reasons for leaving their homeland.

COPING STAGES

There is no doubt that the move to Canada is not unlike other crises that necessitate a rapid mobilization of various coping mechanisms in dealing with stress. Much has been written about the adaptive potential of human beings and their patterns of coping with stressful situations (life threatening situations, severe psychological or physical stresses, psychosocial transitions).[3] The various behaviors utilized in preparing for the move here have been shown to be effective in other contexts;[1,2] for example, seeking and utilizing information about a threatening situation a priori is one of the most important coping strategies in the successful handling of a crisis.[6] While individuals in these situations vary in their behavior, a broad sequence of stages can be observed over a period of time. Similarly with the young draft dodger leaving his home and loved ones, perhaps forever, there seems to be a series of four stages that most of them go through to varying extents in the process of adaptation to their new lives.

The first stage (I) is one of *Disorganization,* in which the individuals may be confused, unprepared, and floundering. Some of the dodgers describe an immediate sense of relief as soon as they get over the border, but this tends to be transient and based more on fantasy and wish fulfillment than reality. It is followed by the sense of isolation, loneliness, and psychic pain that characterizes this period. The length and intensity of this period is dependent on all the factors mentioned earlier, but especially the nature of the people that the individual meets and lives with. A firm sense of belonging in highly valued groups enables more effective coping;[6] this is also the function that communes can fulfill for these young Americans. Adequate preparation beforehand often obviates many of the disabilities of this stage. It is at this time that feelings of ambivalence and guilt (abandoning family, etc.), and suicidal impulses become most manifest. A history of psychoneurotic symptoms predominates in those for whom this stage is most incapacitating.

The second stage (II) is called *Acting Out,* wherein a disproportionate number of these young people drop out of conventional activities, indulge relatively heavily in drugs, and are exploitative and parasitic in their interpersonal relationships; superficiality and non-involvement become the modus vivendi. Destructive relationships with other people, especially heterosexual, are seen and this is the stage in which lovers who have come up together most often separate. It is also during this stage that antisocial activities (stealing, pushing drugs, violence) and consequent trouble with the police become manifest. At this time some individuals may impulsively decide to return to the United States. This stage may be bypassed, but a flicker of it seems to occur fairly early on in the stays of many draft dodgers we have studied. Like the stage previous to it, it is short-lived, and in only a small minority, perhaps less than five percent, does this type of behavior persist in a predominant form. In going over the past histories of this particular core group, it becomes apparent that there were strong hints about this type of behavior before even entering Canada. The anti-social behavior occurs more often among deserters than draft dodgers, perhaps due to their youth and different backgrounds.

The third stage (III) we call *Searching*, in which the individual explores himself and his relationships, looks for meaning in life, and pursues his interests (school, work, commune, etc.). It is as if a psychic brake is put on and the activity becomes less frenetic, less impulsive, and more considered (and considerate). The individual begins to take stock of himself, and continues the process of maturation that was interrupted by the major move away from the States. It is at this time that many of the individuals return to school, develop vocational skills, and are concerned with developing close interpersonal relationships. There is also considerable thought given at this stage to family and friends back in the States, reasons for leaving, and a more realistic appraisal of their lives in Canada. A few even return during this stage to the States, but it is a more reasoned move back, not at all like the impulsive or petulant moves that occured in Stage II.

The least common stage is the fourth (IV), labeled *Adaptation and Integration*. During this period the individual becomes totally involved in his new life style. For the first time he is engaged wholly in being a Canadian, not specifically an American-in-exile. He is concerned about "making it" and about Canadian social problems and politics. He has adapted to his new life in Canada; it is no longer particularly stressful, and he seldom longs for a return to the United States. By this stage the draft dodger or deserter has been in Canada long enough to cement close interpersonal ties, to develop a sense of belonging and a *raison d'etre* here.

There is much variation among the draft dodgers as to the duration and intensity of each of the stages; there is also occasionally considerable overlap or mixing of two (or more) stages. In addition, some of the individuals get bogged down at one of the first three stages. Few of the expatriates I have known or interviewed have fully achieved the level of Stage IV. There are fairly obvious reasons for this, however. The United States engenders in its youth a fierce sense of patriotism; the American young people who have decided to come to Canada usually do so with ambivalent feelings. Even among the radicals there is considerable self chastisement and even guilt about not having gone to jail or gone underground in the United States. They are often still in close touch with their families and friends back home. In addition, American draft dodgers tend to associate with each other, maintain close ties, and live in enclaves or draft dodger communes where most of their close associations are with individuals in the same situation.

This situation is further reinforced by a network of support mechanisms that has evolved on both sides of the border. These mechanisms include communal living arrangements, central meeting places, eating halls, sympathetic Canadian groups making themselves available, indigenous newspapers, employment agencies, counseling services and volunteers. These have made the process of adaptation considerably easier for many, but on the other hand they have ensured a kind of isolation or even elitism (at times it's "cool" to be a draft dodger). Many have retained a wholly American identity, and even criticize or mock Canada for not being quite as good as the United States. There are over 30 draft dodger counseling offices across Canada (War Resister's League, Toronto Anti-Draft Programme, Exile Information Office, Toronto American Deserters' Committee, etc.), and numerous newspapers (AMEX—The American Expatriate in Canada, Guerrilla, Harbinger, Dreadnought, etc.); these sources provide advice on how, where, when, and whom to approach for food, employment, lodging, money, socializing, etc. The clergy has been very active in aiding the new arrivals. The National Council of Churches publishes a newspaper entitled "Contact," as a service to parents in the States of draft dodgers and deserters in Canada. In addition, the World Council of Churches is raising $210,000 to assist these young people in Canada, and has sent a minister from the United States to work specifically with this population.

DRAFT DODGER PATIENTS

Having become known to some of the draft dodger communes and groups as a nonjudgmental psychiatrist, I began receiving referrals from these sources. During a period of over two years a total of sixteen draft dodgers and eight deserters were referred in various states of emotional upset shortly after their arrival in Canada. All those considered here were males, although five females were seen in the same time period, all expatriate Americans who had come up either to be with their lovers or were political dissenters. Even in this highly selective patient sample one recognizes the clear difference between dodgers and deserters.

Six of the patients were age 22 or over, while eighteen were under that age; the deserter patients tend to be younger than the draft dodgers, come from families that are larger and in lower socio-economic strata (most draft dodgers are from middle-class origins), and have poorer educational records (only one of the deserters attended college). It is interesting that very few of the draft dodger and deserter patients had the wholehearted support of their parents: seventeen of the parents disagreed with their sons *and* provided no moral or financial support, three disagreed but supported their son's decision, one agreed with draft evasion but refused to support his leaving the States, and three agreed and supported their sons. (Lack of parental support is prevalent in almost half of the "normal" dodger population.) Of the 24 assessed, only nine had even bothered to seek reliable information about life and conditions in Canada prior to their major move to this country, and what they did do was scanty. The expatriates who were seen by the author for professional reasons were referred during coping Stages I (10), II (10), or III (4). With one exception, the deserters were not referred during Stage III. The diagnoses used ran the gamut of psychiatric (APA) nosology (with all its limitations). Transient situational reaction was the most common diagnosis, used in eight cases. Personality disorder was next in frequency, five having been diagnosed in this way. The diagnosis of psychoneurotic disorder was used five times, and psychotic reaction was diagnosed in three cases.

There was one instance of psychophysiologic disorder in a young man who was also quite neurotic (severe anxiety reaction). In general the more incapacitated young men presented histories and pre-morbid personalities that provided strong hints about prior emotional difficulties before even leaving the United States. These are high risk individuals to be drafted in the first place[9] so that their problems in exile should not necessarily come as a surprise.

The outcome of the 24 cases varied considerably. Some returned to the States, some ended up in the local drug scene, pushing or turning on, and others required psychotherapy. On the other hand, some returned to school, found work, made friends, started a family, or adjusted to life in Canada in other ways. There is no reason to conclude that the draft dodger group is any different from a comparable group in the States in terms of mental health. It is important to note that these 24 draft dodgers and deserters are *not* considered to be representative of the population of exiled Americans who have not sought psychiatric assistance; this is obviously a skewed sample. But we must remember that the age group under discussion is, in general, highly vulnerable to suicidal thoughts, feelings of depression, and emotional liability. In talking with other evaders up here it is clear that those who do not run into serious adjustment problems are a stronger group to begin with. Better educational, social and emotional histories are seen; many had come up here previously to lay the groundwork for their eventual move, and had quite realistic expectations about their life in Canada. Most of the patient group had little parental support, and this remained a crucial factor until the young people joined a group that made them feel welcome and worthwhile. Even in this small patient group, the deserters appear to be more disturbed, with less resources and poorer prognosis than the draft dodgers. To sum up, the most critical factors in determining draft dodgers' and deserters' adaptation to a new life outside their native land are age, education, parental support, pre-existing personality, preparation for the move, contacts in Canada, and probably a great deal of luck (Table 1). It appears from our patient sample that the middle-class young men face fewer difficulties in coping with this major stressful experience; it may be that an individual's previous enriching experiences and stimulation give

Table 1. Factors Affecting Draft Dodger Adaptation

Positive	Negative
Older	Younger
Draft Dodger	Deserter
Parental Support	Parental Rejection
College Education	Poor Academic Record
Preparation	Lower Class
Contacts in Canada	Prior Emotional Difficulties
Ideology	Impulsivity
Money or Job	Unemployed or Unoccupied
Luck	

him a stronger foundation, and ultimately provide him with more alternatives, more degrees of freedom[11] in solving his problems.

PRESENT SITUATION

Two unexpected and important changes have occurred recently in the patterns of American selective service inductees and members who have been coming to Canada. The flow of these young people began in earnest about five years ago when the Vietnam war was moving into high gear. At that time, and for about three years thereafter, a large proportion of these immigrants were middle-class draft dodgers, college educated and politically oriented. For the past year and a half, however, an increasing number of deserters have entered Canada. This latter group is in general younger, more poorly educated, from lower socio-economic strata, and less prepared to make a life here in Canada. They have, as a consequence, been having more troubles with other draft dodger groups, social services, and law enforcement authorities.

A second trend that is much more recent has been a very significant decrease in these types of immigrants. The last two months of 1970 saw the flow of dodgers and deserters slow to a trickle (at its peak in 1969, upwards of 100 a day were arriving in Toronto). There are a number of reasons for this development. There is a growing movement afoot among draft-age young men to "stay and fight" the draft. The lottery system and decrease in draft call numbers have reduced the chances for many individuals to be drafted; some draft boards are reluctant to call vocal, anti-draft, white middle-class youth. But beyond this there is a new ethic to dissent within the system, or to go to jail, or to go underground in the States, rather than to escape. Draft counselors advise eligible youth all over the States on how best to avoid induction, and good lawyers can delay court proceedings for three or more years. The anti-draft movement may have lost some of its clout because of the decreased participation of older, effective members. This has meant especially that the most educated and politically oriented have stopped their exodus across the northern border.

But in addition to these American-originated reasons, there are ample problems facing them in Canada that now cause possible immigrants from the United States to stop and think seriously. For one, the unemployment problem is even worse in Canada than it is in the United States; it has been very difficult for American expatriates to get jobs once here, or offers of employment in advance to help them across the border. Secondly, there are other serious social problems in Canada that are especially threatening to the young idealist who has left his homeland in the expectation of finding a Utopia. Disparities between rich and poor, French-Canadian and English-Canadian, whites and blacks (on a smaller scale) are all too evident here. The recent political kidnappings and murder shocked many Canadians who for long had adopted a "holier than thou" attitude *vis à vis* the States, and exulted in the delusion that "it can't happen here." That it *did* happen, and was followed by the imposition of the War Measures Act, which suspends most civil liberties, was a salient lesson for citizens of Canada, but particularly upsetting to draft dodgers and deserters. In fact, some of the ex-Americans participated in demonstrations against the Act.

Even though it provided considerable potential for abuse, the Act proved very popular in Canada, and Americans who demonstrated against it were castigated. This is fairly representative of growing antipathy to many of these newcomers. The American immigrants were not prepared for Canadian society's apathy and acceptance of the status quo. Canadians are tradition and establishment oriented, and society is still hierarchical and vertically structured.[10] It had been expected that the hostility to draft dodgers would come from conservative circles, and this has been borne out. But some of the most vociferous opposition to them has come from Canadian academics and the New Left. To their spokesman,[8] American draft evaders and deserters represent an extension, albeit subtle, of American imperialism and colonialism, Americans here are accused of taking Canadian jobs, money, places in schools, of denying a Canadian identity, of spreading the American way of life, and of not learning about Canada. Recent headlines in Canada's largest newspaper pointed a finger at perhaps 200,000 Americans here illegally, grossly aggravating the unemployment situation.[14] The draft dodgers are in the news almost constantly, in articles, editorials, radio and television programs. At least one book on the subject will be published shortly.[16]

Assimilation does not appear to be one of the dodgers' top priorities.[13] The whole subject has influenced passions, both supportive and hostile. When one dodger is arrested, the whole group is (metaphorically) indicted. Canadian radicals defend Canada's sovereignty and institutions. The fact that much of Canadian business and industry is American-owned, and that many Americans hold important governmental, academic, artistic, and commercial posts rankles many Canadians. Anti-Americanism is growing, but in a not necessarily healthy direction. It is indeed ironic that some of those American expatriates who most wish to adopt the Canadian way of life are subject to the most violent abuse (especially in universities).

It is my feeling that the current reduction in the influx of draft dodgers and deserters will continue because of the further augmentation of the factors just discussed. At the present time, however, they are an issue to be reckoned with both in Canada and in the United States. They are not paragons of virtue, nor are they all sinners. Many of them represent a distinct loss for the United States and a gain for Canada; others entail just the opposite for their host country and homeland. The majority, however, are young people whose needs must be considered; they are a part of the population, like it or not, that is not yet functioning at optimum potential.

REFERENCES

1. CHODOFF, P., FRIEDMAN, S. AND HAMBURG, D. 1964. Stress, defenses, and coping behavior: observations in parents of children with malignant disease. Amer. J. Psychiat. 120:743–749.

2. Coelho, G., Hamburg, D. and Murphy, E. 1963. Coping strategies in a new learning environment. Arch. Gen. Psychiat. 9:433–443.
3. Coping and Adaptation. 1970. National Institute of Mental Health. U.S. Public Health Service Publication No. 2087.
4. Gaylin, W. 1970. In the Service of Their Country: War Resisters in Prison. Viking Press, New York.
5. Granatstein, J. 1969. Conscription in the Second World War, 1939–1945: A Study in Political Management. Ryerson Press, Toronto.
6. Hamburg, D. and Adams, J. 1967. A perspective on coping behavior: seeking and utilizing information in major transitions. Arch. Gen. Psychiat. 17:277–284.
7. Kline, C. et al. 1971. The Young American Expatriate in Canada: Alienated or Self Defined? Amer. J. Orthopsychiat. 41 (1):74–84.
8. Matthews, R. 1970. Draft dodgers as imperialists. AMEX—The American Expatriate in Canada. 2(4): June–July.
9. Ollendorf, R. and Adams, P. 1971. Psychiatry and the draft. Amer. J. Orthopsychiat. 41(1):85–90.
10. Porter, J. 1965. The Vertical Mosaic: An Analysis of Social Class and Power in Canada. University of Toronto Press, Toronto.
11. Rae-grant, Q. 1970. Personal Communication.
12. Sharp, M. 1970. Press Conference, Minister for External Affairs (Canada) in London, England. November.
13. Toronto Star. 1970. Editorial, December 17.
14. Toronto Star. 1971. Front Page Story, January 25.
15. Turner, W. 1971. (ex-FBI agent; author of Hoover's FBI: The Man and the Myth) on CBC Weekend, January.
16. Williams, R. 1971. The New Exiles: American War Resisters in Canada. To be published.

Developmental Life Transitions: Retirement and Aging

VII

As people move into the last stage of their lives, that of retirement and old age, they encounter new roles that are determined in part by society and in part by their own changing physical condition. The significant roles of the previous thirty to forty years—parent, worker, independent adult—must often be laid aside as grown children begin their own families, retirement occurs, and health and economic status decline. Successful adaptation to these major transitions is highly dependent on the surrounding circumstances as they affect the personal and interpersonal resources on which the aging individual may draw. (See de Beauvoir[4] for an interesting discussion of the issues.)

Two common concepts discussed in current analyses of adjustment to aging are disengagement and activity. The disengagement theory, as developed by Cumming and Henry,[2] suggests that a decrease in social interaction and activity is a natural response to age-related factors like poor health, the loss of peers, the death of older relatives and the consequent shrinking of the social world, and is desirable both for the individual and for society. To be successful, disengagement must be a mutual process; if either society or the individual starts the process before the other is ready, the individual will be unhappy.

According to the activity theory, old people should maintain their middle-age roles and activities as long as possible, replacing those which must be relinquished with new ones (for example, planning travel or developing hobbies after retirement) and resisting as much as possible the contraction of their social and physical "life space." The theory denies old age as a special life stage and fails to take into account the real biological changes of aging. While both theories are useful, neither is sufficient in itself to explain successful aging; further reference to other important factors, such as health, socioeconomic status, education, and family relationships is needed.

Several recent theories have been developed, having in common a basic belief in the continuity of human behavior patterns. Havighurst and others[5] and Neugarten, Havighurst, and Tobin[6] argue that there is no sharp discontinuity at the onset of old age, and that personality characteristics developed over a lifetime will determine behavior and adjustment in a pattern consistent with earlier experience (see also Williams and Wirths[7]). Clark and Anderson[1] take a developmental approach, dividing adjustment to old age into five adaptive tasks: recognize old age and its limitations, readjust physical and social life space to reasonable (that is, controllable) dimensions, substitute alternative and feasible sources of satisfaction, reassess criteria for self-evaluation taking into account changed circumstances, and revise goals and values, if necessary, to fit new life style. The various continuity theories agree that there are many possible patterns for successful aging and that these patterns are generally consistent with the individual's previously developed personality characteristics.

At some point in middle age most adults must relinquish the parenting role as their children leave for college or jobs and eventually get married themselves (see Parts IV and V). In the first article Bernice Neugarten and Karol Weinstein discuss the new role of grandparent that usually follows in the natural progression of events. They interviewed seventy pairs of middle-class grandparents with respect to their comfort in the role, the significance of grandparenthood to them, and their styles of being grandparents. While the majority of those interviewed said they were quite comfortable with their new role, about one-third mentioned having some difficulty (for example, conflict with the parents over child-rearing practices, indifference, or uneasiness with the self-image of grandparent). Several different styles of grandparenting emerged: "formal" (very interested in child but clearly distinguishing grandparent role from parent role), "fun seeker" (informal and playful, mutual satisfaction), "distant figure" (infrequent contact on holidays and ritual occasions), and "reservoir of family wisdom" (authoritarian, dispenser of special skills, deferred to by young parents). Most grandparents find a style that suits them and manage to obtain satisfaction and enjoyment from their new role.

Another major role transition is from worker to retiree, and individual adjustment to this change varies widely. Several factors affect both the timing of, and the successful adjustment to, retirement. Those who elect early or voluntary retirement often have poorer health than their peers, which may make their adjustment more difficult. Financial insecurity can be another stress factor, since most people experience a significant drop in income upon retirement. Planning and preparation is especially important in this area, but it can also be helpful with other aspects of aging and retirement. The abundance of leisure time that retirement brings may or may not be a problem depending in part on the individual's attitude toward work. Those for whom work is a source of interest or a moral issue (one ought to work as long as one is able) may find retirement very difficult.

Another issue of critical importance is loss of independence, the growing need to depend on others for financial support, transportation, medical care, or even daily physical activities (for example, dressing, cooking, or cleaning). Harriet Warren explores this issue in the second article, using a questionnaire to determine health status and ability to perform various daily living tasks. A majority of her sample (invalids and residents of institutions were excluded) considered themselves relatively healthy, and 42 percent were rated completely independent according to her scale. When broken down into age groups, though, the relationship between age and independence becomes apparent; 71 percent of those sixty–sixty-nine years old, 38 percent of those seventy–seventy-nine years old, and 27 percent of those eighty–eighty-nine years old were completely independent. The most commonly reported areas of concern were mobility (getting out and around) and keeping up with housecleaning. These two practical problems are important contributing factors to the larger problems faced by many elderly people: loneliness and loss of dignity and self-respect.

In the third article Elizabeth Hughes provides a vivid example of the difficulties faced by an older person who has become dependent on an institution, and the techniques she uses to cope with these problems (see also Curtin,[3] pp. 74–78, for an example of nursing-home life). Mrs. Mello, still grieving over her husband's sudden death and feeling tired and vulnerable, decided to sell her home and trade all her assets for the security of a retirement home. She was unprepared for the regimentation and the assault on individual dignity that she encountered. When told by the housekeeper that the little personal touches in her room like the bookcase, small rug, and drapes must be removed because they made cleaning more difficult, Mrs. Mello had to appeal to the administrator to relent. Even in her own room she had no authority. Because the home required them to sign over all their assets when they entered, the residents were totally dependent on the staff's good will and helpless in disputes with staff or administration. Adults who had been independent and self-regulating for fifty or sixty years found the rules (which governed life in the home from the 6:30 A.M. rising and the 10:00 P.M. lights-out schedule to the use of communal bathrooms) burdensome and humiliating.

Mrs. Mello fought back against the smothering regimentation, though, getting the record player so she and the others could enjoy music again, taking daily walks away from the home, and challenging the spending of proceeds from the sale of residents' craft projects. An individual can deal with monotony, loneliness, and dependency, but it takes energy and determination. Health professionals and social service workers who work with the elderly can be helpful if they recognize and respect their needs for maintaining a sense of individuality and dignity. From a position of inner strength people can cope with retirement, declining health, loneliness, financial insecurity, and all the other changes and problems old age brings.

REFERENCES

1. CLARK, M., AND ANDERSON, B. G. *Culture and aging.* Springfield, Ill.: Charles C. Thomas, 1967.

2. CUMMING, E., AND HENRY, W. E. *Growing old: The process of disengagement.* New York: Basic Books, 1961.
3. CURTIN, S. Aging in the land of the young. *Atlantic,* July 1972, 230, 68–78.
4. DE BEAUVOIR, S. *The coming of age* (P. O'Brian, tr: ns.). New York: G. P. Putnam's Sons, 1972.
5. HAVIGHURST, R. J.; NEUGARTEN, B. L.; AND TOBIN, S. S. Disengagement and patterns of aging. In B. L. NEUGARTEN, ed., *Middle age and aging: A reader in social psychology.* Chicago: University of Chicago Press, 1968.
6. NEUGARTEN, B. L.; HAVIGHURST, R. J.; AND TOBIN, S. S. Personality and patterns of aging. In B. L. NEUGARTEN, ed., *Middle age and aging: A reader in social psychology.* Chicago: University of Chicago Press, 1968. (Originally published 1965).
7. WILLIAMS, R. H., AND WIRTHS, C. *Lives through the years.* New York: Atherton, 1965.

Article 18   The Changing
American Grandparent

BERNICE L. NEUGARTEN

KAROL K. WEINSTEIN

Despite the proliferation of investigations regarding the relations between generations and the position of the aged within the family, surprisingly little attention
has been paid directly to the role of grandparenthood.[1] There are a few articles
written by psychoanalysts and psychiatrists analyzing the symbolic meaning of
the grandparent in the developing psyche of the child or, in a few cases, illustrating the role of a particular grandparent in the psychopathology of a particular
child.[2] Attention has not correspondingly been given, however, to the psyche of
the grandparent, and references are made only obliquely, if at all, to the symbolic
meaning of the grandchild to the grandparent.

There are a number of anthropologists' reports on grandparenthood in one
or another simple society as well as studies involving crosscultural comparisons
based on ethnographic materials. Notable among the latter is a study by Apple[3]
which shows that, among the 51 societies for which data are available, those societies in which grandparents are removed from family authority are those in
which grandparents have an equalitarian or an indulgent, warm relationship with
the grandchildren. In those societies in which economic power and/or prestige
rests with the old, relationships between grandparents and grandchildren are
formal and authoritarian.

Sociologists, for the most part, have included only a few questions about
grandparenthood when interviewing older persons about family life, or they have
analyzed the grandparent role solely from indirect evidence, without empirical

BERNICE L. NEUGARTEN, PH.D. is Professor of Human Development, Department of Behavioral
Sciences, University of Chicago.

KAROL K. WEINSTEIN, M.A., is with the Psychology Department, Roosevelt University,
Chicago, Illinois.

Source: Journal of Marriage and the Family, Vol. 6, No. 2 (May 1964), 199–204. Copyright
1964 by National Council on Family Relations. Reprinted by permission.

data gathered specifically for that purpose.[4] There are a few noteworthy exceptions: Albrecht[5] studied the grandparental responsibilities of a representative sample of persons over 65 in a small midwestern community. She concluded that grandparents neither had nor coveted responsibility for grandchildren; that they took pleasure from the emotional response and occasionally took reflected glory from the accomplishments of their grandchildren.

An unpublished study by Apple[6] of a group of urban middle-class grandparents indicated that, as they relinquish the parental role over the adult child, grandparents come to identify with grandchildren in a way that might be called "pleasure without responsibility."

In a study of older persons in a working-class area of London, Townsend[7] found many grandmothers who maintained very large responsibility for the care of the grandchild, but he also found that for the total sample, the relationship of grandparents to grandchildren might be characterized as one of "privileged disrespect." Children were expected to be more respectful of parents than of grandparents.

THE DATA

The data reported in this paper were collected primarily for the purpose of generating rather than testing hypotheses regarding various psychological and social dimensions of the grandparent role. Three dimensions were investigated: first, the degree of comfort with the role as expressed by the grandparent; second, the significance of the role as seen by the actor; and last, the style with which the role is enacted.

The data came from interviews with both grandmother and grandfather in 70 middle-class families in which the interviewer located first a married couple with children and then one set of grandparents. Of the 70 sets of grandparents, 46 were maternal—that is, the wife's parents—and 24 were paternal. All pairs of grandparents lived in separated households from their children, although most lived within relatively short distances within the metropolitan area of Chicago.

As classified by indices of occupation, area of residence, level of income, and level of education, the grandparental couples were all middle class. The group was about evenly divided between upper-middle (professionals and business executives) and lower-middle (owners of small service businesses and white-collar occupations below the managerial level). As is true in other middle-class, urban groups in the United States, the largest proportion of these families had been upwardly mobile, either from working class into lower-middle or from lower-middle into upper-middle. Of the 70 grandparental couples, 19 were foreign born (Polish, Lithuanian, Russian, and a few German and Italian). The sample was skewed with regard to religious affiliation, with 40 per cent Jewish, 48 per cent Protestant, and 12 per cent Catholic. The age range of the grandfathers was, with a few exceptions, the mid-50's through the late 60's; for the grandmothers it was the early 50's to the mid-60's.

Each member of the couple was interviewed separately and, in most instances, in two sessions. Respondents were asked a variety of open-ended questions regarding their relations to their grandchildren: how often and on what occasions they saw their grandchildren; what the significance of grandparenthood was in their lives and how it had affected them. While grandparenthood has multiple values for each respondent and may influence his relations with various family members, the focus was upon the primary relationship—that between grandparent and grandchild.

FINDINGS AND DISCUSSION

Degree of Comfort in the Role

As shown in Table 1, the majority of grandparents expressed only comfort, satisfaction, and pleasure. Among this group, a sizable number seemed to be idealizing the role of grandparenthood and to have high expectations of the grandchild in the future—that the child would either achieve some special goal or success or offer unique affection at some later date.

At the same time, approximately one-third of the sample (36 per cent of the grandmothers and 29 per cent of the grandfathers) were experiencing sufficient difficulty in the role that they made open reference to their discomfort, their disappointment, or their lack of positive reward. This discomfort indicated strain in thinking of oneself as a grandparent (the role is in some ways alien to the self-image), conflict with the parents with regard to the rearing of the grandchild, or indifference (and some self-chastisement for the indifference) to caretaking or responsibility in reference to the grandchild.

The Significance and Meaning of the Role

The investigators made judgments based upon the total interview data on each case with regard to the primary significance of grandparenthood for each respondent. Recognizing that the role has multiple meanings for each person and that the categories to be described may overlap to some degree, the investigators nevertheless classified each case as belonging to one of five categories:

1. For some, grandparenthood seemed to constitute primarily a source of *biological renewal* ("It's through my grandchildren that I feel young again") and/or *biological continuity* with the future ("It's through these children that I see my life going on into the future" or, "It's carrying on the family line"). As shown in Table 1, this category occurred significantly less frequently for grandfathers than for grandmothers, perhaps because the majority of these respondents were parents, not of the young husband

Table 1. Ease of Role Performance, Significance of Role, and Style of Grandparenting in 70 Grandmothers and 70 Grandfathers

	Grandmothers (N = 70)		Grandfathers (N = 70)	
	N	Per Cent	N	Per Cent
A. Ease of role performance:				
1) Comfortable/pleasant	41	59	43	61
2) Difficulty/discomfort	25	36	20	29
(Insufficient data)	4	5	7	10
Total	70	100	70	100
B. Significance of the grandparent role:				
1) Biological renewal and/or continuity	29*	42*	16*	23*
2) Emotional self-fulfillment	13	19	19	27
3) Resource person to child	3	4	8	11
4) Vicarious achievement through child	3	4	3	4
5) Remote; little effect on the self	19	27	20	29
(Insufficient data)	3	4	4	6
Total	70	100	70	100
C. Style of grandparenting:				
1) The Formal	22	31	23	33
2) The Fun-Seeking	20	29	17	24
3) The Parent Surrogate	10*	14*	0*	0*
4) The Reservoir of Family Wisdom	1	1	4	6
5) The Distant Figure	13	19	20	29
(Insufficient data)	4	6	6	8
Total	70	100	70	100

* The difference between grandmothers and grandfathers in this category is reliable at or beyond the .05 level (frequencies were tested for differences of proportions, using the Yates correction for continuity).

but of the young wife. It is likely that grandfathers perceive family continuity less frequently through their female than through their male offspring and that in a sample more evenly balanced with regard to maternal-paternal lines of ascent, this category would appear more frequently in the responses from grandfathers.

2. For some, grandparenthood affords primarily an opportunity to succeed in a new emotional role, with the implication that the individual feels himself to be a better grandparent than he was a parent. Frequently, grandfatherhood offered a certain vindication of the life history by providing *emotional self-fulfillment* in a way that fatherhood had not done. As one man put it, "I can be, and I can do for my grandchildren things I could never do for my own kids. I was too busy with my business to enjoy

my kids, but my grandchildren are different. Now I have the time to be with them."

3. For a small proportion, the grandparent role provides a new role of teacher or *resource person*. Here the emphasis is upon the satisfaction that accrues from contributing to the grandchild's welfare—either by financial aid, or by offering the benefit of the grandparent's unique life experience. For example, "I take my grandson down to the factory and show him how the business operates—and then, too, I set aside money especially for him. That's something his father can't do yet, although he'll do it for *his* grandchildren."

4. For a few, grandparenthood is seen as providing an extension of the self in that the grandchild is one who will *accomplish vicariously* for the grandparent that which neither he nor his first-generation offspring could achieve. For these persons, the grandchild offers primarily an opportunity for aggrandizing the ego, as in the case of the grandmother who said, "She's a beautiful child, and she'll grow up to be a beautiful woman. Maybe I shouldn't, but I can't help feeling proud of that."

5. As shown in Table 1, 27 per cent of the grandmothers and 29 per cent of the grandfathers in this sample reported feeling relatively *remote* from their grandchildren and acknowledged relatively *little effect* of grandparenthood in their own lives—this despite the fact that they lived geographically near at least one set of grandchildren and felt apologetic about expressing what they regarded as unusual sentiments. Some of the grandfathers mentioned the young age of their grandchildren in connection with their current feelings of psychological distance. For example, one man remarked, "My granddaughter is just a baby, and I don't even feel like a grandfather yet. Wait until she's older—maybe I'll feel different then."

Of the grandmothers who felt remote from their grandchildren, the rationalization was different. Most of the women in this group were working or were active in community affairs and said essentially, "It's great to be a grandmother, of course—but I don't have much time. . . ." The other grandmothers in this group indicated strained relations with the adult child: either they felt that their daughters had married too young, or they disappoved of their sons-in-law.

For both the men and the women who fell into this category of psychological distance, a certain lack of conviction appeared in their statements, as if the men did not really believe that, once the grandchildren were older, they would indeed become closer to them, and as if the women did not really believe that their busy schedules accounted for their lack of emotional involvement with their grandchildren. Rather, these grandparents imply that the role itself is perceived as being empty of meaningful relationships.

Styles of Grandparenting

Somewhat independent of the significance of grandparenthood is the question of style in enacting the role of grandmother or grandfather. Treating the data inductively, five major styles were differentiated:

1. The *Formal* are those who follow what they regard as the proper and prescribed role for grandparents. Although they like to provide special treats and indulgences for the grandchild, and although they may occasionally take on a minor service such as baby-sitting, they maintain clearly demarcated lines between parenting and grandparenting, and they leave parenting strictly to the parent. They maintain a constant interest in the grandchild but are careful not to offer advice on childrearing.

2. The *Fun Seeker* is the grandparent whose relation to the grandchild is characterized by informality and playfulness. He joins the child in specific activities for the specific purpose of having fun, somewhat as if he were the child's playmate. Grandchildren are viewed as a source of leisure activity, as an item of "consumption" rather than "production," or as a source of self-indulgence. The relationship is one in which authority lines —either with the grandchild or with the parent—are irrelevant. The emphasis here is on mutuality of satisfaction rather than on providing treats for the grandchild. Mutuality imposes a latent demand that both parties derive fun from the relationship.[8]

3. The *Surrogate Parent* occurs only, as might have been anticipated, for grandmothers in this group. It comes about by initiation on the part of the younger generation, that is, when the young mother works and the grandmother assumes the actual caretaking responsibility for the child.

4. The *Reservoir of Family Wisdom* represents a distinctly authoritarian patri-centered relationship in which the grandparent—in the rare occasions on which it occurs in this sample, it is the grandfather—is the dispenser of special skills or resources. Lines of authority are distinct, and the young parents maintain and emphasize their subordinate positions, sometimes with and sometimes without resentment.

5. The *Distant Figure* is the grandparent who emerges from the shadows on holidays and on special ritual occasions such as Christmas and birthdays. Contact with the grandchild is fleeting and infrequent, a fact which distinguishes this style from the *Formal*. This grandparent is benevolent in stance but essentially distant and remote from the child's life, a somewhat intermittent St. Nicholas.

Of major interest is the frequency with which grandparents of both sexes are either Fun Seekers or Distant Figures vis-a-vis their grandchildren. These two styles have been adopted by half of all the cases in this sample. Of interest, also, is the fact that in both styles the issue of authority is peripheral. Although deference may be given to the grandparent in certain ways, authority relationships are not a central issue.

Both of these styles are, then, to be differentiated from what has been regarded as the traditional grandparent role—one in which patriarchal or matriarchal control is exercised over both younger generations and in which authority constitutes the major axis of the relationship.

These two styles of grandparenting differ not only from traditional concepts; they differ also in some respects from more recently described types. Cavan, for example,[9] has suggested that the modern grandparent role is essentially a maternal one for both men and women and that to succeed as a grandfather, the male must learn to be a slightly masculinized grandmother, a role that differs markedly from the instrumental and outer-world orientation that has presumably characterized most males during a great part of their adult lives. It is being suggested here, however, that the newly emerging types are neuter in gender. Neither the Fun Seeker nor the Distant Figure involves much nurturance, and neither "maternal" nor "paternal" seems an appropriate adjective.

Grandparent Style in Relation to Age

A final question is the extent to which these new styles of grandparenting reflect, directly or indirectly, the increasing youthfulness of grandparents as compared to a few decades ago. (This youthfulness is evidenced not only in terms of the actual chronological age at which grandparenthood occurs but also in terms of evaluations of self as youthful. A large majority of middle-aged and older persons describe themselves as "more youthful than my parents were at my age.")

To follow up this point, the sample was divided into two groups: those who were under and over 65. As shown in Table 2, the Formal style occurs significantly more frequently in the older group; the Fun Seeking and the Distant Figure styles occur significantly more frequently in the younger group. (Examina-

Table 2. Age Differences in Styles of Grandparenting*

	Under 65 (N = 81)	Over 65 (N = 34)
The Formal:		
Men	12	11
Women	13	9
Total	25 (31%)	20 (59%)
The Fun-Seeking:		
Men	13	4
Women	17	3
Total	30 (37%)	7 (21%)
The Distant Figure		
Men	15	5
Women	11	2
Total	26 (32%)	7 (21%)

* These age differences are statistically reliable as indicated by 2×3 chi-square test applied to the category totals (P = .02).

tion of the table shows, furthermore, that the same age differences occur in both grandmothers and grandfathers.)

These age differences may reflect secular trends: this is, differences in values and expectations in persons who grow up and who grow old at different times in history. They may also reflect processes of aging and/or the effects of continuing socialization which produce differences in role behavior over time. It might be pointed out, however, that sociologists, when they have treated the topic of grandparenthood at all, have done so within the context of old age, not middle age. Grandparenthood might best be studied as a middle-age phenomenon if the investigator is interested in the assumption of new roles and the significance of new roles in adult socialization.

In this connection, certain lines of inquiry suggest themselves: as with other roles, a certain amount of anticipatory socialization takes place with regard to the grandparent role. Women in particular often describe a preparatory period in which they visualize themselves as grandmothers, often before their children are married. With the presently quickened pace of the family cycle, in which women experience the emptying of the nest, the marriages of their children, and the appearance of granchildren at earlier points in their own lives, the expectation that grandmotherhood is a welcome and pleasurable event seems frequently to be accompanied also by doubts that one is "ready" to become a grandmother or by the feelings of being prematurely old. The anticipation and first adjustment to the role of grandmother has not been systematically studied, either by sociologists or psychologists, but there is anecdotal data to suggest that, at both conscious and unconscious levels, the middle-aged woman may relive her own first pregnancy and childbirth and that there are additional social and psychological factors that probably result in a certain transformation in ego-identity. The reactions of males to grandfatherhood has similarly gone uninvestigated although, as has been suggested earlier, the event may require a certain reversal of traditional sex role and a consequent change in self-concept.

Other questions that merit investigation relate to the variations in role expectations for grandparents in various ethnic and socioeconomic groups and the extent to which the grandparent role is comparable to other roles insofar as "reality shock" occurs for some individuals—that is, insofar as a period of disenchantment sets in, either early in the life of the grandchild or later as the grandchild approaches adolescence when the expected rewards to the grandparents may not be forthcoming.

When grandparenthood comes to be studied from such perspectives as these, it is likely to provide a significant area for research, not only with regard to changing family structure but also with regard to adult socialization.

REFERENCES AND NOTES

1. Most studies of family relationships of older people have given no specific attention to grandparenthood. Several of the most recent examples are: ETHEL

SHANAS, *Family Relationships of Older People* (Research Series 20), New York: Health Information Foundation, 1961; ARTHUR J. ROBINS, "Family Relations of the Aging in Three-Generation Households," in *Social and Psychological Aspects of Aging,* ed. by Clark Tibbitts and Wilma Donahue, New York: Columbia University Press, 1962, pp. 464–474; and MARVIN B. SUSSMAN, "Relationships of Adult Children with Their Parents in the United States," paper given at the Symposium on the Family, Intergenerational Relationships, and Social Structure, Duke University, 1963.

The fact that grandparenthood has not yet come within the central focus of research in social gerontology is evidenced also by examining the three major volumes that have appeared since 1960 in the field of social gerontology: (1) Tibbitts and Donahue, *op. cit.,* which comprises one part of the proceedings of the Fifth Congress of the International Association of Gerontology; (2) Clark Tibbitts, ed., *Handbook of Social Gerontology,* Chicago: University of Chicago Press, 1960; and (3) Ernest W. Burgess, ed., *Aging in Western Societies,* Chicago: University of Chicago Press, 1960, the companion volume to the *Handbooks.* In the first of these three volumes, the topic does not occur at all; in the other two, it is treated on only a few pages of text. In none of the three is "grandparenthood" an entry in the index to the volume.

2. KARL ABRAHAM, "Some Remarks on the Role of Grandparent in the Psychology of Neurosis" (1913), in *Clinical Papers and Essays on Psychoanalysis,* I, New York: Basic Books, 1955, pp. 47–49; SANDOR FERENCZI, "The Grandfather Complex" (1913), reprinted in *Further Contributions to the Theory and Techniques of Psychoanalysis,* New York: Basic Books, 1952, pp. 323–324; ERNEST JONES, "The Fantasy of Reversal of Generations" (1913) and "Significance of the Grandfather for the Fate of the Individual" (1913), in *Papers on Psychoanalysis,* New York: William Wood, 1948, pp. 519–524; MAUREEN BOIRE LABARRE, LUCIE JESSNER, AND LON USSERY, "The Significance of Grandmothers in the Psychopathology of Children," *American Journal of Orthopsychiatry,* 30 (January 1960), pp. 175–185; ERNEST RAPPAPORT, "The Grandparent Syndrome," *Psychoanalytic Quarterly,* 27 (1958), pp. 518–537; and ELSIE THURSTON, "Grandparents in the Three Generation Home: A Study of Their Influence on Children," *Smith College Studies in Social Work,* 12 (1941), pp. 172–173.

3. DORRIAN APPLE, "The Social Structure of Grandparenthood," *American Anthropologist,* 58 (August 1956), pp. 656–663. See also S. F. NADEL, *The Foundations of Social Anthropology,* Glencoe, Ill.: Free Press, 1953; A. R. RADCLIFFE BROWN, *African Systems of Kinship and Marriage,* London: Oxford University Press, 1950.

4. RUTH SHONLE CAVAN, "Self and Role in Adjustment During Old Age," in *Human Behavior and Social Processes,* ed. by Arnold M. Rose, Boston: Houghton Mifflin, 1962, pp. 526–536; HANS VON HENTIG, "The Social Function of the Grandmother," *Social Forces,* 24 (1946), pp. 389–392; M. F. NIMKOFF, "Changing Family Relationships of Older People in the United States During the Last Fifty Years," *The Gerontologist,* 1 (June 1961), pp. 92–97; WILLIAM M. SMITH, JR., JOSEPH H. BRITTON, AND JEAN O. BRITTON, *Relationships within Three Generation Families,* Pennsylvania State University College of Home Economics Research Publication 155, April 1958;

GORDON F. STREIB, "Family Patterns in Retirement," *Journal of Social Issues,* 14 (No. 2, 1958), pp. 46–60.

5. RUTH ALBRECHT, "The Parental Responsibilities of Grandparents," *Marriage and Family Living,* 16 (August 1954), pp. 201–204.
6. DORRIAN APPLE, "Grandparents and Grandchildren: A Sociological and Psychological Study of Their Relationship," unpublished Ph.D. dissertation, Radcliffe College, 1954.
7. PETER TOWNSEND, *The Family Life of Old People,* London: Routledge and Kegan Paul, 1957.
8. Wolfenstein has described fun morality in childrearing practices as it applies to parenthood. Perhaps a parallel development is occurring in connection with grandparenthood. As Wolfenstein has delineated it, fun has become not only permissible but almost required in the new morality. MARTHA WOLFENSTEIN, "Fun Morality, An Analysis of Recent American Child-Training Literature," in *Childhood in Contemporary Cultures,* ed. by Margaret Mead and Martha Wolfenstein, Chicago: University of Chicago Press, 1955, pp. 168–173.
9. Cavan, *op. cit.*

Article 19 Self-Perception of Independence Among Urban Elderly

HARRIET H. WARREN

While most older people prefer to live independently as long as they can manage for themselves, coping becomes increasingly strenuous with advanced age. The ease with which the older person adjusts to age-related changes or stress varies as does his ability to accept alternate means for carrying out daily tasks. The value placed on his health and his physical limitations will influence an older person's ability to cope and the priority of needs. It is important to know how older people assess their state of functional health, that is, the degree they can manage on their own or are restricted physically, and what they consider problems in their daily needs.

A substantial part of the population is now more than 65 years of age. The number of older people has increased from three million, in 1900, to twenty million, and the number will continue to increase.[24] Contrary to popular opinion, the majority of the older population do not live in institutions but in a variety of independent living arrangements within the community.[30] Communities are responding today to the needs and interests of older people by providing a variety of services. Too often the person most in need is unaware of the services, cannot arrange to receive the services at the appropriate time and place, cannot afford them, or finds them unacceptable in the manner offered. The effectiveness of any support system requires not only a concentrated effort toward continuous interagency communication and coordination, but also a commitment to seek out and respect the input of potential consumers. Users of services should not only determine priority of needs but should also be involved in the planning process for the development of such services. As Blenkner points out in her comments on societal solutions for the normal dependencies of aging, no one has demonstrated that the average client cannot do as good a job as the average

HARRIET H. WARREN, M.S., O.T.R., is presently Director of Occupational Therapy at New Hampshire Hospital, Concord.

Source: The American Journal of Occupational Therapy, Vol. 28, No. 6 (1974), 329–336.

caseworker in deciding what the client's problem is and which service would be most effective in helping to solve it.[1] Too often older people have been stereotyped as a homogeneous group with most programs and services planned within that context. Life styles vary as needs do. It is important that occupational therapists and others working in the community recognize this and encourage planning *with* as well as *for* older people.

Dialogue with older people in the environment in which they function provides the most appropriate guidelines for establishing services that would be most meaningful to them in terms of living on their own. Hasselkus and Kiernat, in "Independent Living for the Elderly," found that when the therapist was actually part of the elderly person's home environment the elderly person was able to approach each situation with much greater sensitivity and in the proper perspective than he did in an artificial setting.[10] Implicit in one's approach to a target population is an accumulation of supportive data to justify an occupational therapy service. The substance of this investigation lies within this framwork of data collecting. This investigator posed three questions: How do older people rate their general health? How do they perceive their level of independence in caring for their daily needs? What tasks concern them most about being able to manage on their own? Initially, in directing responses to these questions on health maintenance through the use of a descriptive survey interview, it was important to identify perceived capabilities and needs, since health status influences almost every aspect of an older person's life.

Assessment of Health

There is no true index of health. The measurements of physical and mental health are difficult to differentiate and too complex to be reduced to numbers. Aging is variable. To use chronological age as a basis for the comparison of different functional capacities leads to confusion in interpretation of results. Training, habits, education, stressful use or disuse of joints, hereditary constitution, work roles, and other factors must be considered. The fact that a person has managed to become "old" speaks well for his capacity to cope with the demands of living.

The importance of obtaining accurate information about health status, level of independence, and needs of older people in their environmental settings is well documented in the gerontological literature. One of the clearest and most recurrent findings is the mutual interdependence of the physical state, adaptive behavior, and the emotional state.[6,8,15,19,31]

METHODS OF ASSESSMENT

Numerous methods to assess health are needed, each to be used for a different purpose. The physician and other professionals evaluate health in order to maintain or to improve the health and well-being of a patient.[3,7,11] Similarly, those who assess the health of groups in the population do so to identify problems in terms of prevention, medical care, and rehabilitation programs that can be initiated or expanded.[2,16,25,33] A composite of methods, such as examinations, performance tests, observation of behavior, or the client's subjective judgment, would give a clearer profile of health. One's self-concept is a factor to be considered in aging, which is often a more important index of a person's behavior than is his chronological age, the biological changes, or the crises that have occurred. Subjective self-evaluations of health have been used by investigators as health indices and correlated with other variables such as age, morale, income, employment, socioeconomic status, and functional ability.[4,14,21,26] Studies indicate that most self-ratings of health show congruity with objective measurements, although a tendency toward optimism is seen in the "old" older population.[9,18,32] Pertinent to the focus of this study, Shanas found *perceived health status* to be highly correlated with restrictions in mobility, sensory impairments, and overall incapacity scores.[29]

With advancing years most people require some modification in their pattern of living because of changes in functional capacities, but the process of aging does not move forward at the same speed for everyone. In order to provide a picture of the extent to which an older person can function adequately in his environmental setting, various types of functional assessment techniques have been used. While occupational therapists have long used activities of daily living scales to rate level of independence in self-care, the literature reveals few efforts to adapt and test such measuring devices for use with healthy, older people living on their own in the community. The standardized Index of Activities of Daily Living by Katz,[13] developed for studying hospitalized old people with fractured hips, has been used in the evaluation of healthy, older persons in the general population as well as of more than 1,000 older people in institutional settings.

Concentrating on indices that measure physical functioning of people in their natural settings, Shanas[29] used an Index of Incapacity in a cross-national comparison of the ability of older people in Denmark, Britain, and the United States to function independently. The Index was a modification of one used by Townsend[34] in his study of an older population in England. The Townsend scale included personal care, sensory functions, and the ability to perform

Table 1. Frequency Distribution of 65 Elderly Subjects Perceived Health Status Compared with Others Same Age

	Self-Perception of Health								
	Ages 60–69			*Ages 70–79*			*Ages 80–89*		
	Good	Fair	Poor	Good	Fair	Poor	Good	Fair	Poor
Better	5	5	0	9	3	1	6	7	1
Same	1	1	1	4	6	3	3	1	3
Worse	0	1	0	0	1	2	0	0	1
Total	6	7	1	13	10	6	9	8	5

certain physical tasks necessary for living independently. Shanas reduced the number of variables and tailored the interview schedule to six questions about bathing, grooming, dressing, and ambulation. Shanas found that the proportion of older people who are bedfast, housebound, limited in mobility, and required a broad program of services was roughly 24 percent in Denmark, 21 percent in Britain, and 14 percent in the United States. In each country the most incapacitated were single or widowed older women who were also among the poorest persons in the elderly population.

Lawton[16] has named eight tasks of functional capacity as the Instrumental Activities of Daily Living that, operationally defined, mean those tasks most relevant to the living of a minimally adequate social life. The eight tasks include telephoning, shopping, cooking, housekeeping, laundry, mode of transportation, responsibility for one's own medication, and handling finances. Although not required to perform all these tasks to maintain independent living, it is expected that the individual will seek various sources of help as these tasks become more difficult to accomplish. Although the scale has had little empirical testing, occupational therapists could find it a useful assessment technique with an elderly population, together with measuring devices designed for other personal characteristics and activity patterns.

Rosow and Breslau[27] developed a Guttman scale of functional health suitable for social research limited to interview data. Respondents can be classified by the progression of consecutive, positive responses to attitude statements. In a study of two social classes of older people drawn from purposive sampling within a large metropolitan community, findings indicated that 86 percent of the respondents reported no constraints in their social functioning, which included going to church, movies, meetings, or visiting. Whereas about one-half of the sample was free of any current illness or debilitating condition, only 17 percent reported their health as poor. However, two-thirds of the sample were under the age of 75.

METHODOLOGY

Sampling Procedure

Through the use of a table of random numbers, subjects were selected from an occupancy listing of the elderly residents in an urban housing development.[20] Persons under 60 years of age or, according to visiting nurse records, those who were nonambulatory, too ill, or unable to respond to questioning were also excluded from the sample. Twenty-seven persons were replaced to account for those not at home when called, 27 refused to be interviewed or had no telephone, 5 were dismissed through exclusion criteria, and 65 affirmative responses accounted for 124 randomly selected residents. Of 92 potential subjects, 70 percent or 65 subjects represent the study population. In comparing the sample against the criteria of age and sex through the use of a t-test, no significant difference was found between those who were interviewed and those who refused to be interviewed. Although the 30 percent not included in the study appeared to be similar with respect to age and sex to the 70 percent interviewed, they may have differed because of intervening variables, for example, illness. The sample, therefore, was composed largely of fairly healthy older individuals.

Site Selection

The study population was drawn from a low-income housing project in a metropolitan section of a large northeastern city. According to 1970 United States Census Bureau statistics, the population of this section was 62,800, of which 42,000 were nonwhite.[23] From numerous ethnic groups, twenty-one percent of the individuals in this section were of foreign stock and included a large number of Irish and Canadians. These people were either foreign born, or had at least one parent who was foreign born. More than one-half of the section's residents did not complete high school. The largest employment category was clerical, and one-half of the workers were in white-collar occupations. Median family income for this section was $4,099, whereas the city's median was $9,133.

The housing project included 176 units in six buildings for elderly residents. Two of these were designed specifically for older people. Four adjacent buildings for use of an age-integrated population were added later. These four additional buildings have no laundry area and elevators stop on alternate floors. Organizations that provide a variety of services to the elderly included a tenant management organization primarily concerned with housing and maintenance, a visiting nurse program providing health care and linkage with medical facilities, a senior citizen drop-in center focusing on activity programs, and a golden age club organized and directed by residents. A neighborhood health center, offering maternity services and serving children from infancy through age 21, is located within three blocks of the housing units for the elderly. However, most of the older residents preferred the medical resources of the visiting nurse program, private physicians, or local hospitals.

Test Instrument

A two-part interview schedule, consisting of 119 questions, was constructed for data collection. Part A emphasizes the general health status and level of independence in the ADL tasks. Part B, with emphasis on activity, leisure, and the meaning of each to older people, appears on page 337 of this Journal.* A fact sheet of general information was also included. Operationally defined, older people included persons who were 60 to 90 years of age, ambulatory, and physically and mentally capable of responding to questions.

Part A consisted of 32 questions. The first two questions inquired about the subject's assessment of his general health status. Health was defined as how an older person felt about himself in terms of function or degree of fitness. He

---

* *American Journal of Occupational Therapy*, 1974, 28, 337–45.

was asked to rate his present state of health as good, fair, or poor and to report whether he considered his health better, the same, or worse than that of others his age. The next two questions inquired about auditory and visual problems.

The remaining questions in Part A concerned the subject's perception of his level of independence in caring for his daily needs. Level of independence meant performance category on stipulated ADL tasks according to a four-point scale: does not perform any part of the activity; performs part of the activity but needs someone to assist; performs the activity but needs someone there for safety or supervision; performs the activity without assistance or supervision. Daily needs were defined as ADL tasks, or specifically, bathing, grooming, dressing, meal preparation, laundry, housecleaning, and the ability to get out and around.

Table 2. Percentage Distribution of 65 Elderly Subjects by Performance Scores in ADL Tasks

ADL Tasks	Performance Score*								TOTAL
	1		2		3		4		
	N	%	N	%	N	%	N	%	N
	Bathing and Grooming								
Shampoo hair	3	4.6	3	4.6	0	0	59	90.8	6
Shave/Fix hair	1	1.5	2	3.1	0	0	62	95.4	3
Use tub/Shower	2	3.1	3	4.6	0	0	60	92.3	5
Cut toenails	17	26.2	4	6.2	0	0	44	67.7	21
	Dressing/Fasteners								
Undershirt/Bra	0	0	0	0	0	0	65	100	0
Corset/Girdle	0	0	0	0	0	0	65	100	0
Shirt/Blouse	0	0	0	0	0	0	65	100	0
Slacks/Dress	0	0	0	0	0	0	65	100	0
Socks/Stockings	1	1.5	0	0	0	0	64	98.5	1
Shoes	1	1.5	0	0	0	0	64	98.5	1
Zipper	0	0	0	0	0	0	65	100	0
Buttons	0	0	0	0	0	0	65	100	0
Shoelaces	1	1.5	1	1.5	0	0	63	96.9	2
Hooks & Eyes	1	1.5	2	3.1	0	0	62	95.4	3
Pins	0	0	2	3.1	0	0	63	96.9	2
Buckles	0	0	0	0	0	0	65	100	0
Meal Prep	1	1.5	9	13.8	9	13.8	46	70.8	19
Laundry	4	6.2	8	12.3	10	15.4	43	66.2	22
Housecleaning	1	1.5	14	21.5	10	15.4	40	61.5	25
Get out/around	6	9.2	4	6.2	18	27.7	37	56.9	28

* Scale: Dependence (1) to Independence (4)

Twenty variables on bathing, grooming, and dressing were set up in check-list form. Some questions required double items for men and women such as "undershirt/bra" (see Table 2). Four questions related to the tasks of meal preparation, laundry, housecleaning, and the ability to get out and around. The lead for each of these questions was "Which one of the following statements best describes your ability to . . . ," followed by the name of the task and four statements regarding degree of dependence. It was necessary throughout this section to make allowances for the individual for whom the activity did not apply. For example, in the question on laundry the first statement reads: "Do all my washing and ironing/Could do but is not required of me." Subjects were also asked a question on mode of transportation and on assistance required in getting out and around. The final question on Part A asked what task concerned them most in taking care of their daily needs. Six tasks were listed, plus an additional "No concerns" category; the respondents were asked to name only one task.

Scoring

For the question on level of independence, a score of 84 indicated complete independence, as operationally defined in this study. A four-point scale was used to indicate performance level for each of the 21 stipulated variables using one point at the "dependent" end and up to four points at the "independent" end of the scale. The possible range of scores was from 21 to 84, since the most dependent respondent would receive a score of 1 for each of the 21 variables. A score of 4 for each variable would indicate that the subject could perform the task with complete independence. Sources used to develop Part A were adapted in part from Kutner, Katz, and Lawton.[13,14,16]

For a pilot study twelve graduate students in the same program as the interviewer completed Parts A and B of the interview schedule and provided a critique relative to clarity and relevance. For a further refinement of the instrument for its length and the timing, three elderly residents of another housing project were interviewed and the results reviewed.

Data Collection

The project proposal and consent forms were channeled through the health center and the tenant management organization, which served as primary sources for authorization of the study. Subjects were called by telephone for appointments, usually scheduled for the following day. All data collecting took place during May 1973. At the scheduled time, one of the two interviewers asked the questions on both Parts A and B of the interview schedule. Responses were recorded during the interview. If the respondent seemed confused by the question, the question was repeated. The interview time averaged 45 minutes. When all interviews were completed, a cross-check of inter-rater reliability was performed using the Spearman correlation formula. A correlation coefficient of 0.73 indicated consistency for the results of two interviewers.

Subjects

The subjects consisted of 9 males and 56 females between 60 and 90 years of age, with a median age of 76.2. This sample included 6 single, 11 married, 2 who were separated, 1 who was divorced, and 45 widowed males and females. Length of residency in the housing project ranged from less than 1 year to 19 years; 39 percent for 4 years or less, 18 percent between 5 and 9 years, 23 percent between 10 and 14 years, and 20 percent between 15 and 19 years. Forty percent were foreign born, and the largest percentage of these were from Ireland. Religious preferences included 51 percent Catholic, 46 percent Protestant, and 3 percent Jewish.

Age groups included 14 individuals between the ages of 60 and 69, 29 between 70 and 79, and 22 between 80 and 89. Sixty-three percent had reached the educational level of 8th grade or less, 26 percent had completed 12th grade, 8 percent had attended technical schools, and 3 percent had attended college. Occupational history showed 41 percent in each of the skilled and unskilled categories, 11 percent housewives, 5 percent in professions, and 2 percent in clerical-secretarial work. One person, in the 60- to 64-year-old group, reported full-time employment; all others were retired. Seventy-four percent of the subjects lived alone. Modal interviewees were widows between the ages of 70 and 79.

RESULTS AND DISCUSSION

Health Rating

Regardless of age, 43 percent of the subjects rated their health as good, 38 percent as fair, whereas only 19 percent considered their health poor. Results tend to support findings of Kutner,[14] Shanas,[29] and others studing community-based populations. These authors conclude that there is little difference in perceived health status as age increases, despite the prevalence of chronic conditions.[14,22,29] Caution should be used in interpreting the term "fair" as an indicator of health status, which could signify the presence of a major health problem, or suggest a preference for "taking the middle ground." By adding a fourth indicator of "very good" to the three that were used, the "fair" indicator may have elicited more selective information. Some studies indicate that older people, particularly those in low-income groups, seem to display reality-oriented judgements in their subjective evaluation of health, even though "old" older people tend toward optimism. In a large, longitudinal study of noninstitutionalized, ambulatory elderly subjects, for example, Maddox[18] found that two of every three respondents expressed self-evaluations congruent with medical evaluations. Age alone does not seem to affect one's perception of his health status or his ability to manage on his own. Rather, it is the insults of life stresses incurred and the manner in which he has been able to cope with them that determine the state of well-being. Most older people seem to have managed with whatever means they possess and with standards acceptable to them.

Compared to others their age, 57 percent of the subjects felt that their health was better, 35 percent considered their health about the same, whereas only 8 percent thought that their health was worse than others. When the subject's self-evaluation of health was contrasted with how they saw themselves with respect to others their age, few considered their health worse than others, particularly at the "good" end of the spectrum (Table 1). Kutner, in his New York City study of 500 older people, suggested that many older people regard their chronic illnesses, disabilities, and incapacities more or less "natural" consequences of old age.[14] He found that, even in the large sample of high- and low-income groups with a fairly even age distribution, only 25 percent mentioned health as a problem. Instead, the question of health seemed to be embedded within the context of other life problems of equal significance. His adaptability and his personality structure constitute major factors in his adjustment as the years advance, and are involved in how the older person manages problems in his lifetime.

The findings indicated that the sample of older people involved in this study was relatively healthy in comparison to other groups of a similar age. This sample also indicated general optimism about health in comparison to other groups of older people. Although age and sex of those excluded from the sample were determined comparable, information on health was unavailable and, therefore, was not included as a variable. This factor may represent a limitation in the present study, although the health findings corroborate those of most researchers who have studied community-based populations.

Independence Rating

In response to a series of questions on ADL tasks required for independent living, 27 or 41.5 percent of the subjects considered themselves completely independent, whereas the remaining 38 or 58.5 percent reported some limitations within the six stipulated tasks. Since the mean for the total number of subjects was 79.6 of a possible score of 84 for complete independence as defined in this study, findings revealed that subjects perceived themselves as a fairly independent group of older people (Table 2).

In a chi-square analysis of the relationship between independence level and age the statistically significant difference was at the 0.05 level, indicating that dependency is associated with advanced age (Table 3). Results generally supported the findings in the cross-national study by Shanas.[29] Using an Index of Incapacity, which covered most of the same tasks, Shanas found that 50 percent of the sampling in Denmark, Britain, and the United States were able to function with no limitations and 25 percent with only minimal limitations. She also found that higher degrees of incapacity were related to advanced age. Shanas reported a higher proportion of the American aged as having no limitations or only slight difficulties in the performance of such tasks. In discussing the differences between the American and the European aged, Shanas reasoned that older people in the United States, more so than in Europe, seem to feel that it is psychologically wrong to admit to illness or incapacity, that it is necessary to be active and

Table 3. Chi-Square Analysis by Independence in ADL Tasks for 65 Elderly Subjects

Independence Level in ADL	Age		
	60–69	70–79	80–89
Complete independence	10	11	6
Degrees of dependency	4	18	16

$\chi^2 = 7.149$; $df = 2$; $p < .05$

self-sufficient irrespective of age and infirmity.[29] Such a viewpoint seems somewhat misleading, however, unless by "American" aged Shanas meant to imply second or third generation American aged, since the United States has such large and varied ethnic groups within its population. Kutner's survey of older people in New York City revealed that those from Britain, Ireland, and German-Austria had a similar bent toward self-sufficiency and personal independence.[14]

To summarize findings relative to perceived independence level on the ADL tasks, dressing presented no problem to subjects, with 91 percent reporting complete independence, and one reporting a need for help to put on socks and shoes. Sixty-one percent reported complete independence in bathing and grooming activities, whereas the two personal care items causing the most difficulties for the remaining 39 percent were the tasks of cutting toenails and shampooing hair. Twenty-one of the 65 respondents reported degrees of dependency in toenail cutting, and 17 individuals or 26 percent were completely dependent. Seventy-one percent managed their own meals, including shopping and food preparation; an equal proportion of the remaining 29 percent reported difficulty in shopping for groceries or were temporarily receiving a daily meals-on-wheels dinner. Approximately an equal number of subjects handled all laundry (66 percent) and housecleaning (61 percent) by themselves. Those who reported some limitation still continued to do hand laundry and light-housekeeping tasks. In four of the six buildings there were intermittent elevator stops and no laundry facilities, which may have accounted for many of the subjects' reported difficulties. In mobility, or the ability to get out and around, 57 percent reported complete independence, an additional 29 percent left their apartments regularly to do errands or visit, but limited their walking to one or two blocks. Most of the remaining 14 percent left their apartments only on rare occasions, depending upon friends or relatives to do the errands.

It is doubtful that a single assessment device could differentiate behaviors as actually observed by the interviewer. The check-list style used to assess physical self-maintenance tasks could not measure personal appearance and cleanliness nor could it differentiate between those who required more time than others did to accomplish the same task. Had the time element been included, more information may have been learned about the meaning of such activities to the older person. How an individual uses time and energy, and how he values the expenditure, has special significance in the gradual slowing down process. Hans Selye, in *The Stress of Life*, assists those in the helping professions to understand and appreciate the influence one has over the use of his adaptation energy in responding to everyday demands of living.[28]

Three out of every five older subjects perceived themselves as relatively free from limitations in performance of ADL tasks usually required for independent living. One of every five professed to have difficulties in performing one or more tasks to the extent that they could not function without assistance. This investigator did not attest to actual performance of a given task.

Areas of Concern

Subjects were asked to state what tasks concerned them most in managing on their own. While 63 percent presently expressed no concern about any task, the fact that 58 percent had previously reported varying degrees of dependency in ADL tasks would suggest that many who reported no concerns had adapted to their limitations, had developed an attitude of resignation, or a seeming indifference that masked their real concerns from the casual observer. Of those who did report concern, 37 percent were about evenly divided between those who worried about their ability to get out and around and those who were concerned about being able to continue housecleaning activities. These findings parallel those reported by Shanas in respect to mobility and other studies focused on the capabilities of community-based populations of older people.[5,17,29]

Since residents in this study were afforded the convenience of homemakers, meals-on-wheels, a drop-in center, and a visiting nurse, there seemed to be a certain security in knowing that there were easy accesses to services for stressful situations.

SUMMARY AND CONCLUSIONS

In summary, subjects perceived themselves as a relatively healthy and fairly independent group of older people, with particular limitations in mobility and the more physically taxing activities of housecleaning. The site selection of special housing units offered a more protective environment than usual for older people who continue to maintain independent living arrangements.

ACKNOWLEDGMENTS

Appreciation is extended to Dr. Beverly Bullen for technical assistance; to Dean Bernard Kutner, whose research and interest contributed to this study; to Eleanor Nystrom, co-investigator; to the staff of organizations within the Bromley Health Housing Project; and to the many men and women who contributed the data for this study. Additional statistical data of this study can be obtained upon request from the Journal office.

REFERENCES

1. BLENKNER, M. The Normal Dependencies of Aging. In Kalash, R. A. (ed.). *The Dependencies of Old People*, Ann Arbor, Institute of Gerontology, 1969, p. 35.
2. BLOOM, M., AND BLENKNER, M. Assessing Functioning of Older Persons Living in the Community. *Gerontologist*, 10:31–37, 1970.
3. BRODMAN, K., ERDMAN, A. J., LORGE, I. et al. *The Cornell Medical Index*, an adjunct to medical review. *J.A.M.A.*, 140:530–534, 1949.
4. BULTENA, G. L., OYLER, R. Effects of Health on Disengagement and Morale. *Aging Hum. Devel.*, 2:142–148, 1971.
5. CLARK, M. Patterns of Aging Among the Elderly Poor of the Inner City. *Gerontologist*, 11:58–65, 1971.
6. DOVENMUEHLE, R. H. Health and Aging. *J. Health Hum. Behav.*, 1:273–277, 1960.
7. DOVENMUEHLE, R. H., BUSSE, E. W., AND NEWMAN, E. G. Physical Problems of Older People. *J. Am. Ger. Soc.*, 9:209–217, 1961.
8. EISDORFER, C. The Implications of Research for Medical Practice. *Gerontologist*, 10:62–68, 1970.
9. FRIEDSAM, H. F. AND MARTIN, H. W. A Comparison of Self and Physicians' Health Ratings in an Older Population. *J. Health Hum. Behav.*, 4:179–183, 1963.
10. HASSELKUS, B. R. AND KIERNAT, J. M. Independent Living for the Elderly. *Am. J. Occup. Ther.*, 27:181–188, 1973.
11. HORVATH, S. M. AND BENDER, A. D. Physical Capacity of the Aged. *J. Am. Ger. Soc.*, 9:247–252, 1961.
12. KATZ, S., DOWNS, T. D., CASH, H. R. et al. Progress in Development of the Index of ADL. *Gerontologist*, 10:20–30, 1970.
13. KATZ, S., FORD, A. B., MOSKOWITZ, R. W. et al. The Index of ADL, a Standardized Measure of Biological and Psychological Function. *J.A.M.A.*, 185:914–919; 915; 1963.
14. KUTNER, B., FANSCHEL, D., TOGO, A. M. et al. *Five Hundred Over Sixty*. New York: Russell Sage Foundation, pp. 148; 153; 172; 270–293, 1956.
15. LAWTON, M. P. Assessment, Integration and Environments for Older People. *Gerontologist*, 10:38–46, 1970.
16. LAWTON, M. P. The Functional Assessment of Elderly People. *J. Am. Ger. Soc.*, 19:465–481; 170, 172; 1971.
17. LAWTON, M. P. AND KLEBAN, M. H. The Aged Resident of the Inner City. *Gerontologist*, 11:277–283, 1971.
18. MADDOX, G. L. Some Correlates of Differences in Self-Assessments of Health Status Among the Elderly. *J. Geront.*, 17:180–185, 1962.
19. MADDOX, G. L. Self-Assessment of Health Status: A Longitudinal Study of Selected Elderly Subjects. *J. Chron. Dis.*, 17:449–460, 1964.
20. McCALL, R. *Fundamental Statistics for Psychology*. New York: Harcourt, Brace and World, Inc., 1970.
21. PALMORE, E. AND LUIKART, C. Health and Social Factors Related to Life Satisfaction. *J. Health Soc. Behav.*, 13:68–80, 1972.
22. PIHLBLAD, C. T. AND McNAMARA, R. L. Social Adjustment of Elderly Peo-

ple in Three Small Towns. In Rose, A. M., et al., p. 55. *Older People and Their Social World*. Philadelphia: F. A. Davis, Co., 1965.

23. Research Department Committee: Four Areas of Boston, 1970: A Report based on the 1970 Census, IV, Center City, Massachusetts, United Community Services of Metropolitan Boston, 1973.
24. RILEY, M. W. AND FONER, A. *Aging and Society*, Volume One. *An Inventory of Research Findings*. New York: Russell Sage Foundation, 1968, p. 21.
25. ROSENCRANZ, H. A. AND PIHLBLAD, C. T. Measuring the Health of the Elderly. *J. Geront.* 25:129–133, 1970.
26. ROSE, A. M. AND PETERSON, W. A. *Older People and Their Social World*. Philadelphia: F. A. Davis Co., 1965.
27. ROSOW, I., AND BRESLAU, N. A Guttman Scale for the Aged. *J. Geront.* 21:556–559, 1966.
28. SELYE, H. *The Stress of Life*. New York: McGraw-Hill Book Co., 1956.
29. SHANAS, E., TOWNSEND, P., WEDDERBURN, D., et al. *Old People in Three Industrial Societies*. New York: Atherton Press, pp. 26; 27; 29; 46; 53, 1968.
30. STOTSKY, B. A. *The Elderly Patient*. New York: Grune and Stratton, 1968, p. 71.
31. STOTSKY, B. A. Social and Clinical Issues in Geriatric Psychiatry. *Am. J. Psychiatry*, 129:31–40, 1972.
32. SUCHMAN, E. A., PHILLIPS, B. S., AND STREIB, G. F. An Analysis of the Validity of Health Questionnaires. *Social Forces*, 36:223–232, 1958.
33. TOWNSEND, P. *The Last Refuge*. London: Routledge and Kegan Paul, 1962.
34. TOWNSEND, P. Measuring Incapacity for Self Care. In Williams, R. H., Tibbitts, C., Donahue, W. (eds.), *Processes of Aging*, Volume Two. New York, Atherton Press, 1963.

Article 20 Angry in Retirement

ELIZABETH Z. HUGHES

I never thought that my life would end with such strong feelings of hopelessness and loneliness. Living—or should I say dying?—in a place that resembled an institution or a home for the aged had never entered my mind.

Most of my life, for about 50 years, everything just went along. My life was centered around my husband, Armando. He actually designed and built our two-story, gray shingled house. Both of us were particularly proud of the rose garden in our backyard. Armando made four beautiful fountains similar to those in his Italian, boyhood home. I can still hear him saying, "Angelina, can we spare a few coins for a bag of cement?" We had only pennies in our pockets during the depression, but I couldn't deny Armando his pleasure. I knew that a bag of cement meant more to him than bread. Our garden became almost too elaborate for the neighborhood; yet it was so beautiful to us that it became a part of our very life.

Armando had his work at the newspaper and his garden. I had the care of two children, church work and some part-time translating jobs. I became involved in many of the projects that social workers do today.

Time saw us facing the early death of our only son. Our daughter married and moved halfway across the country. Generally, though, my church work, cooking and caring for a large house kept me very busy. Admittedly, the house and garden consumed much of our energy. We talked about selling our home. After all, the two of us really didn't need so much room. A small apartment with the comforts of home in our own neighborhood was what we were looking for— close to our friends, our church and the stores we had traded in for years. But we just couldn't find anything we liked. With our retirement budget, we

ELIZABETH Z. HUGHES, PH.D., is associate dean at the University of Maryland's School of Nursing.

Source: Human Behavior, September 1974, pp. 56–59. Copyright © 1974 Human Behavior Magazine. Reprinted by permission.

couldn't seem to find an apartment that would give us the comforts of our own home.

I had never given any thought to how I would live if Armando should die. He said that I should remarry, but it was usually said in jest. I responded by saying that one good man in my lifetime was enough! Then suddenly my life changed. The safe, warm feeling I had with Armando stopped. His death was sudden; it still hurts to talk about it. I felt drained and so empty inside.

All my friends wanted to help in some way. Some brought home-cooked food and others gave their opinions about what I should do. They kept insisting that I couldn't manage by myself. The house was just too big, and I needed a place where someone would take care of me. I felt so very tired and weak at that time that I found myself tending to agree with them. If I could only have resisted their suggestions and waited awhile, maybe I wouldn't be here today. One of my friends offered me a room in her apartment, but she talked constantly and I knew I couldn't live with that! She would just sap all my energy. In looking back, I really needed more time to heal the hurt I felt with the loss of Armando. I felt pushed into making decisions; decisions had always been Armando's responsibility. It was easier now for me to let others make them.

So the house went up for sale and was sold immediately, one month after Armando's death. I was hoping that it would remain on the market for a while, but the first people who looked at the house bought it. Although the settlement day was two months away, I had to decide just where I would go. I really wanted to stay in my own neighborhood.

A friend of mine had lived in a home for old people for many years; in fact, she still lived there. I had visited her several times, and, to me, the home looked clean and pleasant. My minister took me to the home for one quick visit, and within a few weeks I was on my way to becoming a resident of the institution. I knew that my way of living would change. How could I not know that when I compared my lovely 10-room house and my garden to one small, cream-colored, cement-block room?

The administrator at the Home told me that in exchange for lifetime residency, I would have to relinquish all of my assets, which included the proceeds from the sale of the house, all of my life savings and my monthly social security check. Yes, I knew that I was allowed to bring only a few of my lifetime possessions. So I chose to bring my bookcase filled with some treasured books, my favorite chair, a small dressing table, a small television set and my sewing machine.

There were a few rules, of course. The use of communal bathrooms and my monthly spending allowance of $20 were explained to me. I had been accustomed to my private bath, and as for my "allowance," by the time I bought a few personal articles and a bit of fresh fruit, the $20 was gone. The dreadful loneliness and confinement were to become a part of my daily agony; I was unaware of these feelings until I entered the institution.

I felt so anxious that first day. How could it be that Angelina was going to be in a place like this "for good." The tightness in my throat wouldn't go away. I could hardly swallow and my mouth was too dry to speak. I kept wondering about the irreversible decision I had made.

I can still remember walking up the two cement steps of the big red brick building and through the heavy wooden doors of my "new home." I looked around to see if someone might be there to greet me. I really wanted someone to say welcome. But both the big parlor on the left-hand side of the main corridor and the dining room on the right-hand side were empty. As I came toward the end of the corridor, I noticed a small white-haired woman quietly sitting and staring at the floor. Maybe it was my need for a welcome of some kind. Anyway, I walked up to the crouched woman and touched her bent shoulder. The woman peered up at me; her lips formed a faint smile. I felt some warmth when she smiled.

After signing a big book on the administrator's desk, I took the elevator to my room. The room seemed so small! There was one window and one closet. A feeling of warmth came again, though, at the sight of my furniture from home. Just about every piece of wall space was covered. In fact, the room looked cluttered. But how do you squeeze 57 years of living in a 10-room house into a tiny box-shaped room? The long, narrow closet was filled with clothes for all seasons. The institution had no place for trunk storage, so winter and summer wear hung side by side. Maybe I had no need to be worried about clothes since the nurse's first question was, "Mrs. Mello, where do you intend to be buried?" Should I tell her I came here to live?

To make a homier room, friends made a pair of drapes and brought a scatter rug. The floral drapes hanging on either side of the only window and the small rug at the foot of the bed made the head of the housekeeping department very angry. How could the cleaning people wash the windows with those fancy drapes? How could they mop the gray rubber-tile floor with that scatter rug there? And the bookcase with all those books was in the way. Besides, no one had a bookcase in her room. Why should Mrs. Mello be different?

Within two weeks, the administrator, a member of the board of directors and Mrs. Pierce, the head of the housekeeping department, came single file into my room. Mrs. Pierce began to search the closet. She pointed to the drapes and rug and told the two men that these items hampered her cleaning.

I can still feel the anger rage inside me. It was hard for me to keep my chin up and maintain my pride. In my own home I was queen, but here I felt like a child. I told Mr. Ford, the administrator, that I had done nothing wrong. The bookcase did clutter the room, but I didn't want to give up my books. Without them I would die! After looking around the room, Mr. Ford nodded his head in approval, and the inspection party left. Fighting back the welling tears, I sat rigid on the edge of my bed. I resented being put into such a humiliating position. Maybe I did regard myself as being more "cultured" than the rest. But without a feeling of self-worth, I would never survive.

Yet, I also knew that I had to be careful now. If I complained or found fault, Mrs. Pierce would tell me to get out and find another place. But where could I go with no money? Being so helpless in this place made me feel angry and desperate. I tried to control myself by saying, "Angelina, behave." It's so hard to keep fighting when you are alone and existence seems suddenly without purpose.

Mrs. Pierce and her staff called us "old ladies" or kids. Women who had not paid much money to the Home obeyed Mrs. Pierce and kept quiet. They dared not retaliate. It became increasingly apparent to me that we were at the mercy of the institutional staff. They controlled whether we got an extra piece of toast for breakfast or a clean towel each day. But even more important, they controlled how we felt about ourselves.

As each day unfolded, I became more and more aware of the demanding lifestyle I had hastily chosen. My first day in the craft room was humiliating. The craft instructor gave me a small piece of unfinished knitting. She said it had been started by someone who had died last week and that if I did well, I could have some unused yarn. Looking at the tattered, jumbled four inches of knitting, which resembled the beginnings of a scarf, I could feel myself getting hot with anger. I had knitted sweaters and capes before. I felt the instructor was testing me; if I passed the test, I could use new yarn. Finishing about four inches of knitting was a matter of minutes. My reward was immediate; a new skein of yarn was mine.

Craft activities scheduled for three mornings a week tended to fill the hours adequately for some of the other residents. They made pincushions and wastepaper baskets made from straws. I found myself barely tolerating the craft sessions. I came to them because I didn't want to be known as a troublemaker, one who did not conform to the institution's expectations. But one day, after an hour of making mice out of gray foam rubber, I had to leave the room. My head was pounding; I felt tight inside. After years of music, gardening and lectures, I was now putting yarn whiskers on sponge rubber mice, something any three-year-old could do. The craft instructor's plan was to sell these mice at the institution's fall bazaar. But, really, who would buy them? What purpose would they serve? Buying them would just be patronizing.

I felt good about the bright red cape I had completed for the fall bazaar. But the price tag placed on my piece of handicraft was only $2.50! The low price angered me. I had spent three consecutive days knitting. Was this the value of my time and energies?

I knew that all the bazaar moneys went to the institution. And then what happened to the moneys? Did the residents share in the bazaar proceeds that came from their individual pieces of craft? When asked this question, the craft instructor looked a bit perturbed. To be queried by an old lady was certainly unexpected. There was no direct answer to my question. Now a scheme began to slowly unfold in my mind.

Although I had been in the Home for seven months now, I still retained some of my identity. Angelina I was, and no one was going to force me to conform. I enjoyed spending a few days visiting friends in their homes, but it was not a very pleasant feeling to come back. I missed not being able to make a cup of coffee when I felt like it. And I used to enjoy making fudge or a batch of cookies. It would be nice to have a kitchen; we could cook some of our favorite dishes for each other.

Mealtime tended to be each day's highlight. Every able person queued up at least a half hour before serving time. But once everybody was in line, there was a funeral-parlor silence. Each person moved always in her own personal orbit. In fact, to me, the total day-to-day silence of the Victorian parlor across from the dining room was deafening! Maybe music would bring people closer together. Right now, the place was like a mortuary!

Within a day or two, I asked Mr. Ford if the Home happened to have a record player. He indicated that one had been put in storage several years ago. He would hunt it up for me. I was excited because, to me, where there is music, there is life. That evening I announced at dinner that Mr. Ford was trying to locate a record player. Records would be needed. So if anybody had ways of getting some, I would start stacking a small pile in my room.

Each evening for about a week, one or two residents, with record in hand, would knock on my door. Some of the residents thought the record player was a foolish idea. They had existed for 10 years without music, and they didn't want to get involved. But within a month or so, I had a stack of records on top of my bookcase. And we also had the record player to be used in the parlor.

Since my first days at the institution, I had felt like a troublemaker. The feeling started with the drapery-rug episode and was increased by my daily walks. I started to recruit others to take daily walks with me regardless of the severity of the weather. On cold days, my head, neck and hands were wrapped in my own handknitted cap, scarf and gloves. I soon noticed that a few of the residents were beginning to join me on my walks.

A place to sit outside would be nice. A porch would be really nice. When they build the next home, I would like to be here to be sure they do it right! Daily walks became my way of letting off steam. Living by so many rules was new to me, and I found myself resisting them.

The institutional staff awakened residents at 6:30 every morning, and lights had to be out by 10 p.m. I enjoyed reading or watching the 11 o'clock news on television, but I had to seek permission to keep my lights on until after the news was over. I must go ahead and live!

Certainly the occasional support from Mr. Ford seemed to help in maintaining my basic need for activity and involvement. I had been very active in my neighborhood before I came to the Home. And now I started poetry-reading sessions and planned a few violin concerts.

A few of the residents were on my side. However, most of them tended not to accept my rebellious ways. "Why does Mrs. Mello have to change things?" they would mutter.

The residents dared not risk supporting any change that had even a hint of being contrary to the staff's demands. Residents who had bequeathed very limited moneys to the institution felt particularly vulnerable and hesitated to reveal their true feelings. After all, they should feel grateful to have a roof over their head, and three meals a day. Dared they expect anything more? So they conformed to the expectations of the staff.

Almost a year has passed. Is this it? There seems to be nothing else I can do. I've become less rebellious. Maybe there is a need to calm down and be tamed. It takes so much energy to penetrate and be heard. I feel choked; I can't express myself, and I could contribute so much. As long as this is my home, I want to feel a part of it in some way.

The music in the parlor is beautiful. We love it. The monotony has lessened. We talk and the music plays softly. Maybe there will be more than pincushions.

Developmental Life Transitions: Death and Bereavement

VIII

It has often been said that death is more painful for the survivors than for the victim. Although death is a natural and inevitable part of the life cycle, it is a taboo subject that people avoid talking or thinking about. As a result, they are seldom prepared for the death of a loved one or for their own death. Adults have many different strategies for coping with this shock (see Part III for children's responses). These strategies are determined by the circumstances of the death (unexpected, or after a long illness, for example), by personal characteristics (such as age), and by cultural expectations.

People generally accept the fact that someday they will die, but they are comforted by the vagueness and uncertainty of the timing and the means. For the very old whose health begins to fail, and for those who have a terminal illness, this comfort is gone and they must face the reality of their own death. Kübler-Ross[7] has outlined five stages that dying people experience: denial and isolation (an initial refusal to accept the reality of their impending death), anger (at fate, but directed at anyone nearby), bargaining (for brief respites), depression (both reaction to and preparation for loss), and finally acceptance. Not everyone will have sufficient time, strength, or external emotional support to achieve acceptance and peace.

The last stage of old age is a time to look back over one's life and evaluate its meaningfulness. One's conclusions may lead one to a general sense of either "integrity" (satisfaction) or "despair" (see pp. 10–11). This life review process involves the use of reminiscence to gain a new perspective on one's life. Through one's memories (perhaps selectively chosen), one can decide what was of real significance and recall one's achievements. The outcome of this process has a significant impact on the ability to handle the approach of death. For old people who realize that the significant people and things of their lives are gone, death can be viewed without anxiety.[4] Even younger people can face death with equanimity and peace of mind if they can find some purpose or accomplishment in their lives.

Although there is a growing interest in the problems of those who know they are dying, traditionally interest has focused on the survivors. The articles in this section deal with how people mourn and adjust to the death of important people in their lives. The shock of such a death provokes common physical and emotional reactions. Lindemann[7] describes some physical symptoms associated with acute grief: sighing, muscular weakness, tightness in the throat, emptiness in the stomach. Sleep disturbance, loss of appetite and weight, and increased dependence on alcohol, tobacco, and sedatives are also common.

Emotional reactions generally go through four phases. The initial response is one of numbness, denying all feeling and detached from reality. This serves to cushion and buffer the shock. As feeling returns, the bereaved person tends to focus on his own loss and to be preoccupied with the deceased person, a phase Parkes[8] describes as yearning for the lost one. This is a time of emotional disorganization and inability to function, when feelings of depression, anger (at the deceased or others), restlessness (as normal activity patterns are disrupted), and guilt threaten to overwhelm. Gradually people learn to accept their loss and break their ties with the dead person, until finally they are ready to build new relationships to replace the lost ones.

In the first article Avery Weisman examines the problems of people who have experienced the untimely death of a loved one and the ways outsiders can ease their adjustment. He outlines three types of untimely death: premature (early in the life cycle), unexpected (sudden or unpredicted), and calamitous (unpredicted, violent, and demeaning). For those whose bereavement is unpredicted, there is no chance for anticipatory grieving, a phenomenon found by Chodoff, Friedman, and Hamburg[3] among the parents of leukemic children. Even while the child was still alive, parents experienced the physical symptoms of grief described earlier, the total preoccupation with the ill child, and finally a sense of detachment. When the child did die, the mourning was less intense than for those parents who had had no such preparation.

Weisman suggests that some intervention can ease the shock of an untimely death. Such intervention should be aimed at transforming a distressing untimely death into a more acceptable form and at fostering mourning that is appropriate for the individual and for society. The goals of coping with an untimely death, as Weisman sees them, are to make the shock tolerable, to keep the survivor accessible and open to the emotional and practical support others may offer, to make the death acceptable, and to realize and accept that death itself can come at any time.

The second article is Bard Lindeman's personal account of the untimely death of his wife and his painful adjustment to his loss. His wife's death, premature and unexpected, left him alone with their three dependent children. He experienced many of the feelings considered a normal part of acute grief: numbness, anger, unreality ("Strangely, I felt I was losing my contact with reality and reason."). He was lucky, though, in having his friend and minister offering compassionate emotional and practical support. His children also proved to be a source of support and strength. Because his eight-year-old daughter had some trouble understanding her mother's death and expressing her feelings about it (see Part III on childhood bereavement), Lindeman worked to develop a special close relationship with her that turned out to help them both. Lindeman found that with time the wounds started to heal and gradually he was able to pick up the threads of his life. (For a poignant and articulate account of a young widow with small children, see Lynn Caine's *Widow;*[2] also on widowhood, Glick, Weiss, and Parkes.[5])

In 1972 there were 9.6 million widowed women and 1.8 million widowed men in the United States, a large and vulnerable population for whom preventive intervention could be significant. Evidence of significant deterioration in the general health of young widows and widowers has been found, and the mortality rates for widowers in their first six months of bereavement are unusually high for their age groups. In the third article Phyllis Silverman discusses one such preventive project, the Widow to Widow Program, which is primarily designed to reach younger widows before they have serious problems, and to help them in making the difficult adjustments necessary to their new status.

The widow "caregiver" can offer the newly widowed woman comfort and understanding in a way nonwidows seldom can. The new widow tends to feel isolated and somehow stigmatized, constantly called on to reassure the friends and family who should be supporting her. From the program aide women can accept reassurance, encouragement, and realistic advice. The real coping tasks for widows are changing one's self-image from "married" to "widowed" (which may also mean learning to make decisions), learning to be alone, and making new friends and enjoying the company of others. These are similar to Lindemann's[7] grief work: break the ties, readjust to a world without the deceased, and make new relationships. Silverman found that most women do respond (for example, get jobs and learn to handle finances) and make the adjustment satisfactorily.

In other societies there may be features built into the culture to ease the trauma experienced by the bereaved. In the fourth article Joe Yamamoto, Keigo Okonogi, Tetsuya Iwasaki, and Saburo Yoshimura examine cultural differences in the mourning process of Japanese widows. In Japan both culture and religion sanction the practice of ancestor worship, the use of ancestor altars, and the implied presence of the deceased relative. Unlike Western ideas—that the bereaved must break their ties with the dead—the Japanese encourage widows to maintain their ties (for example, by tending the altar each day). Yamamoto and his colleagues felt that these religious customs were helpful in the period of acute grief, giving comfort and easing the loss by offering some form of substitute relationship with the deceased.

Ablon's[1] study of an American-Samoan community showed a totally different cultural approach to bereavement. Supports are built into the social structure so that a wide circle of family and friends converge to offer full emotional and practical support; the whole community donates money for the elaborate funeral rites that tradition demands. Their attitudes—that death is a natural event and that family members can and should be replaced when lost—play a significant role in the adaptation of the bereaved. The widowed tend to remarry and be quickly reintegrated into society. These cultures offer two very different examples of ways in which society can alleviate the emotional distress of bereavement and facilitate coping and adaptation.

REFERENCES

1. ABLON, J. Bereavement in a Samoan community. *British Journal of Medical Psychology*, 1971, 44, 329–37.
2. CAINE, L. *Widow*. New York: William Morrow, 1974.
3. CHODOFF, P., FRIEDMAN, S. B.; AND HAMBURG, D. A. Stress, defenses and coping behavior: Observations in parents of children with malignant disease. *American Journal of Psychiatry*, 1964, 120, 743–49.
4. DE BEAUVOIR, S. *The coming of age* (P. O'Brian, trans.). New York: G. P. Putnam's Sons, 1972.
5. GLICK, I. O.; WEISS, R. S.; AND PARKES, C. M. *The first year of bereavement*. New York: Wiley, 1974.
6. KÜBLER-ROSS, E. *On death and dying*. New York: Macmillan, 1969.
7. LINDEMANN, E. Symptomatology and management of acute grief. *American Journal of Psychiatry*, 1944, 101, 141–48.
8. PARKES, C. M. The first year of bereavement: A longitudinal study of the reaction of London widows to the death of their husbands. *Psychiatry*, 1970, 33, 444–67.

Article 21 Coping with Untimely Death

AVERY D. WEISMAN

Ecclesiastes tells us that "All things have their season, and in their times all things pass under heaven . . . a time to be born, and a time to die; . . . a time to keep, and a time to cast away. . . ." While we might all agree in the abstract that there is a time to die because death, after all, is inevitable, most people in our society also seem to believe that with few exceptions, every death is unnecessary and preventable, and therefore untimely.[16] Furthermore, at the critical moment of their grief, few mourners can be expected to achieve the perspective and serenity conveyed by Ecclesiastes. Death is more than a biological fact: it is a psychosocial crisis for which all societies have established rituals and expectations about how and when survivors should mourn.

It is a basic paradox to maintain that death is inevitable, while clinging to the illusion that it can be postponed indefinitely.[24] But there is nothing paradoxical about finding man's attitudes paradoxical. If there is a general notion that most deaths are untimely, then it becomes especially important for mental health professionals to deal effectively with people who suffer acute bereavement without the ameliorating advantage of anticipatory grief.

We are a generation of survivors, afflicted by constant reminders of sudden death. We can scarcely pretend to understand why some people must die brutal, meaningless, and arbitrary deaths, long before their time—whenever that is— while others go on, risking their lives, wasting their years, but surviving.

Earlier generations used to ask, "What shall we do to be saved?" This threatened generation might find it more appropriate to change the question to,

AVERY D. WEISMAN, M.D., is Associate Professor of Psychiatry, Massachusetts General Hospital, Harvard Medical School.

Source: Psychiatry, Vol. 36 (November 1973), 366–78. Copyright by The William Alanson White Psychiatric Foundation, Inc. Reprinted by special permission of The William Alanson White Psychiatric Foundation, Inc.

"What shall we do to survive?" Leaving aside religious differences, there is a secular distinction between being saved and surviving. Furthermore, in this era of computerized anonymity, we should ask still another question: "What shall we do *to survive with significance?*" Significant survival is not measured in years, but by how competently we have coped with problematic situations and how well we have lived, within the scope of our values and ideals.

We hear a great deal about the desirability of "living our own life." I am not always sure what this phrase means. But since death is inexorable, and dying is a part of being alive in the first place, one of our central objectives should be to overcome the anonymity that surrounds us, and to *die our own death*, whenever it occurs, however untimely it seems.

Without enumerating various theories[5] that define or, more often, deny death, most of us fear death because it seems to be the universal refutation of everything we find desirable in life. Concepts of death are usually extensions of what we dread most about living, filtered through our efforts to mute, mitigate, and deny the pains and frustrations we have experienced. On the whole, fears of death and dying are decided not by nebulous notions of the "Unknown," but by how we *systemize our sufferings and shortcomings*.

THE MEANING OF UNTIMELY DEATH

Every generation survives the last, and as long as we are alive, we are survivors who seek to deny the reality of personal extinction. When the sequence of birth, childhood, maturation, adulthood, senescence, and death is interrupted before running its natural course, we are usually shocked by being confronted with the most ultimate of facts: death and dying. Fears of shock and suffering usually prevent us from realizing that living and dying are phases of the same process. Death is not a mere end product of having lived—a tiny exit at the distant end of a long corridor, open only to those who have nowhere else to go, nothing else to do, nothing else to become. Aging reminds us of death's inexorability, so we shun the aged, disenfranchise them, render them obsolete and disengaged, as if the purposelessness of old age makes death less certain for those who are younger. Mortality tables fractionate the certainty of death into a variety of separate causes, tacitly suggesting that if we could prevent any or all of these deaths, we might postpone our own demise indefinitely. But the statistics about death itself are easy to remember—100 percent. Someone recently asked me about our death research. I told him we found that the longer one lives, the older one gets—but that as scientists we couldn't prove it, because we had no control groups!

Of course, we must acknowledge that there are many very untimely deaths which are capricious, ironic, ugly, and unnecessary. But not all untimely deaths are alike, and our horror at untimely death, especially our own, should not lead to the conclusion that nothing can be done about the victims or the survivors.

Although there can be combinations, it is useful to distinguish three kinds of

untimely deaths: (1) premature deaths, (2) unexpected deaths, and (3) calamitous deaths.

Premature death is the demise of a person in the earlier phases of human
development. Logically, the earlier the death, the more premature it is. In actuality, however, the prototype of premature death is not a newborn, but a child.
This image of untimely death is painful and poignant, not merely because the
young of any species evoke tenderness, but because the child has already become
a person. Consequently, a child's death damages our own social and emotional
reality, including our belief in a future in which potentials become fulfilled. The
blamelessness, trust, and innocence which we identify with childhood makes
adults feel as if they had failed to protect the child through its vulnerable years.[18]

Unexpected death is sudden, unpredicted death. It can occur at any age,
among the healthy or the sick. An aged woman who is struck and killed by an
automobile, or a middle-aged man who has a fatal coronary occlusion while
seated at his desk are both examples of unexpected death. The emotional impact
upon survivors is gauged and sharpened by how independent, autonomous, and
distinctive the deceased happens to be at the time of death. It is also measured
by how drastically the death violates our inner timetable of expectation. The
sudden transition from being very alive to being very dead shocks the sense of
reality of even professional observers, let alone those whose lives will be seriously
changed.

While cases of unexpected, premature deaths are common, there are also unexpected and postmature deaths. Glaser and Strauss described certain hospital
patients who were expected to die but failed to do so. When a death does not
happen on schedule, both the family and hospital staff may become very upset,
impatient, even angry and punitive. Expectation of death is at least as much a
social judgment as an expression of medical prognosis. Unexpected untimely
deaths, therefore, are apt to be conspicuous because that person is not only active
and visible at the time of demise, but is a significant force in the life of survivors.

Calamitous death is not only unpredicted, but violent, destructive, demeaning, and even degrading. Murder is calamitous; for the most part, so is suicide.
We commonly observe that certain people who attempt suicide are not seeking
death, but rather searching for ways to make life more tolerable.[6] In fact, they
may repudiate violent death as vigorously as they repudiate help. Recently, one
young woman overdosed herself with sleeping pills, turned on the gas in the
oven, and lay down, expecting to die. Before she dropped off to sleep, however,
she suddenly imagined the scene after death: strangers would break into her
apartment, uncouth men would handle her roughly, and perhaps even undress
her! So she telephoned a friend, who summoned help, and was brought to the
hospital. Similarly, suicides may strew their postmortem path with co-victims,
because survivors may always wonder if a rescue might have been effected, had
circumstances been different.[4]

To cite cases of premature, unexpected, and even calamitous deaths that take place in hospitals would belabor the obvious. In this respect we have an embarrassment of riches—one need only glance at the morning newspaper!

TIMELY DEATH

We are brought up to believe that death is almost always deplorable, even when the prevailing creed provides for some form of postmortem survival. So we naturally assume that when someone we care about dies, it is almost always untimely, and the only timely death happens to someone who, in our opinion, richly deserves it.

But there are timely deaths, and if we consider the topic with a minimum of preconception, it will be clear that we cannot cope effectively with the realities of untimely deaths without a concept of timeliness. We hear about people who are said to be living "on borrowed time." Borrowed from whom? It is at least as reasonable to say that all of us die "within allotted time."

A timely death is one in which *observed survival* equals *expected survival*. There are, in fact, three modes of timely deaths, but only one is familiar. These modes are (1) actuarial deaths, (2) survival quotient deaths, and (3) appropriate deaths.

Actuarial deaths are the familiar realities we meet when we buy insurance policies. Like it or not, our premiums are off-track bets, where the agent gives odds on our survival. It is a highly impersonal transaction, advertising notwithstanding. Odds are determined precisely, and become less favorable with each passing year, illness, or deviation from the norm, until the company finally refuses our bets altogether. In the actuarial sense, timeliness is a matter of dollars and cents. There is a time to die, and the occasion is a problem of probability.

Survival quotient deaths refer to mathematical formulas that help determine the average time of death, or, more euphemistically, the average survival for patients with different kinds of cancer.[28] These formulas have been developed at the Massachusetts General Hospital by Project Omega, as we call our investigation of suicide and terminal illness. Briefly, various contributing factors in survival are weighed. These include anatomical sites, staging, sex, diagnostic criteria, clinical complications, and therapeutic measures. Then, using computerized information about dead patients,* multiple regression formulas are calculated for each site. These formulas are based upon available data, and, naturally, will be sharpened with accumulation of further clinical and research information. Very simply, when the observed survival time for an individual patient is compared with the average expected survival time, the result is called a "survival quotient." The more conventional five-year survival periods are largely for the convenience of cancer statisticians, and do not accurately reflect the biological activity or even therapeutic effectiveness of various treatments.[7,21]

---

* This information was made available by the Massachusetts Tumor Registry.

Project Omega is primarily a psychosocial study. But we are also concerned with organic factors in order to help answer questions about the relation of personality and social factors to length of survival. The survival quotient *does not*, of course, determine who gets a cancer, nor the kind of treatment given. No mathematical formula tells us who shall live or die, or when death will occur. That would be science fiction. The survival quotient simply offers an approximation about the timeliness of death from certain malignancies. These are average expectations for average afflicted populations. High probability is one thing, but the reality of individual sickness and timely death of an individual patient is another.

Appropriate deaths are the third form of timely deaths, and the only mode that directly pertains to the individual. In appropriate death, there is correspondence between observed survival and expected survival, and also patients themselves seem to find death *acceptable* at that moment.[26] This does not mean that anyone afflicted with a serious illness is able to "will" death, independent of physical factors. In no sense should the concept be misconstrued as a "psychogenic death." An appropriate death, furthermore, is not a death which some outside observer decides is fitting and proper. Many aged patients in nursing homes, for example, are treated less than vigorously simply because the physician, with the assent of the family, decides that the patient has lived long enough. Finally, appropriate death should be emphatically distinguished from suicidal death, in which a potential victim, seeking to survive, "appropriates" fatal means of self-cure.

What, then, is an "appropriate death"? Who decides that a death is timely and appropriate? There are concepts and constructions in both psychoanalysis and existentialism which view death as a concomitant element of life. Freud's famous postulate of the "death instinct" was largely based upon a speculation that all organic activity seeks the quiescence found in consummated instinctual behavior.[22] Heidegger and other philosophers,[12] representing an "existential" viewpoint, also see death as depletion of desire, the fulfillment as well as the exhaustion of potentialities. Hence, death is both a "not yet" and a "here and now," just as any potentiality is present in a specific tendency.

Appropriate deaths are those in which suffering is at a low ebb, conflict is minimal, and behavior has been maintained on as high a level as is compatible with physical status. Moreover, the dying patient indicates that what has already been done corresponds to what he expected of himself, of the people who matter most, of those to whom he turned for relief, and, finally, of the world in general. Literally and metaphorically, it is time to die. Relief of anguish and resolution of remaining conflicts join in a harmonious exitus. The patient both accepts and expects death, and is willing, albeit ruefully, to die. It is the ultimate of successful coping. From the general viewpoint of conventional values, this may not be a desirable death, but, after all, few deaths are regarded as desirable, and appropriate death recognizes no age limitations. However, death at any age can still be compatible with everything that the person stood for during life. The phrase, *stood for,* can be construed in several senses: (1) what a person *represented* in terms of an ego ideal, (2) what a person *achieved,* or became, and, finally, (3) what he or she *endured.*

We can cope effectively with death as we cope with every other problematic situation, using every practical means at our disposal to bring about *relief, reward, quiescence,* and *equilibrium.* The concept of a "death instinct" is superfluous, since the repetition compulsion, upon which death instinct depends, is simply Freud's name for efforts to cope with recurrent conflicts in more or less uniform ways. Heidegger is practically useless as a clinical guide. Among other well-known existential authors, only Boss and Minkowski seem to have the necessary clinical experience to make their observations about dying patients fully credible. Sartre, however, has recognized the two functions of consciousness—namely, to be *conscious of events* as we are during their perception, and to be actively *conscious for a purpose.* This insight is exceedingly relevant to what we expect of dying patients and of their survivors. We expect them to perceive immutable facts and then, so far as possible, to act upon them for the purposes of resolution and relief.

The essence of timeliness in death can be epitomized. In the course of fatal illness, we proceed from the *probability* of dying, should treatment fail, to the *necessity* of death, as a personal event. Then, should social, physical, and psychological forces converge, the *obligation to die,* which belongs to the human condition itself, becomes the *freedom to die our own death.*[24]

DYNAMICS OF UNTIMELY DEATH AND BEREAVEMENT

Most of us have a poorly developed sense of reality for death. Although we recognize the universality of death, even our personal obligation to die simply because we were born, we keep death at a distance, remote and impersonal. Freud's dictum that in the unconscious there is no concept of death because we cannot imagine ourselves dead, really makes no more sense than to claim that the unconscious has no concept of being born, simply because we cannot imagine a state of being "unborn," prior to our mother's pregnancy and parturition.

Whatever we consider true or real in the world depends upon our sense of reality for our own personal presence and consciousness. Objects, people, and things are real only insofar as they belong to lived-in experience, or to our libidinal field.[23] Reality consists of whatever confers a sense of being alive and can be acted upon. Physical organs and physiological functions are obviously necessary to sustain organic life. But propositions and perceptions about them are believed in, acted upon, and accepted as *real* only to the degree that they participate in purposeful activity. Similarly, the distinction between brain and mind is not limited to the boundaries of the skull. The brain is an organ, but the mind is everything that we are mindful of and to which we can assign a meaning.

Deaths that occur at a distance from lived-in experience, outside of our libidinal field, are simply not real. These impersonal deaths can be read about, even conceptualized, but not experienced, because such deaths really don't matter. Sentimentality aside, it is not true that "Every man's death diminishes me." If it were, we would dwindle into nothingness in the course of a day. Untimely deaths of people far distant, emotionally and geographically, are hardly deaths at all. Our libidinal fields have not been damaged, our psychosocial network is intact, we have not been injured, and we are not bereaved.

Untimely deaths do matter when our very existence is traumatized; our name for this process is *bereavement*. There are several stages of bereavement,[15] which need only be summarized here. Because some untimely deaths are not necessarily unexpected, there may be a preliminary phase of anticipatory grief, followed by mourning, depression, sense of loss. In most instances, however, when anticipatory grief is absent or greatly abbreviated, untimely death usually evokes numbness and disbelief about the immediate fact of death itself. Terrible anxiety and anguish may precede dejection and awareness of loss. The key survivor undergoes a painful period of crying, searching, craving for the deceased—often with intermittent anger, even traces of delusions, depersonalization, and hallucinations. The co-victim not only is exceedingly vulnerable and lonely, but also feels hopeless, helpless, and sometimes worthless, as well.

When an untimely death penetrates our sense of reality, we often respond like a micro-organism punctured by a painful injury. We withdraw, finding a psychological counterpart to physical avoidance in *denial*. By denying a painful portion of immediate reality, we protect our own presence and consciousness. We simply disbelieve, repudiating anything that would threaten our enduring assumptions about reality and its key relationships.[27] Denial is an emergency response that closes off further perceptions by means of physical and psychological numbness. Quite literally, we are *wounded*, having come to grief before we are prepared to mourn.[25]

The process of bereavement is a problem of coping with a wounded sense of reality. Its aim is to facilitate *healing*—namely, resolution, relief, and restoration of a corrective equilibrium between reality sense and reality testing. Because untimely death is so often regarded, first, as unreal, complete coping requires that we gradually reunite reality testing with the unalterable facts of death.

It is obvious that our sense of reality for a death diminishes according to its distance from our personal field. Conversely, grief increases in proportion to the intimacy of our relationship with the deceased. Intimacy does not necessarily correspond to kinship or marital formalities. Rather, it depends upon the key function that the deceased serves in present, conscious, lived-in experience. There are many people, for example, who appropriately grieve more for a pet than for their siblings. Loss of a parent means something less to an adult than to a child, who still depends upon its parents and perceives its own reality according to their presence. Deaths of childhood friends may be painful only to the degree that our own mortality is presaged.

COPING PROBLEMS IN UNTIMELY DEATHS

Just as most wounds heal and bone fractures reunite, leaving scars but permitting restoration of function, so, too, most people get over bereavements. We may not, of course, wholly forget that a death happened, nor can the loss be totally obliterated. But the process of bereavement is complete when the afflicted survivor becomes operational once again. In the same way that a doctor treats a wound or fracture, our professional task is to provide emergency service, keep complications to a minimum, and then, during convalescence, prevent delayed reactions, invalidism, and secondary benefits.

The more closely the death of another person resembles our own—in fact, *is* our death—the more severe the damage inflicted. Coping problems differ in both timely and untimely deaths, according to our vulnerability. But coping is also apt to be different in the three modes of untimely deaths: premature, unexpected, and calamitous.

How long is "normal" bereavement? This common question is really a pseudoquestion, like "How high is up?" I find it significant that we are seldom asked: How short is normal bereavement? There are different cultural, subcultural, and community expectations which determine not only the *form* of mourning, but its *duration*. Mandelbaum has described antithetical funeral and mourning practices in different societies. Some groups emphasize brief ceremonies and short bereavements. Others seem to turn the observance of death into a central preoccupation of the clan. And there are still others that adopt a mixture of love, hate, fear, appeasement, and sacrifice.

Every group has its own version of rites of passage. But these passages are two-way streets: one is for the deceased, who is properly conducted from life to death; the other is for the key survivors, who must undergo certain restrictions before being readmitted to regular society. Cultural customs regarding death epitomize the social values of the group, and emotional responses are at least partially influenced by group expectations and local practices.[11] Any close death is a wound, but the method of healing is either prescribed by society or at least presided over by some established authority. The key survivor is usually under some kind of taboo, simply by virtue of his or her proximity to the deceased or nearness to death. Community pressure modifies personal behavior.

I recall, as a boy, overhearing adults talking about how much a widow or widower grieved after the death of her or his spouse. I got the impression that the depth of affection and the openness of mourning were synonymous, as if decibels of moaning and drops of tears adequately measured love, suffering, and loss. The adults also implied that there were two kinds of open grieving, genuine and hypocritical. Genuine mourning was said to signify genuine loss; hypocritical mourning meant that the survivor suffered from guilt, rather than loss. In any case, if the key survivor did not mourn conspicuously, it was clear-cut evidence that he or she did not care. True enough, some funeral practices accentuate catharsis. But others can convey great loss through maintaining strict control.

There are traces of this attitude of "social desirability" in certain psychiatric

circles today. It is often reported that a patient "never really grieved" for someone now dead. Delayed grief is certainly a clinical entity, in that numbness and
closure to the sense of loss may be unduly prolonged. But I do not believe that
therapeutic reactivation of memories and unsettled conflicts about a long-dead
person requires copious shedding of tears and open mourning before we can
believe that bereavement is at an end.

People mourn in different ways: many suffer silently, others preserve de-
corum and control, some react soberly, and a few even respond with forced
frivolity. By observing the behavior of people at funerals, we can usually form
a good idea of what the group expects of the key mourners themselves. Consequently, many coping problems in untimely deaths are manufactured by differences between the individual pacing of bereavement and the prevailing cultural
and community expectations. Bereavement runs its own course, and there is
ample room for individual variations within the norm. For example, adolescents
follow a growth curve that is more or less identical. But some adolescents grow
faster than others at times, and more slowly at other times, before reaching maturity. While the shape of the curve is similar, not every growth pattern is identical.

Cultural expectations also lead us to assume that untimely deaths are more
traumatic than anticipated deaths. As a result, it is expected that mourning will
be more conspicuous and that more serious psychological complications will
develop. While anticipatory grieving may modulate mourning by preparing key
survivors for the shock of actual death, there is no evidence that the bereavement
which follows is necessarily more benign or short-lived.[20] Indeed, some people
who have cared for a relative over a long period, expecting, yet seeking to postpone inevitable death, suffer far longer with acute bereavement symptoms than
people who lose someone very close in an untimely manner. There is even something to be said in favor of the quick, clean death. It spares survivors from the
extended anguish of a deathwatch over the inexorable deterioration and decline
of a loved one.

Although the relation of physical and psychiatric illness to grief has been
described by many investigators,[17] we have not yet determined the effect of different kinds of untimely deaths upon survivors. Many factors need to be considered: age, kinship, intimacy, impairments, complications, treatments, religious
and cultural norms, and personal vulnerability. We do know that many psychiatric patients suffering from so-called "reactive" depression are not those who
have been bereft by an untimely death.

PRINCIPLES OF PSYCHOLOGICAL INTERVENTION

There are no ready-made, cookbook recipes for coping with untimely death.
However, there is a range of appropriate interventions, timed to the phase of
bereavement, adjusted to the individuality of the key survivor, and modified
according to our own attitude toward death.

Our goal cannot be to restore the deceased, nor to deny that death occurred.

Interventions have only one practical objective: to promote the process of bereavement so that the pacing of resolution and relief corresponds with psychosocial customs and expectations. Coping should terminate in relief, reward, quiescence, and equilibrium. This means restoration of health and recovery of reality. In other words, psychological intervention in cases of untimely death has two parts: (1) fostering a timely and appropriate bereavement process, analogous to an appropriate death for the individual patient, and (2) transforming the most malignant untimely deaths into more acceptable forms.

More specifically, our task is to change calamitous deaths into unexpected deaths, unexpected deaths into premature deaths, and premature deaths into appropriate deaths. If we believe that every death is preventable, and that militant denial is the only available method of containing grief, then psychological intervention is an exercise in futility. On the other hand, if we truly accept death as a part of living, then untimely death can be construed as an allotment, not simply a tragic stroke of fate.

As professionals, we are sometimes called upon to "break the news." Whom should we tell? Who is the key survivor? Who is the most vulnerable in the immediate family? Who is the strongest member? Whose support can be enlisted for the emergency? It is a sad truth that only in recent times, and perhaps not yet in many instances, have hospitals recognized that families need bereavement rooms so that the news of death can be presented in privacy. Sympathy is not enough, and a set speech obviously benefits the professional more than the survivors.

The task of breaking bad news and supplying emergency support for key survivors occurs frequently in hospitals and emergency rooms. It is reasonable to expect that in the future certain professionals will identify this as one of their functions, and will be trained accordingly. Until that time, however, the task should not be delegated to another person, unless the replacement carries at least equal authority and social esteem. It often helps if two people in authority —say, the physician and nurse, the chaplain and social worker, or any other supportive combination—can be present and share the ensuing events.[1,2]

There is no scenario enabling us to break bad news painlessly. Key survivors will be injured, but will survive. Informants should expect disbelief, shock, or bitter antagonism, because, after all, they are the bearers of sad tidings, even the psychological perpetrators of an injustice. Sometimes initial responses are rage, confusion, or aimless, wordless bustling about the room. This may be followed by accusations, self-directed or at the authority whom the family has trusted until that point. In contrast, early numbness and incredulity may give a false impression of calmness and control. Therefore, during these very early moments, following simple but mitigated communication of basic facts, the professional should be prepared to stand by, absorbing various responses, without flinching, without hastening to ameliorate through intellectual explanations, without being defensive, guilty, or hurried, without arbitrary acts of personal contrition.

It is difficult to take individual examples out of context in order to illustrate how to break bad news and to initiate the bereavement process. However, consider the following case in which a woman was crudely told about her husband's death in an automobile accident. She answered the telephone. An unidentified man's voice said, "Mrs. A, this is the police. About an hour ago your husband was killed by a truck. Can you come down to headquarters and identify the body?"

Her reaction was bewildered disbelief, followed quickly by panic and then denial. "Are you serious?," she exclaimed. "Is this a crank call?" An interchange followed, and she was finally persuaded that the call was legitimate. She then waited alone for the police to arrive and take her to the mortuary.

Obviously, this was a calamitous death. But her response could have been somewhat different had the call been less impersonal, cruel, and itself calamitous. Perhaps a professional, even a policeman, might have come to her house, introduced himself, and done something more to mitigate the impact and to prepare her for the unexpected death. Conceivably, he might have said something along these lines: "I am here because there was an accident, a bad accident. Some cars collided—your husband was in his car and was injured very badly. We took him to the hospital right away—but the doctors found that he was hurt too seriously —and I must now tell you that he didn't survive."

This is, of course, a rather brutal example of "hard tell." Long after the woman recovered from her bereavement, she dwelt upon the calamitous way in which the news broke up her life. Not only did she repeat that moment over and over in her mind, but she hated the nameless voice that first called her.

Tact can be taught, not just to make bitter news sound hypocritically sweet, but to change the context in which the information is given. Note that in the version of how she might have been told, the violence of the fatal calamity was changed into an unexpected accident. Her husband's injuries in such a bad accident were depicted as *necessarily* severe. He was not left to die alone, anonymously, but was given appropriate *care*, with an implication of personal concern. Telling the woman that her husband was hurt too seriously to survive, especially after a medical examination, shows that suffering was at a minimum, that experts were involved, and that the arbitrary chaos and violence had been diminished. Even the indirect statement, "He didn't survive," muffles without hiding the fact of death.

Untimeliness, by definition, carries a shock. But studies have shown[9] that even the bereavement of parents who lose a child because of leukemia can be ameliorated over time with appropriate care, communication, and compassion.

In the aftermath of untimely death, beyond the point of "breaking the news," turmoil and disorganization continue. Because the professional is likely to leave for other duties very soon, it is necessary to enlist the practical support of others. In most cases, survivors must depend upon each other. We look for a significant other—friend, neighbor, relative—who can preserve equanimity and make immediate decisions. Funeral arrangements provide a socially necessary task, as well as being a ritual which brings people together in a show of solidarity.

But key survivors often need practical assistance about living arrangements, care of children, telling distant relatives and friends, and so forth. The significant other who is specifically charged with these responsibilities can be quite young; the main requirement is the ability to see what needs to be done and do it. The opposite and undesirable attitude is that of the well-meaning, but helplessly confused bystander who pleads, "What can I do?," hoping not to be asked.

Survivor guilt has no place in the bereavement process. It should be *forcibly* discouraged whenever it is detected. Survivors sometimes derive an unwitting secondary benefit from the "widowed" or "sick" role. Therefore, the professional who helps in the bereavement process should also know when to withhold help, when to insist upon reactivation, and when to terminate grief. Sooner or later, bereavement comes to an end. The reality of the deceased becomes a significant memory, nothing more. Resolution requires that survivors find a certain reward, as well as relief, in knowing that the deceased had been so important to them. It is not intended to be a consolation when we accept the ongoing reality of a tender memory, even if the dead person lived and died prematurely. Rather, it is a sign that grief has come to its expected conclusion.

If weeping is prescribed by the culture, it should be encouraged. If strict emotional control has more value, this should be supported. Recollections are necessary at any stage, but if emotion threatens to become too intense in later phases of bereavement, we should intercede. People tend to idealize the dead, often giving them larger-than-life virtues. Occasionally, idealization beyond reality may deprive key survivors from mourning the person they actually lost! However, if there were genuine virtues, good works, tender recollections that deserve to be kept alive, then the death itself was timely enough.

Man *must* reminisce and remember, as he would be remembered. Retaining a link with the past is a way to imagine a link with a significant future. Timely reminiscences do not inevitably belong to the process of bereavement, but to the ongoing equilibrium that finally arrives. When coping is completed, untimely death acquires the solemnity befitting a timely, appropriate death. The overpowering truth is that by living our own life, the best we can hope for is to die our own death, not to live in the vain illusion of endlessness. However brief the life, unless our entire existence is a search for the ineffable, the very last chapter may be why the first chapters were written.

CONCLUSION

The purpose of professional intercession in untimely deaths is to permit bereavement to run its natural course. To do so, we must understand that not all untimely deaths are alike, but that each of the three kinds—*premature, unexpected,* and *calamitous*—may require a different method of management.

How we interpose our influence depends upon appreciating that while bereavement has predictable phases, coping with untimely death is partially determined by cultural expectations, social values, and community practices. We can best deal with untimeliness, however, by recognizing the reality of timely deaths, as well. Then we match our interventions with the stages of bereavement. Our first aim is *tolerability*, to reduce the shock and to protect the wound. The second aim is *accessibility*, which means that significant others are present and available, both for solidarity and for practical support. The third aim is *acceptability* of death. The course of coping entails gradual relief of distress, resolution of problems, and restoration of reality testing. The fourth aim is *expectability*. Because death may come at any time to anyone, our basic orientation is that living and dying are concomitant phases of the same process. Consequently, death is always an allotment necessitated by being alive, and not wholly accidental. While our conventional belief is that we must survive as long as possible, we must also tolerate the unsummoned brevity of survival itself. Therefore instead of asking for indeterminate survival, we can ponder the question of how short life can be and still be conducted as if there were an appropriate time to die.

REFERENCES

1. ABRAMS, RUTH. "The Responsibility of Social Work in Terminal Cancer," in B. Schoenberg et al. (Eds.), *Psychosocial Aspects of Terminal Care*; Columbia Univ. Press, 1972.
2. BEACHY, WILLIAM. "Assisting the Family in Time of Grief," in D. W. Montgomery (Ed.), *Healing and Wholeness*; Richmond, Va.; John Knox Press, 1971.
3. BOSS, MEDARD. "Anxiety, Guilt and Psychotherapeutic Liberation," *Review Existential Psychology and Psychiatry* (1962) 2:173–195.
4. CAIN, ALBERT C., AND FAST, IRENE. "The Legacy of Suicide," *Psychiatry* (1966) 29:406–411.
5. CHORON, JACQUES. *Modern Man and Mortality*; Macmillan, 1964.
6. FARBEROW, NORMAN L., AND SHNEIDMAN, EDWIN S. *The Cry for Help*; McGraw-Hill, 1961.
7. FEINSTEIN, ALVAN. "A New Staging System for Cancer and Reappraisal of 'Early' Treatment and 'Cure' by Radical Surgery," *New England J. Medicine* (1968) 279:717–753.
8. FREUD, SIGMUND. "Thoughts for the Times on War and Death (1915) II. Our Attitude Towards Death," *Complete Psychological Works*, Vol. 14; Hogarth, 1957.
9. FRIEDMAN, STANFORD B., et al. "Behavioral Observations of Parents Anticipating the Death of a Child," *Pediatrics* (1963) 32: 610–625.
10. GLASER, BARNEY, AND STRAUSS, ANSELM J. *Time for Dying*; Aldine, 1968.
11. IRION, PAUL. *The Funeral: Vestige or Value?*; Abingdon Press, 1966.
12. KAUFMAN, WALTER. "Existentialism and Death," in H. Feifel (Ed.), *The Meaning of Death*; McGraw-Hill, 1959.
13. MANDELBAUM, DAVID. "Social Uses of Funeral Rites," in H. Feifel (Ed.), *The Meaning of Death*; McGraw-Hill, 1959.

14. MINKOWSKI, EUGENE. "Findings in a Case of Schizophrenic Depression," in R. May et al. (Eds.), *Existence; Basic Books,* 1958.
15. PARKES, COLIN MURRAY. *Bereavement: Studies of Grief in Adult Life;* Internat. Univ. Press, 1972.
16. PARSONS, TALCOTT, AND LIDZ, VICTOR. "Death in American Society," in E. Shneidman (Ed.), *Essays in Self-Destruction;* Science House, 1967.
17. REES, W. DEWI. "Bereavement and Illness," in B. Schoenberg et al. (Eds.), *Psychosocial Aspects of Terminal Care;* Columbia Univ. Press, 1972.
18. RHEINGOLD, JOSEPH C. *The Mother, Anxiety, and Death: The Catastrophic Death Complex;* Little, Brown, 1967.
19. SARTRE, JEAN PAUL. *Existential Psychoanalysis,* trans. H. Barnes; Philosophical Library, 1953.
20. SCHMALE, ARTHUR. "Coping Reactions of the Cancer Patient and his Family, Workshop V. Grief and Bereavement, in *Catastrophic Illness in the Seventies;* New York, Cancer Care, Inc., Natl. Cancer Foundation, 1971.
21. SCHULTZ, MILFORD. "Clinical Staging of Cancer and End Result Reporting," in *Cancer—A Manual for Practitioners* (4th Ed.); Boston, Amer. Cancer Society, 1968.
22. SCHUR, MAX. *Freud: Living and Dying;* Internat. Univ. Press, 1972.
23. WEISMAN, AVERY D. "Reality Sense and Reality Testing," *Behavioral Science* (1958) 3: 228–261.
24. WEISMAN, AVERY D. *On Dying and Denying: A Psychiatric Study of Terminality;* Behavioral Pubs., 1972.
25. WEISMAN, AVERY D. "On the Value of Denying Death," *Pastoral Psychology* (1972) 23:24–32.
26. WEISMAN, AVERY D., AND HACKETT, THOMAS. "Predilection to Death," *Psychosomatic Medicine* (1961) 23:232–255.
27. WEISMAN, AVERY D., AND HACKETT, THOMAS. "Denial as a Social Act," in S. Levin and R. Kahana (Eds.), *Psychodynamic Studies on Aging: Creativity, Reminiscing and Dying;* Internat. Univ. Press, 1967.
28. WORDEN, J. WILLIAM, JOHNSTON, LEE C., AND HARRISON, ROBERT. "The Survival Quotient as a Method for Prediction of Cancer Death," unpublished.

Article 22 Widower, Heal Thyself

Early into the night that was September 8, 1971, my strong and spirited wife died. She was 40 years old, and in the 16 years of our marriage she had been sick only once.

Without a clear warning, without even a foreboding, I was suddenly alone. Rather, I was *alone* with our three children.

In this terrible moment, I brought them together in our family room. I heard myself say, "Mom didn't make it. She died."

The youngest, a girl, was 8. She had only a small comprehension of this dread announcement. The boys were 12 and 15 and I shall never forget the single, piercing cry of anguish from the older son as he took this news into his mind and heart.

In the two and a half years since that night, we four have lived through our grief: months of deep, numbing, painful hurt. We have drawn closer as a family and found strength in one another. I shall share some of that time and its challenges to us, but what I write is not meant to be maudlin. Neither do I want any reader to be uncomfortable with these narrative moments. My purpose is to speak about *The Widower* . . . his children . . . and his problems.

At a juncture when he is asking if life is indeed worth living, the widower must give his attention to an unnerving succession of all new problems; and when he turns for support and counseling, he discovers there are few who can, and will, start him along the way back.

There are 1,834,000 widowers in the United States. If they all lived together, their colony would be the fifth largest American city. Yet, there is not one professional agency which regularly, intelligently reaches out to these confused, lonely men.

No one is standing ready to teach the widower that he is a convalescent and

---

Source: Today's Health, May 1974, pp. 48–52, 68–71. Reprinted by permission of the author.

should get by one day at a time. Nor is he told that he must begin by helping himself. Only then can he reasonably expect volunteer help from others.

The widower with young children has extra lessons to learn. A pathetic, middle-aged widower said to me, "If only I didn't have the children. I think I could make it . . ."

He was far off the mark. With encouragement, the widower's children will leap forward in maturity and responsibility. From their example, the rational widower can derive pride and strength.

On the cruel September night 32 months ago, I had no understanding of these things. I was still very much the prisoner of shock—not only from the trauma of death, but additionally from the pain and anger which accompany bafflement.

You see, I had no explanation as to why my wife was dead. I could not tell my children why their mother wasn't coming home, as we had planned. I had only the empty words of a contrite surgeon who, in asking for the right of autopsy, said: "This should never have happened."

Hours later, when the prayers had been offered and the house was quiet, I let myself out to walk the streets. In the black stillness of this mild autumn night, I walked and ran and searched for reason to fill my mad void.

Thus, I began my long journey: one which ultimately would teach me that the nobility of my children was a testament to the woman I had lost too soon.

*   *   *

We had met at Middlebury College, on a New England campus which, for most of the year, is spectacular for its natural beauty. We discovered we were both sensitive, eager for new friends, and that we came out of middle-class Eastern families where the children were held to be special.

It was 1949 and my senior year. She was a sophomore dating the football captain and, in this carefree time, our fall weekends held a particular magic. We did a most natural thing then—we fell in love. Six years later, following military service in Japan and some early years in wire-service journalism, I married Adele Kathleen Mullen. The quiet ceremony was performed in her home on a sunny Saturday in the spring of 1955.

As our partnership began, Del was an executive secretary in an advertising agency. She worked days. I was a wire editor for the Associated Press and worked three different night shifts. We were like ferry boats, passing at the slip. We had an hour together in the evening and almost two hours every morning. But she never complained, and we endured this giddy life for 10 months. Our home then was a three-room apartment in Queens. We lived next in suburban New Jersey, first in an inherited home and later a split-level which Del chose for its quiet good taste. Our fourth address was Oak Park, Illinois: an old, buff-colored brick house with broad rooms and high ceilings. She left her mark on all our homes because she was, proudly, a homemaker and a parent. She also was a subtle shaper of character. Her qualities were: selflessness, an abiding concern for people, and a sense of responsibility that was, in a word, uncommon.

Moreover, she had conditioned those who knew her to believe she wasn't subject to incapacity. And in the fall of 1971 when she signed herself into a local hospital for "exploratory surgery," a search for the cause of what was euphemistically called stomach distress, she masked her private fears while admonishing the family not to worry.

The morning of her operation, the patient did two things which speak of her character. First, she refused to remove her wedding rings, throwing the rules-enforcers into confusion. Despite petty threats, the rings stayed in place.

Additionally, there was a reassuring note for her very uptight husband:

"Relax and stay cool," it began. "Everyone is more than considerate. My thoughts, as always, are all of you. Rest in the security of my two full hands—the one grasping yours, the other holding onto the Lord's.

"Who could be more blessed?" she asked. "I've got it made."

The surgery required three hours. All the while, I sat in her room, alone, nervous, and feeling terribly inadequate. Toward the end of the wait, I gave in to anxiety and placed a long-distance call to her twin sister in Swarthmore, Pennsylvania.

I had only begun to unburden myself when a nurse interrupted to say that the operation was over, and successful. Minutes later I learned from my wife's physician that the gallbladder was judged healthy and left intact. However, the surgeon had discovered a blockage in the pyloric canal, or opening, which leads into the intestines. With the aid of a small, crude pencil-drawing, the woman-physician explained how the pylorus had been cut, widened, and then stitched. She said the next 72 hours were certain to be hard ones for us, but there no longer was any need to worry.

I did little else but worry as my wife now fought against giant, wracking waves of abdominal pain and showed only small improvement.

On the sixth day after her operation, I believed I saw color in her face. But on the following day, the eighth of September, I was told on arrival at the hospital, "She doesn't look so good tonight. Her color is bad, and she's got some gas."

Throughout this interminable week, I had reassured myself that my wife was strong. Yet, when I saw her now, with her clear skin turned sallow and the unmistakable mark of sickness in her eyes, I felt a terror deep inside me. I had cause to be terrified for, unknown to me then, the so-called "gas" distending her stomach was, in fact, peritonitis.

Outside in the corridor, where I had been asked to wait for a moment, I was approached by a small, brusque, and embarrassingly pregnant nurse.

"You'd better stand by," she said.

"Stand by for what?" I demanded.

"Things don't look so good," she said, lowering her head.

I swung around then and punched hard against the wall across from my wife's room. I wouldn't realize it for another week, but I had just fractured the fifth knuckle on my right hand. The pain was immediate, sharp, and constant. In my pain, however, I somehow felt closer to my wife.

As I struggled to focus then, I saw Del's physician coming toward me and, for the first time, she admitted that her patient was seriously ill. I heard her say, "There is a 50–50 chance . . ."

These virtual strangers, the nurse and the physician, guided me toward a desk where I sat and leaned forward to put my head down. In the dark space between my folded arms, I prayed for a miracle to happen inside that beleaguered hospital room just 25 feet away and housing the very core of my life.

Strangely, I felt I was losing my contact with reality and reason. My senses, however, remained sharp. I listened to the calls over the intercom and recognized the urgent injunction for a breathing machine—for my wife's room. I heard people walking in the corridor, and I vividly recall an elderly man who padded back and forth in soft slippers. During my blackest, most terror-filled moments, I heard the whisper of his feet. I remember thinking that he had come through his trial. He was practicing to walk away from this hell.

"He's going home," I thought. "Why not Del? . . ."

It was 8:40 p.m. when death claimed my wife, and while it mercifully ended her pain, it unloosed the terror inside me. Now, I had no choice except to deal with this emotion—and I wasn't sure how to begin.

I am not pious, nor even religious in the conventional sense, but I soon found myself on my knees in prayer. I was put there by a kindly shove from my friend and minister, Dean Lueking, who understands that when death reaches deeply into a man's life it brings with it a momentary touch of madness. From that hour forward, this genuinely good man took my arm. His calm, deliberate, and intelligent manner, as he tended to my needs and those of my children, was my example. I felt compelled to behave in a comparable way. Thus, I was able to think, to sort out priorities and act upon them.

I have talked since with other widowers. Some were frozen in shock. One confessed that for weeks he was able to do only small, mindless chores. He chose to dust and vacuum, again and again. Another was unable to answer his seven children when they asked, "Where's mommy?"

Surely, all children react differently to this monumental trauma. I suspect the shock can be softened by two things: a faith, and the quality of the love bond between mother and child. The stronger that bond the greater the loss but, importantly, the stronger the bond the greater the realization in later months and years that the memory of the mother is good and strong and, therefore, sustaining.

When I sorely needed it, my teenage son sent me a sign that it was going to be all right with us. This was the eve of the funeral service and we were sitting alone, in the hushed parlor. I was prepared to speak quietly to him, hoping to bring a small reassurance. Instead, he took the forum.

"You know, dad," he began in a steady voice and with his clear blue eyes as dry as bone, "if we think long enough and hard enough about this, I suppose someday it will make sense."

* * *

The nine-room house was empty. The grandmothers and aunts all had gone and the children were back in school. I had all day, and every day, to myself. In the army, far from home in strange, spartan barracks, surrounded by alien faces and hostile noises and smells, I had known loneliness. Yet nothing in my life's experience had conditioned me for these first days and weeks truly alone.

I wandered about, looking into drawers, opening closets, cabinets and, naturally, there were reminders of her.

I remember, too, great waves of fatigue. I remember my confusion and that I ceased to care. My mind was like a pair of binoculars whose threads had become worn. No matter which way, or to what degree, you turned the range finder, you could not bring a clear picture into my eyepiece and thus, I had almost no attention span.

"Work is therapy," a friend advised. Another said, in a way calculated for me to overhear: "He's got to get out of that house."

But a friend who truly understood wrote that he knew where I was at. He had come through a bitter, snarling divorce action and empathized: "A house without a woman, is a home that is boarded up . . ."

I discovered some small peace in running. One of my early passions, dating back to adolescence and a campaign against corpulence, was athletics. Now I found a natural release—an escape, actually—by running a mile and a quarter through neighborhood streets and around a park near our home.

Along the far edge of the park is a flat, open piece of ground, perhaps 100 yards long. I would push myself to sprint this course, believing that if I was able to run fast enough down the straightaway that I might call her back for a time. Forcing my body to run faster and faster—moving ever closer to exhaustion—I bargained with the imaginary timer.

As bizarre as this may sound, it is a normal emotion, part of the accepted practice of saying "good-bye." I learned this in reading of the three stages of grief. The first stage is shock, the time of numb disbelief when the widower invariably says: "Oh, no. It can't be . . ." The middle stage is the period when memories and images are called back, often at the slightest provocation. Psychiatrist Robert B. White, M.D., of the University of Texas, has written of this period: "Many spouses have illusions of seeing or feeling the presence of the dead partner. The widower may hear a door slam, or the floor creak and, for a moment, believe his wife is there. These things are all part of the slow process of saying good-bye, because the ultimate, total separation is too overwhelming to comprehend or accept all at once."

In the final stage, the widower will experience periodic sadness but is now able to remember and talk about his wife, with naturalness. He is in control of his emotions. Eventually, he will speak of her with pleasure, freely using her name.

In midafternoon of my first routine day as a widower, I heard the back door open. This was my alarm: I must now pick up the mantle, for a subdued, blue-eyed Jan Lindeman was coming home from the third grade, beginning her new life as a latchkey child. We spent the remainder of the sleepy September afternoon at the public library, where Jan sat still as a log on the floor in the children's room, browsing inside an estimated 67 books, and rejecting every one.

With the experience of that afternoon, I declared to myself, and perhaps to little Jan, that we were going to move on ... to get along ... to make it. That night I hung a new watercolor in the hall downstairs, and in my bedroom I tied one of Jan's classroom creations to the overhead light fixture. It was a misshapen fish, made of a wire coat hanger and a discarded silk stocking. It hung down, I believe, as a symbol of life.

I also cluttered up the kitchen bulletin board with school notices and schedules, making the space a future book. I wanted to say, often and with enthusiasm, that to the limit of my ability I was going to try to be two parents or, more realistically, a constant parent.

Of course, we were innocents. I knew this those first evenings when reality was dinnertime or, more accurately, *supper*time. Someone had to get a meal. That someone was me, a kitchen novice who, literally, had difficulty operating an electric can opener on his first several assaults. Someone also had to see about the laundry, do the shopping, balance the checkbook, make three sets of school lunches, drive the two younger children to school, pay the insurance, the real estate and income taxes, the mortgage, take the cat to the vet's, have the lawn-mower fixed, argue with the roofer, attend the PTA meetings, sign the report cards, see about winter coats and boots, dental and doctor checkups, urge letters to grandmothers, and bedevil three children who weren't so confused or grief-stricken that they didn't know their own minds when it came to making beds, hanging up pajamas and clothes, and eating green vegetables. Given the opportunity, they would vote negative every time.

I clearly recall the satisfaction and shameless, inordinate pride that came with the accomplishment of small household tasks. Once I had managed to get the supper served and eaten, the dishes done, the garbage carried out, the homework secured, and the lunches made, I felt in a rank with Caesar Augustus. I had conquered a new country and was looking about for my next challenge. However, this euphoria came only after we had put some time between us and the night of our loss. We began slowly and erratically living through a period of trial and error.

On one of the frequent mornings that fall when we awoke to find hard rain washing across the streets and walks, I was openly angry to further discover that neither Paul nor Janet had chosen to wear rubbers or boots. I noisily plunged into a small storage room off our garage and returned with two sets of almost new galoshes.

"Here, these belong to someone," I said, dropping them before two anguished-looking children.

Paul did nothing and he said nothing. Janet said: "I can't buckle them."

"I'll buckle them."

This response forced the truth from her: "But the kids will all laugh at me wearing galoshes."

Paul quickly joined her defense, telling me that "kids don't wear galoshes any more."

I instinctively knew that I was losing the argument. I was also losing my cool. I began shouting that they didn't need me, or any father. They knew it all. I told them then that I was going to quit, to resign, that they should find someone else to kick around. I screamed that when people walked in the rain with "nothing on their feet" they caught colds.

"Do you know what happens when you get sick? You go to the hospital," I shouted, "and when some people go to the hospital, they die ..."

There it was. No matter how hard I wished I hadn't said it, it was too late. I couldn't get it back.

That night, I apologized and was forgiven. When I asked Paul, he explained that rubbers and galoshes were *out*. Shoe-boots were *in*. At dinner, I kept the talk light and pleasant. Everyone pretended that nothing had happened.

I became obsessed with the idea that our evening meal be an event, a time when we could pour out our concerns and grow strong again through mutual love. I insisted on candles and I poured wine for all who wanted it, even little Jan. Like a prosecutor with his first jury trial, I probed for some happy event out of the school day. But in those first weeks, we were all leaden and emotionally drained.

Still, I plunged ahead, hoping to will some joy into our midst or, if need be, impose it by paternal decree. I was wrong, of course. It was too soon. We all needed time for the wounds to close. Children are the most natural creatures, and my three were merely exhibiting their true feelings. As I listened to the silence of our dinner hour, with its notable absence of laughter, I vowed that one day in the future I would fill my home with good music. I knew even then that I would —indeed, I *must*—marry again, and as soon as I believed them strong enough for this news, I delivered it to the boys. I believe that untruth and hypocrisy are two sins of the first magnitude, and a part of our regeneration as a family came through an open sharing of all subjects.

To make things happen, to put more time behind us, I knew we must build new experiences. So, after a sabbatical, Paul returned to his guitar and Jan began taking piano lessons again. Moreover, Jan and I shared time together every Monday night. While she took an hour's swimming lesson, I got my exercise swimming laps. I also took some of the strain out of our life through the hiring of a housekeeper.

In our home today, mention of the word housekeeper is a source of high good humor, yet there was nothing particularly funny about the situation as we lived through it. In summary, there was a rather melancholy parade. The first woman was, in her own words, "too nervous." I shall never forget the night she interrupted my viewing of the six o'clock news to announce that she had "lost" the steak in the broiler. Pathetically, it was true. She had accidently forked the meat off the broiler, lodging it in a heretofore unknown crevice at the rear of the stove. It took a great negative dexterity to accomplish this. To retrieve the meat required consummate manual skill. With far less aplomb than was demanded to survive this night, and bowing to the children's urging, I immediately hired housekeeper number two. She proved a dour woman who attempted to sue me for an alleged fall in the basement, an accident which no one saw or heard about until a registered letter arrived from her lawyer.

The third woman persisted in telling me of her hard life as a widow and how much orange juice my boys consumed in a 24-hour period. The fourth candidate, a Slavic émigré, was worth the waiting. Cheerful, indefatigable Maria Perucki has made the home run again and, through time and love, has become an integral stripe in our family fabric. Thus, days slipped into weeks and, little by little, the scar formed. I went back to work, we began joining other families for supper in their homes, and the children again watched television, even medical and hospital programs. All of us progressed into the latter half of the second of grief's stages.

We did not—for we could not—mention Del by name, nor were we able to talk about what had happened to us. However, I was able to scrawl silly sayings on each of the lunch sacks: "I'm a Jan Fan," "Tall Paul has it all," "You can do more with Les . . ." and I was able, once more, to laugh at myself.

I remember in particular a Friday evening in the late fall. I had sent Les out back to mow the lawn. Now I sat alone, working on a bourbon highball and waiting for the six o'clock news on television. Suddenly, I felt a terrible loneliness. I called Les in to "keep me company."

"Let's talk," I said. He nodded yes.

"What's new?" I began.

"Not much."

I switched on the television then, hoping for another discussion topic. When I next looked down the couch, my six-foot, 180-pound co-captain had slumped over and was fast asleep.

I had to laugh over my clumsy attempt to press my taciturn teenager into being my partner. He already was my lieutenant, invariably my rudder when a temper storm swept over me and, in my absence, a surrogate father. Yet, he was still a boy, and I must guard against rushing him through his already ruptured innocence.

*   *   *

Of the three children, the youngest was least able to cope. Often, Jan was an angry eight-year-old woman who neither understood nor could adequately ventilate her emotional upset. She was easily offended by any of the three of us males, and once wronged, in her eyes, she would harangue us for not caring or being sufficiently loving toward her. A too-frequent scenario would consist of: (1) An insulted Janet storming from the dinner table, slamming her bedroom door. (2) Brother Paul dispatched as conciliator. (3) Return of Janet, arm around her brother, no longer the malefactor.

There were, as well, some transient problems at school. I learned that once or twice she lay down in a classroom aisle, disrupting the story hour. I was more amused than concerned, for while this was a negative sign, there were also positive ones and they predominated. To begin, Janet got herself up, washed, and dressed every morning without fuss or complaint. A half-dozen times I helped comb the snarls from her hair, or told her to wear the purple pantsuit, but beyond that she was on her own. When you consider the cold, dark, and stark discomfort of a winter morning in Illinois, Janet has to tally extra points. Moreover, in my mind's eye I would see other eight-year-olds eating a breakfast of oatmeal, eggs, bacon, and hot cocoa. Then I would find Jan alone in the kitchen nook, quietly spooning her glass of cold instant breakfast. That was her reality.

Another of my concerns for Jan was her tendency to suffer stomachaches and low-grade fever when a business trip kept me out of the home overnight. I decided to obviate this by taking her along, and so we traveled together to Coral Gables, Florida; Hazleton, Pennsylvania; and my hometown, New York. With each passing month, Janet and I were growing closer, and more than ever—more than I had ever imagined—I liked being her companion and father. I think she understood.

*   *   *

On the second day following his mother's death, I took Paul for a walk. We kicked through fallen leaves on neighborhood sidewalks, taking our time, going nowhere in particular. I meant only to see if this lively 12-year-old, who up to this hour had given no quarter to his grief, had any questions. Seemingly there were none. So I began telling him that it was his special gift to be incandescent.

"You're the spirit of this family," I told him. "Don't stop laughing, and don't stop making us laugh with you. Now, more than ever, we need your laughter, your spirit . . ."

I don't remember that he responded. I know he listened hard and that he hasn't let us down. In time, he again filled the house with his sound and motion which ranges from the rock music of his guitar to the regular rhythm of the basketball bouncing against the floor or the wall or the ceiling in his room. Because he is lithe and swift, fast with his hands, feet, mind, and tongue, he was nicknamed *Pistol*.

One night, when he proved just too much clatter and flip-talk for me, I exploded. It was a routine tantrum with my standard inquiry: "What do I have to do to get some peace around here?"

"But Dad," he said, staring me straight in the eye, "you told me to be the spirit of the family."

Because he is the most natural of the children, I sensed my peripatetic son was going to cope with his loss. What few doubts remained were erased that June as Paul, and two equally combative teammates, led their team of 11- and 12-year-olds to the local Little League championship. It was a beautiful summer evening which we celebrated with a cookout. While Jan and Les set out the potato chips, ketchup, and mustard, the winning pitcher tended to the fire in the grill.

Once the fire was glowing, Paul stretched out on his back, staring up at the sky which held the last light of this exhilarating day. He was pleased and tired and uncommonly quiet. I guessed that he was replaying the games in his mind.

In the best of three series, Paul's team had won two straight. But in the opener, Paul had failed to finish his pitching assignment. He badly wanted to do well in the second game, and as the final inning began, his team led 4-0. However, the "enemy" quickly scored a run and loaded the bases. Two were out. Paul then struck the next batter out with a perfect low-and-away pitch, for a swinging strike.

I wanted now to say that success after adversity is twice as choice but found myself falling into the shorthand of a sports cliche. "You were knocked down, got back up and did your job," I said. "I'm proud of you . . ."

Following his gaze across a stretch of cloudless evening sky, I added: "Your mother would have been too."

He never turned his head. He just slid his eyes toward me, and although he said nothing, I heard him tell me that his thoughts matched mine.

*   *   *

Because Leslie saw no good sense in sitting home, and because he was caught up in a challenge for the quarterback's post on his sophomore team, the oldest child missed only two days of school back in September 1971. I'm sure that in his football he found a natural release. I think we both worked out some problems through football: he as the quarterback, I as the proud, rabid spectator.

Like a character out of a lesser Fitzgerald novel, I sat in the crude wooden high school stands each Saturday and saw a much larger drama than the spectacle of awkward sophomores wrestling each other to the soft earth. I watched my son lead his mixed band of gladiators, most with baby faces, and I thought back to a Sunday morning nine years before. Croup had suddenly threatened his lungs and I was holding him by the heels over a bathtub. Hot water and steam were coming from every faucet in this tiny, locked room and I was working to dislodge the mucus.

I looked down now at the thick, strong calves and the powerful thighs held firm in the football stretch-pants and thought how quickly they go from sickness to laughter; from infancy to adolescence; from helplessness to independence. And wherever this action flowed, I sped my eye to pick out the familiar number 12, wearing the bright green and white.

Whenever I allowed it, my mind kept coming back to the same unspoken thought: *"If only his mother could see him . . ."*

Since that fall of 1971, Les has had good and bad days—one idle year as a reserve quarterback and a strong senior season. He is ready for college and, perhaps, more football. There are other dreams and desires to complement this one, though. Writing is among them. One night, recently, he handed me his own written version of how far we have come as a family. It begins with the night of September 8, 1971:

"I recall Gram walking out of the family room where we all were and I knew then it was gonna be *our* problem. The four of us would have to hack it on our own. There was no substitute for mother; there couldn't be.

"Later on, a couple weeks later I guess, I felt a small excitement at living on our own. I took pride in the four of us, that we could make it without something drastic happening, like us moving. We drew closer.

"Mornings would bring instant breakfast instead of bacon and eggs, but morning would come nonetheless. It was no picnic—don't get me wrong. But at least we could keep on truckin', drawing strength from one another and keeping on the move. That was important to me, and that is what it is all about."

*   *   *

EPILOGUE

This cannot of course serve as a widowers' primer. Rather, it is the way one widower has traveled and I mean to say to other widowers that yes, you will sleep through the night again; your food will taste good again; you will laugh, enjoy music, and watch your children with joy. In time, you will throw off your lassitude and rationalize your anger.

The holidays will be a hard time. Describing ours only makes the obvious point. We spent the first Thanksgiving away from home and decided a traditional Christmas was necessary because of Janet. On the morning of December 25, I came downstairs to find the tree had tipped over, spilling water and staining the rug. The remainder of the day was uneventful. When my mother, a widow for 16 years, left for her home back east, she summed it up: "Thank God, it's over."

A subject which I prefer only to brush is dating, or more candidly, sex. As I told the children, I was going to see and socialize with women—and I did. Having to "date" in the early 40s, or older, and after long years of marriage is, for many, an unnerving experience. I suggest there is something even worse—not getting out and having companionship.

How the widower chooses to begin his new social and sex life depends, of course, on his personality, his philosophy, and the sort of companionship he wants. Some widowers prefer to join Parents Without Partners, where they are sure to meet people with common interests. Like the majority of widowers, I had friends and relatives who put forward suggestions and invited me to dinner parties. I found most of these either tedious or painful. One night the table conversation turned to movies, and my host innocently asked, "Did you see *The Hospital?* It's a great film . . ."

In this way, however, I did meet a beautiful woman. She was a widow, with one child: a teenaged daughter. On June 9, 1973, I married Harriet Meyer, who is a soprano and a teacher of voice.

We are a complete household again and throughout the home there once more is the sound of laughter, of music and, best of all, the rich sound of meaningful conversation between children and their parents.

Article 23 The Widow as a Caregiver in a Program of Preventive Intervention with Other Widows

PHYLLIS ROLFE SILVERMAN

One of the primary problems programs of prevention face is how to seek out people who have not asked for help. There is the question of who is the appropriate caregiver for such a population?[3] The Laboratory of Community Psychiatry confronted this dilemma when it tried to develop a program that would prevent emotional illness in a population of bereaved people. In this instance the target population with the high risk of developing serious emotional distress consisted of younger widowed people. A caregiving group had to be defined that would be acceptable to them. This paper describes the caregiving group chosen, their special qualifications as interveners, and discusses the kind of intervention they provide.

The caregiving group consists of other widows who have recovered from their bereavement. It was hypothesized that if another widow reached out to the new widow she would be accepted as a friend because she was someone who understood since she had been there herself.[2] Does experience bear out this notion that the recovered widow is an appropriate and accepted helper? The material that follows presents data which provide some answers to this question.

THE WIDOW TO WIDOW PROGRAM

The Widow to Widow Program, as this demonstration in preventive intervention is known, has been in operation for three years. Five widows have reached out to over 400 new widows under the age of 60 in this time.[4]

The widow caregivers, called aides henceforth to distinguish them from the widow recipients, all live in or near the community they serve. They have all

PHYLLIS ROLFE SILVERMAN, PH.D., is Lecturer in Social Welfare, Department of Psychiatry, Harvard Medical School, Boston, Mass.

Source: Mental Hygiene, Vol. 54, No. 4 (October 1970), 540–47. Published by Communications Dept. of the National Association for Mental Health, Inc. Reprinted by permission.

287

been widowed about three years, are for the most part in their mid-forties, and have no more than a high school education. While their husbands were alive, they devoted themselves to raising their children and keeping house. Two of them did help their husbands in his business. After his death they all had to think of supplementing their income, which came largely from social security and pensions.

Each of the aides had become involved in community activities subsequent to their becoming widows. It was through our contact with these community organizations that we were able to recruit them. Until now they had never thought of earning their living helping people in this way. All of them could talk about their bereavement, the very difficult time they had and the current problems being widowed still created for them. They had only one reservation about the program as it was described to them. They wanted to be sure that service, not research, was the main purpose of what they would be doing.

In order to assure that their experience as widows would be utilized to the maximum no attempt has been made to supervise their work. At weekly group meetings people they visit are discussed, and they use each other as well as myself for consultation about what they have done and about how they might proceed. Most frequently they chart an independent course of action which seems right for them and the new widow they have visited.

They quickly corrected our notion that within one year, or less, a widow has recovered from her bereavement. They feel that although by then she *may* be over the acute stage of her grief, she is not recovered, and may even be depressed by her growing awareness of what the loss means. They say a widow never recovers but rather learns to adjust to the situation. They thought this took about two years to accomplish. It involves an ability to repattern her life without a husband, to find new friends, new interests, and sometimes a new career. It also means learning to live with loneliness.

INITIAL REACHING OUT AND SERVICE OFFERED

How is their experience translated into their work? Their first task is to establish their credentials as an appropriate caregiver; that is, as someone the widow will accept and see as potentially helpful. The new widow first learns that the aide, too, is a widow in the letter of introduction she receives. The stationery has the names of the three religious groups sponsoring the program on it.* In the letter the aide tells the new widow that she will visit at a given time on a given day. She gives her home phone number and invites the widow to call if for any reason she does not want the aide to visit.

The aides do not feel that a visit before three weeks would be useful to a new widow. At the moment of acute bereavement they do not feel that a new

---

* The program is sponsored by the Archdiocesan Council of Catholic Women, the Mt. Bowdoin YM and YWCA and the Temple Beth Hillel. These are community based agencies traditionally involved at the time of a death in the family.

widow can identify with another widow, because she still thinks of herself as a married woman.

Several things influence the widows' response to this letter. One is their willingness and readiness to consider that they are now widowed. Some women thought this was a "terrible thing to call me" and threw the letter away.* Others thought they were "already on a mailing list, now what do these people want?" Still others were impressed with the fact that someone cared, and were re-assured by the names of the religious organizations on the letterhead. Some of their reactions were colored by who else was available to talk to.† Some women, therefore, called and told the aide not to come; others chose to let her visit be-cause they lacked the energy to call and refuse the visit. Some simply weren't home when she arrived, but most looked forward to the visit. Many widows subsequently became involved with the program although their initial response to the aides' offer to visit was negative.

Once they sensed there was no ulterior motive in the aides' interest, the fact of the aides' widowhood was the important thing that made it possible for them to become involved. The aide mentions it in her letter but it always comes up early in the actual encounter, either in a face to face visit or on the telephone.

The fact of common widowhood is often discussed through the aides' at-tempt to clarify how the widow is managing financially. They talk about social security, VA pensions, and the like, and the aide will describe her own experi-ence and clarify for the widow what benefits she is entitled to and how to be sure of getting them. One widow saw this discussion about money as an "affirmation of life. It makes you think about what is needed to go on living and reminded me that that's what I have to do."

The aide's willingness to answer questions about her own widowhood seems to give the widow permission to unburden herself.

SHARING COMMON PROBLEMS

The fact of the aide's widowhood seems to make it easier for the new widow to accept her, to talk to her, to ask for advice on problems related to her own widowhood, and to feel as if she can still be part of the mainstream of life—that is, she is not so alone and the only one to whom this could happen. Another widow said: "Since you are a widow too, when you said you understand I knew you meant it and that was so important. I can't stand sympathy and that's all anyone else could give me."

Pride and an unrealistic wish to be independent seem to get in the way of a relationship between widow and non-widows. The new widow finds the lat-

---

* This may be why organizations of widowers and widows chose names such as NAIM Conference of Chicago and THEO (They Help Others) in Pittsburgh.

† Many felt that their family and friends sufficed for their current need only to come to an awareness later on that they did not really understand, and were inappropriately impatient with them to recover more quickly than was possible.

ter's efforts to be helpful clumsy. Often they find themselves providing reassurrance rather than being reassured. This does not happen with the aide. The aide is using her own experience as a human being and as a widow to guide her in her encounter with the new widow; she appreciates the real need that exists but never takes the widow's initiative away from her. If the widow becomes dependent on the aide, it does not seem to bother either of them at this point in the encounter. This most often will take the form of frequent phone calls, or the aide will drive the widow to the social security office and the like.

The widow explores the common problems of widowhood with the aide.

One woman was worried about her child, who took out his father's picture and talked to it. And another was upset because her daughter wasn't doing well in school anymore. The aide could talk about not knowing how to help a child, could honestly normalize the behavior in the knowledge that with time the child does make an adjustment, but also recognizing the child's need to mourn which the widow doesn't always see. In the words of one widow:

> I tried not to cry in front of my children. I wanted things to be as normal as possible for them. Then my little one stopped working at school. The teacher said he was depressed. The children felt it wasn't right that things should be the same if their father was dead. They thought I didn't care about him.

As a result of talking to the aide the widow started to show the children her true feelings and her boy's studies began to improve. The aides could talk about their own children, the problems they have now as well as when their husband died and how they saw their husband's death contributing to them. They reported what worked for them and what didn't work, and were receptive of the widow's suggestions for solutions as well. They established with the widow the fact that widowhood is lonely, frustrating, that there is often a bitterness which accompanies it; that you really don't get over it, but get used to it. By so doing they seem to take the fear and worry out of mourning and give the widow a context for her behavior which she can accept and understand. They see normal grief as extending over time, and counsel patience to the widow.

Often, the aides report that the widow tries to be strong and feels she must avoid being excessively dependent and is inadequate if she requires assistance. This feeling is fostered by people such as her doctor to whom she may complain. She is usually told she will get over it and be strong. The aide, on the other hand encourages the widow to return to get a physical check-up to verify that her symptoms are indeed just "nerves." If the doctor prescribes tranquillizers, the aide encourages the widow to use them and not feel that she is weak and defective for needing this "crutch" in order to get through this period until she learns to find her way in her new circumstances. Over and over again the aide is told: "You understand the void in my life." The aide sees this as meaning that it is not always necessary to talk about it because they indeed do know.

The aide is not trying to make the widow into what she is not nor does she want her to act as she, the aide, did, but rather to accept the fact that she is going through hell, to be more accepting of herself and her own needs at this time. This is something that the widow reports she does not learn from her immediate family or friends. If she does, it is because there are widows among them.

LEARNING TO LIVE AS A WIDOW

The aides talk of the needs of the widow to change from seeing herself as married to thinking of herself as widowed. This they see as the first step to recovery.

The aides identify three themes in this process. First is the need to learn to make decisions independently or unilaterally, that is, without the guidance and help of a husband:* "The biggest decision I ever made was what loaf of bread to buy."

The second theme is the need to learn to be alone: "What do I do after the children are asleep? I can't stand the empty silence, and I can't watch another T.V. program."

The third theme follows on these two in that there is a growing need to make new friends and be out with people: "I don't get invited out by our couple friends any more. I'm not always comfortable with them. It makes me feel even lonelier."

As far as decision making is concerned, the aide is primarily helpful in two ways. She is not afraid to give direct advice if it is needed. She seems more willing to do this early in the contact when the widow seems confused and needs direction about for example, money, children, selling the house and so forth. In the latter phase of accommodation she is more apt to offer encouragement, ideas and support; though she is often quite pointed in telling the new widow she needs to act for herself. For a woman who has seen herself, for most of her adult life, as a partner in a marriage, this is not always an easy transition to make. We begin to differentiate between those widows who respond to this encouragement, get a job, learn to drive and begin to repattern their lives, and those widows (a small minority) who have difficulty. This latter group seems to cling to the past and to be searching for a replacement for their deceased husband who will make their decisions for them. Often they try to put the aide in this role but at this point in time she seems to instinctively repel this effort while not rejecting the widow herself. The program is too young to know how the aides can help this latter group of women pass this stumbling block.[2, †]

---

* Mrs. Ruth Abrams, research social worker in the Conjugal Bereavement Study at the LOCP has observed that in the first months after her husband's demise the new widow leans heavily on his wishes and her memories of him, and tries to do things as he would have wanted. Recovery, she observed, begins when the widow can "give up her husband's ghost," which means she is able to learn to make decisions based on her current reality.

† In a survey of psychiatric clinic records I noted that most widows who appeared for treatment, came for the first time two years after the death of their husband. These patients may come from this group, and it may be that the effort now at the end of the first year of bereavement, may have the greatest payoff to prevent a serious emotional disturbance from developing.

To fill the loneliness of any empty evening is not easy. The aide in part helps by being available, if only on the telephone, to talk and to empathize with the problem which she, too, is experiencing. Some widows run away from this by never being home, others put all their energy into taking care of the children. The aide tries to help find a middle road. She acknowledges aloud for the widow that the consequence of running is that one day she will suddenly have no place to go and then she will be really depressed. The problem for the aide is to offer real alternatives for the widow. This brings us to the third theme which involves helping the widow expand her resources and repeople her life differently. Out of this the widow can create for herself alternatives with which to cope with her aloneness and her loneliness.

The aides have helped in two ways. They have helped the widows find other groups of "single" people where they might find common interests. There are several kinds of single groups. One such group is being formed by widows who have been served by the program. They met at several large meetings arranged by the aides to discuss the problems of widowhood. Several cookouts were also planned. These group meetings have attracted women who initially refused to see the aide or spoke to her only on the telephone. Once they came to a meeting they returned because they found:

> It helps to talk to others in the same situation. Sometimes when I get home and think about what other women have said I learn something new about myself. The only reason I came in the first place was because I was embarrassed to refuse (the aide's invitation) again. If I had realized how friendly and nice everyone is I wouldn't have been so reluctant.

Another group of women have sought out single people clubs where it is possible to meet men. These women have relied on the aide to inform them about such groups and have asked her to take them to a meeting to overcome their initial shyness.

The second way of helping has been to get these widows to reach out to other widows in their immediate neighborhood who have refused to see the aide, or who are so physically disabled that they cannot leave home. This provides the widow with an opportunity to do something for someone else as well as to make new friends. Some widows seem ready for this by the end of the first 18 months of their bereavement; and the aides are eager to share with them their role as caregiver.

As the widows move out into the role of caregiver it seems appropriate to consider ways of making them more responsible for the ultimate life of the widow to widow program. The next phase of new activity should involve the widows served in the workings of the program. As they were helped so they can help others and thus give the program continuity and a permanent status in the community. The program is still going on.* We are only beginning to understand

---

* During the first year the program was supported in part by the National and New England Funeral Directors Association and in part under Grant MH-03442 NIMH. This latter support will continue for another 2 years.

the unique role of the widow caregiver. At this point, however, it is possible to say the evidence seems to support our initial hypothesis that another widow is the appropriate intervener in a program of preventive intervention where the client group did not ask for the service.

DISCUSSION

The purpose of any program in preventive intervention is to prevent emotional breakdown in a vulnerable population. While, at this point, we cannot demonstrate that we are achieving this goal in the Widow to Widow Program, it becomes clear that the aides are indeed being very helpful to the widows they reach and that in good part their ability to do so is a consequence of their being widowed themselves.

To better understand the special quality of the aides' helpfulness, two basic problems facing a new widow need examination. The first problem is that of facing the fact of widowhood; that is, accepting their changed marital status and all this involves. The second problem is to learn to manage their own lives, and to demonstrate to themselves and others that they can be and are independent.

Many women see widowhood as a social stigma. They see themselves as marked women, different from everyone else, even carrying this so far as to see themselves as defective, that something must be wrong with them if they lost their husbands. In addition, all widows report they experience a growing social isolation as time passes after their husband's death. They no longer belong with their married friends, who they find gradually withdrawing from them. They can no longer conform to standards which society calls "normal" and they become people in a "special situation," that is, with a stigma. Goffman[1] describes this phenomenon but it is beyond the scope of this paper to explore all its ramifications for understanding the problems of widowhood. He, however, describes the function of the veteran of this role in helping the newly stigmatized person accept his lot. This is exactly the work of the widow aide. Goffman points to the need of the stigmatized individual to feel that:

> he is human and 'essentially' normal in spite of appearance and in spite of his own self-doubts. . . . The first set of sympathetic others is of course those who share his stigma. Knowing from their own experience what it is like to have this particular stigma some of them can provide the individual with instruction in the tricks of the trade and with a circle of lament to which he can withdraw for moral support and for the comfort of feeling at home, at ease, accepted as a person who really is like any other normal person.[1]

Goffman further notes that the veteran serves as an example of someone who can successfully live with his stigma. In addition he functions as a bridge person to the outside world helping them to normalize and be more accepting of people in this category. First, however, they must help the new member accept his own membership in the category. The aide understands instinctively the new widow's reluctance and resistance to accepting this status. The widow is not unique in this. Goffman notes the difficulty the alcoholic, the deaf, and so forth have in accepting their assignment to a special category. However when they do so, their hope for "normalizing" their life is increased and adjustment or recovery can be achieved. In accepting help from another member of the category they take the first step toward accepting their own membership. By the very nature of the problem then, the veteran, in this case another widow, is best equipped to help the new member. She is first a bridge to accepting the role of widow and then to helping the widow find a place for herself in the larger community. In addition, the veteran has a privileged communication with the new member which no outsider can have. The aide can say things about being a member of the category: about feelings (positive and negative), about problems it creates which if mentioned by a nonmember would be considered an intrusion or an impertinence. Intervention becomes the work of the members of the category and we begin to understand the success of such help groups as Alcoholics Anonymous and Parents Without Partners.

There is also a progression in the organization. The members move from initially being recipients of service to becoming providers of service. As a provider of service he develops a sense of independence and adequacy which brings him well on to the road of recovery, accommodation, or adjustment. This is the second need a widow has and as she in turn becomes a caregiver she develops a new sense of independence and worth. Insofar as the Widow to Widow Program can do this, it should be able to accomplish its goal of preventing emotional breakdown in a new widow.

The self help group has several important characteristics. Primary among them are: that the caregiver has the same disability as the carereceiver; that a recipient of service can change roles to become a caregiver; and all policy and program is decided by a membership whose chief qualification is that they at one time qualified and were recipients of the services of the organization. The prototype for self help groups has been Alcoholics Anonymous, run by alcoholics for alcoholics. This program has assiduously remained independent of the formal health and welfare system, using professionals only as occasional consultants, never to make policy or direct a program.

What I am advocating is the development of a self help organization. It may be that this kind of organization is best suited to do the work of preventive intervention. What problems arise for a self help group begun in a Laboratory of a Medical School? Are these problems different for such a program started in a Community Mental Health Center or clinics?

Many mental health agencies have attempted to replicate some aspects of the success of these self help groups by employing so called "non-professional indigenous workers." Unlike A.A. these non-professionals are usually given extensive training and supervision so that they begin to adopt professional values and emulate professional techniques. If they were following the self help model, they should be making policy, developing their own techniques for helping, and the consumer of their services should be able to move into their role of caregiver.

In the average agency setting this would be difficult to achieve since it would mean the professionally trained caregiver could be displaced by his former client. He could also potentially lose control of policy as well as of practice. This would be inappropriate and inconsistent with the mandate an agency has from the community supporting it. The goal, as I see it, should be a partnership between independent self help groups and the formal agency whose special expertise is utilized as needed.

The Laboratory of Community Psychiatry at Harvard Medical School is a research and training center and has no commitment to serve a particular population. Nor is it an agency committed to any particular technology. It does not have a staff who would be offering an additional or competing service and whose position would be threatened if clients became caregivers. It is therefore feasible for the Laboratory of Community Psychiatry to experiment in sponsoring a self help program staffed by non-professionals who meet all the requirements for being potential recipients of the service themselves. Here is a unique opportunity to learn how to stimulate the development of such organizations to do the work of prevention, to learn what form an on-going organization can take in the community, and to experiment with different forms of collaboration between the formal agency and the emerging self help group.

REFERENCES

1. GOFFMAN, ERVING. Stigma. New Jersey: Prentice-Hall, 1963.
2. SILVERMAN, PHYLLIS R. Services for the Widowed During the Period of Bereavement, Social Work Practice. New York: Columbia University Press, 1966.
3. SILVERMAN, PHYLLIS R. Services to the Widowed: First Steps in a Program of Preventive Intervention. *Community Mental Health Journal*, 3:1, Spring, 1967.
4. SILVERMAN, PHYLLIS R. The Widow to Widow Program: An Experiment in Preventive Intervention. *Mental Hygiene*, 53:3, 1969.

Article 24 Mourning in Japan

JOE YAMAMOTO

KEIGO OKONOGI

TETSUYA IWASAKI

SABURO YOSHIMURA

In Japan, religions permit the mourner to maintain contact with the deceased, who become ancestors. This is true both of Shintoism and Buddhism, which include ancestor worship as a part of the respective religious beliefs. Erich Lindemann has stated that the main adaptive purpose in grief work is "to fill the empty space" in the psyche; the religious custom of ancestor worship appears to fulfill this purpose.[6] We therefore decided to study the process of mourning in Japan. There, the lost object is not lost. The mourner can cling to the deceased, who has become an ancestor to be worshipped and fed, and with whom the mourner can share experiences and discuss eventful happenings.

Since Sigmund Freud's "Mourning and Melancholia" was published in 1917, there have been studies of the process of mourning among those bereaving in the community.[4] Because it was our belief that one aspect of the experience of object loss is the religious and cultural context in which the loss occurs, this study is a report of grief where the cultural institution of ancestor worship may greatly alter the process.

According to Oguchi:

> The nearest thing to a universal element in all Japanese religions is the deep-seated regard for ancestors commonly called ancestor worship, which transcends sectarian lines. In other than Christian circles, the dead are normally referred to as *kami* or *hotoke*, Shinto and Buddhist terms respectively, which are used also to designate divine beings. According to traditional beliefs, the spirits of the departed can be called back to this world—usually by shamanistic rites similar to those widespread throughout Asia.[8]

JOE YAMAMOTO, M.D., is Professor of Psychiatry, University of Southern California School of Medicine, Los Angeles.

Source: *The American Journal of Psychiatry*, Vol. 125, 1969, pp. 1660–65. © Copyright 1969 American Psychiatric Association.

The ability to call back the spirits of the departed is a part of the religious beliefs, and it may permit the Japanese mourner to hold on to the deceased, in contrast to the British who have no such beliefs.[5]

There are data concerning bereavement in British widows.[7,9] To compare our data with the studies of Marris and Parkes, we decided to study widows in Tokyo. We wrote to the widows of men killed in automobile accidents, with the hope of interviewing 20 widows between the ages of 20 and 60. We set the upper age limit of 60 because previous studies had suggested that grief may be attenuated in widows over 60.

In our letters to the widows we requested an interview in their homes, suggesting three alternative times, and asked them to return a postal card indicating their first and second choices. The following tabulation describes the returns:

Letters to widows	55
Replies: Yes	23
No	7
No reply	18
Address unknown	7
Widows actually interviewed	20

Three widows who agreed to the interviews were not seen; two were ill and therefore indisposed, and one was not home at the appointed time.

We wanted to see the widows ten to 50 days after their loss. The average actual interval of time between the loss and the interview was 42 days, with a range of 12 to 76 days. The age of the widows averaged 38 years and varied from 24 to 52. They had been married for 14 years on the average, with a range of one year to 26 years. Only one widow had no children; the average widow had two children. One woman had five children, two had three children, and the remaining had either one or two children.

Each widow was seen within the confines of her home. Most often this was a small apartment with a tatami floor. When we were invited in, we would bow low on our hands and knees and present the official gift of a package of incense imprinted with the sign and seal of the chief medical examiner of Tokyo. With the permission of the widow we ritualistically paid our respects to the deceased. Then we asked to tape-record the interview, and all but one widow consented. Without exception, tea was served, and if the widow could afford it she also served cookies or candy.

The religious affiliations of the 20 widows are shown in Table 1. There was one Shintoist, 13 Buddhists, and six who had no affiliation. The Shintoist and Buddhists each had a family altar, a *kamidana or butsudan*, respectively. In addition, four of the six widows who had denied religious affiliations also had such altars. Thus, only two had no altar, and even one of these said she planned to get one soon. It is apparent that the widows are conforming with the cultural

Table 1. Religious Choice of 20 Japanese Widows.

Religion	Number	Butsudan (Family Altar)
Shinto	1	1
Buddhist	13	13
Christian	0	0
None	6	4
Total	20	18

custom of family altars even when they disavow religious affiliation. Only one widow totally ignored this custom.

REACTION TO LOSS

After explaining the purpose of the interview, we asked about the automobile accident and the widow's reaction to it. Table 2 shows the results. Depression or anxiety was experienced by 85 percent, apathy by 55 percent, and insomnia by 70 percent. None of the widows reported cultivation of the idea of the presence of the deceased, but 90 percent reported sensing the presence of the deceased. One widow said that at night she would "wake up and feel he was there." Attempts to escape reminders of the deceased were reported by 55 percent; one widow moved the television out of the living room because she was painfully reminded of how they had watched television together. Sixty percent reported difficulty in accepting the loss, which was typically expressed as an incredulity that the husband was dead. One widow said that at six o'clock she felt he would walk through the door. In addition, all the widows with altars could feel the deceased was there. None blamed herself for the accident, and 60 percent blamed the other drivers.

Although the sample is too small for probability inferences, Table 3 compares the religious and nonreligious widows. The religious widows seemed to be as depressed and apathetic as the nonreligious widows, had more difficulty sleeping, and more often sensed the presence of the deceased. At the same time, they more often attempted to escape reminders of the deceased. However, the fact that they

Table 2. Reactions to Loss of 20 Tokyo Widows.

Reaction to Loss	Percent
Depression or anxiety	85
Apathy	55
Insomnia	70
Cultivation of idea of presence	0
Sense of presence of the deceased	90
Attempts to escape reminders	55
Difficulty in accepting the loss	60
Blames self	0
Blames others	60

Table 3. Comparison of 14 Religious with Six Nonreligious Widows in Tokyo

Reaction to Loss	Religious (Percent)	Nonreligious (Percent)
Depression or anxiety	93	83
Apathy	58	67
Insomnia	79	50
Cultivation of idea of presence	0	0
Sense of presence of deceased	93	83
Attempts to escape reminders	64	33
Difficulty in accepting the loss	50	83
Blames self	0	0
Blames others	57	67

more often sensed the presence of the husband may be a positive sign in helping them adapt to the loss.

Moreover, the religious widows had less difficulty than the nonreligious widows in accepting the loss. They also less frequently blamed others. Perhaps this trend suggests that active religious belief with the rituals and institutionalized belief in ancestors makes the loss less stressful. This and other related issues will be discussed later.

Four widows who said they had no religious preference nevertheless had a *butsudan* at which they made offerings of water, food, incense, prayers, and "goodies" and where related activities occurred. These widows then behaved like the religious widows in following the prescribed rituals, while the remaining two widows without the altars did not.

Table 4 deals with the widows without religious affiliations, comparing the four who had altars to the two without altars. Those without religious beliefs and practices were more troubled, less often sensed the presence of the deceased, more often attempted to escape reminders, had more difficulty in accepting the loss, and more often blamed others.

Table 4. Nonreligious Widows. Comparison of Those with Altars to Those without Altars

Reaction to Loss	With Altars (N = 4) (Percent)	Without Altars (N = 2) (Percent)
Depression and anxiety	75	100
Apathy	50	100
Insomnia	50	50
Cultivation of idea of presence	0	0
Sense of presence of deceased	100	50
Attempts to escape reminders	0	100
Difficulty in accepting the loss	75	100
Blames self	0	0
Blames others	50	100

When the widows in Tokyo are compared with those in London in Table 5, it can be seen that there are variations in the response. In Tokyo 85 percent of the

Table 5. Comparison of Marris' London Widows with 20 Tokyo Widows

Reaction to Loss	Tokyo Group (N = 20) (Percent)	Marris Group[1] (N = 72) (Percent)
Depression or anxiety	85	100
Apathy	55	61
Insomnia	70	79
Cultivation of idea of presence	0	21
Sense of presence of deceased	90	50
Attempts to escape reminders	55	18
Difficulty in accepting the loss	60	23
Blames self	0	11
Blames others	60	15

[1] Reprinted by permission from Parkes, C. M.: Bereavement and Mental Illness, *Brit. J. Med. Psychol.* 38:1–26, 1965.

widows were depressed or anxious, while in London 100 percent were. In contrasting the two groups, 55 percent and 61 percent respectively were apathetic, 70 percent and 79 percent complained of insomnia, and none and 21 percent cultivated the idea of the sense of presence of the deceased.

Marris reports, "A few seemed to cultivate this sense of their husbands' presence. They talked to his photograph and imagined that he advised them . . . ." Ninety percent of the widows in Tokyo and 50 percent of those in London sensed their husband's presence. A British widow whose husband died of pneumonia said, "In the night, I've heard him cough—he used to give a little cough, and he'd get up. One night I even called out to him, 'Tommy, you're coming to bed!' " Fifty-five percent of the Tokyo group and 18 percent of the London group attempted to escape reminders. "Some widows told me the house had become so unbearably charged with memories, that they had spent hours in public parks, or wandering in the street to escape from them." Sixty percent of the Tokyo widows and 23 percent of the London widows had difficulty in accepting the loss. One-half "had illusions of seeing their husbands, or more often hearing his voice or his footsteps after his death." None of the Tokyo widows and 11 percent of the London widows blamed themselves, and 60 percent and 15 percent respectively blamed others.

DISCUSSION

The differences in the pattern of response between the Tokyo and the London widows may be due to the acuteness of the Japanese women's loss, since all were deprived of their husbands due to automobile accidents, an artifact of the experimental design. All of the Tokyo widows were seen during the acute grief phase, that is to say, within 72 days of the death; and thus their data are different from the data of Marris, who saw the London widows from ten to 46 months after bereavement, with an average of two years and two months.

There is an important difference in the two interviewing methods, because we had a schedule on which the responses were listed. Marris listed the widows who had difficulty accepting the loss even though he did not ask each of the 72 widows the same questions. "Altogether, seventeen mentioned this difficulty in grasping the reality of their husbands' death, and fifteen that they continued to behave involuntarily as if he were still alive. The experiences are probably commoner than these figures suggest, since I recorded them only when they were spontaneously mentioned." [7]

There are many other variables, such as the presence of members of the extended family in the Japanese household. We are aware that the presence of relatives or in-laws may be a source of great comfort and may contrariwise be a source of great friction and unhappiness. The so-frequent presence of others in the Japanese household is another cultural variable that complicates the picture by presenting a different life situation and life style. One-half of the widows were living with relatives, either in their own home or in the relative's. The distribution of relatives was of possible importance in that eight of the 14 religious widows had relatives as compared with only two of the nonreligious widows. Of the latter, one widow had a *butsudan*; the other had no altar.

Since both the nonreligious widows without family altars attempted to escape reminders of their husbands, had difficulty in accepting the loss, and blamed others, it might be speculated that they were socially withdrawn and rejecting of help even from the deity. Lindemann has pointed out the tendency of some to withdraw at a time when others would try to be attentive and to comfort the bereaved.[6]

The main point of our research was to observe the natural process of mourning in a culture where the religious beliefs and institutions permit the "cultivation of the idea of the presence of the deceased" as ancestors. If you would for a moment give up your Judeo-Christian beliefs and attitudes about one's destiny after death and pretend to be a Japanese, you might be able to feel how you are in direct daily communication with your ancestors. The family altar would be your "hot line." As such, you could immediately ring the bell, light incense, and talk over the current crisis with one whom you have loved and cherished. When you were happy, you could smile and share your good feelings with him. When you were sad your tears would be in his presence. With all those who share the grief he can be cherished, fed, berated, and idealized, and the relationship would be continuous from the live object to the revered ancestor.

The Work of Mourning

Sigmund Freud's view of the work of mourning was the individual's ability to test reality and to be aware of the absence of the object.[4] This painful change of reality may at the same time exist with fantasy-making. Pollock has suggested that "fantasies and even day dreams concerning the deceased object can interfere with the mourning work, and in instances where the death of the object is not realistically appreciated, the object may continue to exist as an unassimilated introject with whom internal conversations can be carried out." [10] There may then be a problem, Pollock believes, due to the lack of completeness of identification and ego integration of the introject.

Bowlby also expresses the idea that in healthy mourning there is a withdrawal of emotional concern from the lost object and that this prepares the way for new relationships.[1,2] In comparing mourning with the separation of the child from his mother, he describes three phases of mourning: the urge to recover the lost object, disorganization with despair and depression, and the phase of reorganization.

In Tokyo the process of mourning is different due to the cultural beliefs. Ninety percent of the widows worked to maintain ties with the deceased, who becomes an ancestor. This of course is the counterpart of the 21 percent of the British widows who worked to cultivate the idea of the sense of presence of the deceased. The important theoretical difference is that in Tokyo this is acceptable and encouraged by the culture and religious rituals. The London widows would cultivate the idea of the presence of the deceased "by imagining him speaking to them, or in one case by lying on the bed of her dead child and playing with his toys." [9] The Tokyo widows did something like this at least once a day when they offered incense at the family altar. The photograph of the deceased, the urn of ashes, the flowers, water, rice, and other offerings were all for the ancestor. As one widow said, "When I look at his smiling face, I feel he is alive, but then I look at the urn and know he is dead."

A number of widows executed actions previously directed toward the dead persons. For example, one widow would go to the streetcar stop at the hour when the husband previously came home. Another kept his clothes in case he needed them. Still another widow thought it was her husband returning when she heard a motorbike, and she went to the door. Because we neglected to include such acts on our interview schedule, we do not have the number who executed acts associated with the deceased.

Ninety percent of the Japanese widows sensed the presence of the deceased; often this occurred not only in sensing the physical presence of the deceased, but in sensing the presence of the ancestor as well. With this "transitional object," most gained comfort in the rituals around the *butsudan*. The widows often said they had difficulty in accepting the loss, but this was in the sense of feeling that perhaps the husband was merely teasing and would return, for example, at suppertime. One widow was angry at the husband and said that when he returned she would scold him. A large number were hostile toward the people who drove the lethal car or truck.

The Tokyo widows more often wished to escape reminders of their husbands. We will have to allow for the variables of the abrupt death due to the accident and the acute phase during which the women were interviewed. In addition, there were examples of gory reminders that seemed to be related to witnessing the accident scene. One widow described how she walked by the place of the accident and was reminded of the accident scene, and so she escaped this by avoiding that particular street.

Because the widows were victims of automobile accidents, they had no need to blame themselves. The closest any widow came to self-blame was to state that she had suggested that her husband take the streetcar instead of the motor bicycle, and she wished that she had insisted he do so. Another widow explained that she had wanted her husband to wear a helmet when riding the motor bicycle and that he had refused to do so; she thought that she should have suggested it again.

Freud said that the mourner has "to detach the survivor's memories and hopes from the dead." [3] Perhaps there will be a problem ultimately for the Tokyo widows. We do not have data on them six months later, although an attempt at follow-up was made. When letters were mailed to eight widows who had been seen six months previously, only two consented to the second interview. We were able to see just one widow, however; the other one was sick. Five did not reply to the letter, and one had moved. The only widow we saw six months after bereavement was not so depressed or disorganized, and the only residual signs of the loss were mild depression and insomnia. She no longer spoke of sensing the presence of the deceased or of other behavior associated with the deceased. This widow had become a devout Buddhist and felt that the religious beliefs helped.

Our data on the Tokyo widows suggest that at times of death, religious beliefs and customs are a source of comfort and that even when the loss is abrupt and the widow is seen during the acute grief phase, the idea of ancestors substitutes for the "cultivation of the idea of the presence of the deceased." These widows were less accepting of the idea of the loss, which may be evidence of greater difficulty in adapting or may be a function of the abrupt loss due to the accident and the acute grief phase during which they were interviewed. The widow who was seen six months later was no longer having difficulty accepting the loss.

SUMMARY

Because there are quite strong urges to regain the deceased one, it seemed reasonable to assume that religious concepts and practices may have a bearing on the process of mourning. In Japan the deceased become ancestors who are fed, watered, given gifts, and talked to, and so the tie between the widow and the dead husband remains through the concrete medium of the husband's photograph on the family altar. The family altar is almost universal and is a cultural "cultivation of the idea of the presence of the deceased." The rituals appeared to aid the widows, and although they were acutely grieving, they seemed to be adapting to the loss. They certainly required no special fantasy-making since they could "look at the picture and feel he is alive and look at the urn of ashes and realize he is dead."

We have added data on how individuals grieve, specifically those who are not patients but who are experiencing the acute phase of the process of mourning. There is evidence from these data that the feelings of grief are considerable and that when the loss is acute there frequently may be difficulty in accepting it. It appears that mourning requires a major adaptation and that we who help the bereaved need to know what can be expected as a part of the adaptive process and need to understand that religion may play an important role here. Those bereaved without religious support may have greater difficulty in coping with the loss. None of the Tokyo widows worried about their sanity because they felt their husband's presence. Religion aided in this aspect of the grieving, for in the religious beliefs the husband was present as an ancestor.

REFERENCES

1. BOWLBY, J.: Separation Anxiety, Int. J. Psychoanal. 41:89–113, 1960.
2. BOWLBY, J.: Processes of Mourning, Int. J. Psychoanal. 42:317–340, 1961.
3. FREUD, S.: "Totem and Taboo" (1913), in The Complete Psychological Works of Sigmund Freud, vol .13. London: Hogarth Press, 1955, pp. 1–161
4. FREUD, S.: "Mourning and Melancholia" (1917), in The Complete Psychological Works of Sigmund Freud, vol. 14. London: Hogarth Press, 1957, pp. 239–258.
5. HINTON, J.: Dying. Baltimore: Penguin Books, 1967.
6. LINDEMANN, E.: Symptomatology and Management of Acute Grief, Amer. J. Psychiat. 101:141–148, 1944.
7. MARRIS, P.: Widows and Their Families. London: Routledge and Kegan Paul, 1958.
8. OGUCHI, I.: The Religions of Japan—Past Tradition and Present Tendencies. New York: Intercultural Publications, 1955.
9. PARKES, C. M.: Bereavement and Mental Illness, Brit. J. Med. Psychol. 38:1–26, 1965.
10. POLLOCK, G.: Mourning and Adaptation, Int. J. Psychoanal. 42:341–361, 1961.

Coping with Unusual Stress: The Prisoner of War

IX   In every war soldiers have had to face the possibility of being taken prisoner by their enemy. Over the last thirty-five years large numbers of American servicemen have experienced life as prisoners of war. From those who survived we can learn about the effects of their experiences and their means of coping with this severe and unusual stress.

Soldiers captured in the Orient by the Japanese in World War II, or in Korea or Southeast Asia in later wars, have had to withstand particularly harsh conditions. Cultural differences in diet and the general treatment of prisoners contributed to an excessive mortality rate. Thirty-five percent of Americans captured by the Japanese and 38 percent of Americans captured by the North Koreans died as prisoners, as compared with 1 percent of Americans held in Nazi prisoner of war camps.[5] Inadequate food and shelter, physical abuse, disease, forceful indoctrination, and uncertainty about their fate were the primary stresses. Wolff [5] found that ignorance about what to expect and fear of the unknown made some American prisoners more susceptible to Chinese and Korean brainwashing and general management techniques. Some foreknowledge of POW camp conditions was helpful in coping with the frequent cycles of anxiety and temporary reassurance and relief. Among prisoners of the Japanese and the North Koreans many died from what was essentially a failure to cope; deep depression, withdrawal, and giving up led to a speedy death (similar to the "Musselmann" stage among concentration camp inmates, Part X).

In the first article Raymond Spaulding and Charles Ford examine the reactions of the men taken prisoner by the North Koreans in an incident with the U.S.S. *Pueblo* in 1968. During their eleven months of captivity the men evolved a variety of techniques for dealing with the stresses of their situation. These included religious faith, faith in their government, denial, rationalization, humor, and successful reality testing (that is, interpreting clues about their situation accurately). Those who coped poorly used few of these adaptive techniques and tended to be passive-dependent and to need a great deal of group support.[1] Those who could maintain a degree of detachment and insensitivity, who could find some relief in the use of fantasy, or who could keep busy, withstood POW captivity better. Unlike the experience of earlier captives, the *Pueblo* crew found that the North Koreans permitted, and even encouraged, the emergence of natural leaders and the development of group stability and loyalty. In fact, frequent group discussions among roommates served to relieve anxiety and depression.

Post-release reactions of Americans held in Asian camps have been surprisingly uniform. For the first few days they were dazed, apathetic, and in no particular hurry to get home. The dull monotones in which they spoke were characteristic of the subdued, even bland, quality of their emotions. Then tension, anxiety, restlessness, and defensiveness about their behavior as POWs began to appear. Many felt isolated and had difficulty relating to non-POWs. They found some comfort in a strong identification with their fellow prisoners.

After the Korean War, released POWs were returned home by ship to provide them with a brief interlude in which to adjust to their new status and to work through some of their feelings before they rejoined their families.[2] The men of the *Pueblo* were flown directly home for a last-minute Christmas reunion with their families, leaving them little time to make the tremendous transition from captive POW to freed man. This abruptness, and the Navy investigation into the incident, may account for the delayed upsurgence several months later of expressions of active hostility toward the Navy and fellow *Pueblo* crew members, and general acting-out behaviors (such as abuse of alcohol and drugs). A transitional interlude with counseling available, such as Lifton described, might have been helpful in working through the feelings associated with the POW experience, and in dealing with the stresses of repatriation.

In the second article Edna Hunter compares one hundred men who were prisoners of war in Vietnam for five years or more on the basis of the amount of time they had spent in solitary confinement. In psychiatric interviews shortly after their release some significant differences were found between those with high and those with relatively low amounts of time in solitary confinement. The Hi Solo Group received more frequent abnormal ratings on guilt, on resistance to suggestions, on overdeveloped superego, and on unrealistically high need for achievement. The men in the Hi Solo Group were generally subjected to harsher conditions than the men in the Lo Solo Group, that is, lack of adequate shelter and clothing, reduction or withdrawal of food and water rations, and physical torture. Dr. Hunter raises an interesting question about the extent to which the traits characterizing this group might have preexisted their capture and determined (or provoked) in part the treatment the prisoners received. The Hi Solo

Group were of higher rank generally, which may be related to the personality traits found to characterize this group. On the other hand, the captors may have deliberately singled out higher-ranking officers for more severe treatment, both solitary confinement and physical abuse.

Colonel Risner,[3] in his personal account of his seven years as a POW, reports that officers were isolated more often. In his own experience, though, much of the harsh treatment he received was punishment for his efforts to communicate with and organize the other prisoners, an effort he persisted in because he felt it was his duty as a senior officer. When he was put in solitary confinement in total darkness for ten months, he staved off panic and insanity by keeping physically active (running in place and doing push-ups), sometimes screaming into a towel, crying, doing complex math problems, and frequent praying. Religion was a source of comfort and strength for many of the men.[3,4]

The experiences of American servicemen captured by Asian adversaries over the last thirty years offer much information on the effects of stresses like poor diet, exposure to the elements, social isolation, and calculated attempts at external thought control. There is also evidence of the use of coping mechanisms, such as denial, intellectualization, and faith in religious and patriotic principles, and the importance of these processes for physical and emotional survival.

REFERENCES

1. FORD, C. V., AND SPAULDING, R. C. The *Pueblo* incident: A comparison of factors related to coping with extreme stress. *Archives of General Psychiatry*, 1973, 29, 340–43.
2. LIFTON, R. J. Home by ship: Reaction patterns of American prisoners of war repatriated from North Korea. *American Journal of Psychiatry*, 1954, *110*, 732–39.
3. RISNER, R. *The passing of the night: My seven years as prisoner of the North Vietnamese.* New York: Random House, 1973.
4. ROWE, J. N. *Five years to freedom.* Boston: Little, Brown, 1971.
5. WOLFF, H. G. Every man has his breaking point—(?) The conduct of prisoners of war. *Military Medicine*, 1960, *125*, 85–104.

Article 25 The Pueblo Incident

Psychological Reactions to the Stresses of Imprisonment and Repatriation

RAYMOND C. SPAULDING

CHARLES V. FORD

The return to the United States of the 82 surviving *Pueblo* crew members in December 1968 made possible the systematic evaluation of a group of men imprisoned for 11 months in a hostile country. The backgrounds of the crew were highly varied but the stress was fairly uniform; this permitted consideration of how differing personality types tolerated the adverse situation, as well as a comparison of the ego defense mechanisms utilized. The readjustment time following repatriation was a stressful situation and was probably associated with more psychiatric symptoms than the imprisonment. The information obtained has historical and practical significance in relationship to previously reported studies of prisoners of war in the Orient.

A REVIEW OF THE LITERATURE

A number of papers have been published that describe the unique circumstances of being a prisoner of war in the Orient. Cultural differences, including the treatment of prisoners as well as the different diet, are viewed as making this a more stressful situation than being imprisoned by a Western country. Mortality rates among American prisoners of war were exceptionally high: approximately 60 percent for prisoners of the Japanese during World War II [10] and approximately 35 percent for those of the North Koreans during the Korean conflict.[8] Factors cited as etiological include malnutrition, physical maltreatment, and "give-up-itis." The last is certainly of psychiatric concern since it represents a malignant de-

RAYMOND C. SPAULDING, CAPT. MC., USN, is Chief of Psychiatric Service, Naval Regional Medical Center, San Diego, California.

CHARLES V. FORD, M.D., formerly Staff Psychiatrist at the Naval Hospital at San Diego, is now Assistant Professor of Psychiatry, Harbor General Hospital, Torrance, California.

Source: The American Journal of Psychiatry, Vol. 129, No. 1 (July 1972), 49–58. Copyright American Psychiatric Association.

pressive reaction. Also of particular interest to psychiatrists are the efforts made during the Korean conflict to indoctrinate Americans with the political beliefs of the captors. These efforts have been described as "thought reform" and are colloquially known as "brainwashing."

Prisoners of the Japanese studied immediately after repatriation were described by Wolf and Ripley[13] as being comparatively seclusive and taciturn. Their resentment was directed toward superiors and associates rather than toward their captors. This was interpreted as being a displacement, since hostility overtly expressed toward the Japanese was dangerous, if not suicidal, because treatment during captivity had often included capricious cruelty. Multiple vitamin deficiencies and severe weight losses contributed to numerous somatic complaints, neurological difficulties, and alterations in thinking. There was agreement among the survivors that there had been a general lowering of moral standards and a preoccupation with food, which was often obtained by devious means. Sexual drives were absent except when food was plentiful.

Among the survivors two personality types seemed to predominate: those with features of psychopathy and exceptionally mature individuals. The first group survived by their manipulations and the protective quality of their emotional shallowness, but the second group had a surprisingly large number of neurotic symptoms, particularly conversion reactions (e.g., anesthesia), which seemed to have survival value. The best-adjusted individuals helped themselves survive by keeping busy and productive; this was particularly true of the physicians and chaplains. There was practically unanimous opinion that those prisoners who became depressed and anorexic died.

Brill[3] reported on a very large series of repatriated prisoners of war (POWs) of the Japanese and commented that the men displayed a mixture of superficial well-being and optimism, with little spontaneous expression of hostility. They were not compulsive in their living habits. This group appeared to be of above-average intelligence and stability, which could be attributed to "survival of the fittest." Factors that seemed to be correlated with survival were "courage," emotional detachment, belief in one's superiority over the enemy, and a refusal to give up hope. An ability to eat anything and passively accept hostility were also attributes associated with survival. Religious beliefs, social background, and education were not considered important variables. Many men commented that they had gained from the experience and had gained maturity.

Nardini,[10] himself a prisoner of the Japanese, analyzed the qualities of the emotional stresses and factors associated with survival or death. It was his opinion that emotional shock and reactive depression contributed much to the massive death rate: "Self-pity . . . was highly dangerous to life." Survival factors included the "will to live," which was dependent upon the effectiveness of the ego defenses. Such defenses include fantasy and the ability to repress and suppress an awareness of ever-present death. Cleverness, adroitness of thinking, and general cunning were factors contributing to survival, whereas pure intelligence unrelated to interpersonal dealing was less advantageous. A sense of humor was regarded as a definite asset. Other factors contributory to survival included a previously stable life adjustment and the personality characteristics of being emotionally unresponsive or insensitive. Factors that predisposed to high mortality included youth, immaturity, marked dependency, and sensitivity. Men of very low intelligence fared poorly because of poor planning, poor self-care, and similar causes.

IMPRISONMENT IN KOREA

In addition to the stresses listed above, imprisonment by the Communists during the Korean conflict involved increased exposure to inclement weather and the unique stress of "brainwashing." American POWs repatriated after the Korean conflict were described by Segal [12] as bland, apathetic, and retarded. Their talk was shallow, often vague with a lack of content, and was marked by large memory gaps. This "zombie reaction" typically cleared in about three days and was replaced by a mild euphoria. Yet there was no pressure to return home and the men seemed suspended in time; they showed increasing anxiety and tension. They expressed hatred for the North Koreans but surprisingly seemed to think well of the Chinese, who "treated us the best they could." It was learned that they had been treated in a manner that had the objective of attempting to change their political beliefs and attitudes. The techniques were fairly standardized and included assurance of good treatment in a manner that combined subtle hints of torture or exile to Siberia.

They were denied all internal leadership, since officers and noncommissioned officers were segregated in other camps. Natural leaders who emerged were also isolated or removed and the prisoners were divided instead into squads under the leadership of a Chinese Communist squad leader who carried on constant spying. Informing on other group members was encouraged. Confessional autobiographies were written and rewritten, with the emphasis being that one had committed a crime against the people rather than against the individual. Letters from home were carefully screened and only those with pessimistic news were delivered. Cooperative prisoners were rewarded with extra comforts and were used to "sell" other prisoners. Segal did not describe the characteristics of those men who were best able to cope with the stresses, but the implication was that the Chinese thought the more passive and less educated men were the most susceptible.

Lifton [7] reported his experiences and observations of 442 repatriated American enlisted men who were prisoners of war following the Korean conflict and who were returned to the United States by ship. It was hypothesized that providing an "interlude" would permit the men to more effectively integrate the realities of repatriation and to bridge the emotional gap between prison camp and home.

The majority of these men had been prisoners for more than two years. Initially they were mildly confused and surprisingly unenthusiastic about their release. Although they were relieved about being back in American hands, there was no hurry to get home now that "I'm back on this side of the line." Their demeanor was markedly restrained and phlegmatic, except for anger at those who had collaborated with the enemy. The men made little attempt to mix with crew members of the ship, since they felt no one could understand them who had not experienced the incarceration himself.

After a few days at sea, a gradual but definite change began to take place in their behavior. They became increasingly reactive, but in general belligerent, irritable, and critical. While they had previously expressed little but praise for the treatment that they had received since their repatriation, they now began to complain petulantly to both psychiatrists and compartment commanders about the facilities of the ship and the military routine. During psychiatric sessions they were much more direct in expressing their overt hostility. At the time of their climactic arrival in San Francisco, they again behaved with great emotional reserve. During the debarkation scene, observers were extremely moved by the sight of mothers literally reaching up to the ship for their sons, which was in marked contrast to the blandness and lack of outward emotion displayed by the repatriated group.

Group therapy sessions held during the crossing allowed the men to talk about their fears associated with return to the United States. They spoke of their tremendous feelings of isolation, of their inability to communicate, and of the anxiety about the future that all were experiencing. They perceived the homecoming itself as a particularly threatening experience and the prospective fanfare stimulated their feelings of guilt and unworthiness: "That first week is really going to be rough; big parties with relatives and all that. Makes me feel funny because I know I don't deserve it. . . . I sure hope they don't make a lot of fuss over me. I'd rather be just left alone, maybe take a fishing trip for a few days."

The dangers of this unknown world were vividly expressed, and all the men felt left behind and out of things: "We're not familiar with things back home anymore. We have a lot of catching up to do—I've never seen television. I don't even know where our new house is." Their common identity as repatriated American POWs was the only strong group tie available to fill the emotional vacuum that had been created in them.

The transition from apathy to hostility in these men was seen as a form of verbal muscle-flexing and as an attempt by the men to again assert themselves after the humiliation and impotence experienced during imprisonment.

The delayed homecoming, with the opportunity to live together and form these group ties, was nevertheless of definite value. In addition to the much-needed interim support, it offered the men both a necessary working through period for reality testing and a protective form of initial social exposure to "outsiders."

THE PUEBLO INCIDENT: HISTORICAL SUMMARY

Details of what has become known as the *"Pueblo* incident" have been widely reported in the lay press, and numerous crew members have written books or articles giving their personal and somewhat differing accounts.[2,4,6,9,11] It is not the purpose of this paper to recount in detail that information, but a summary of the information generally agreed upon by all crew members is given in order that reference can be made to specific incidents.

On January 23, 1968, the small intelligence ship U.S.S. *Pueblo* (AGER-2) was attacked in international waters off the Pacific Coast of North Korea. One crew member was killed; two others sustained substantial injuries. The ship was then boarded, captured, and sailed into Wonsan Harbor. The crew was then transferred to Pyongyang by railroad, where they were imprisoned in a building known to the crew as "the barn."

The commanding officer was isolated from the others, who were quartered three or four to a room. Crew members were threatened with death and interrogated; some were severely beaten. "Confessions" about "criminal aggressive acts" were obtained from all crew members as a result of these threats and ill-treatment. The capture of the men's service records gave the North Koreans a strong position in their interrogations. Later, in March, the crew was transferred to other accommodations near military installations, where officers had their own rooms while enlisted men generally were housed eight to a room.

Their treatment by the North Koreans varied, but in general the living quarters, sanitation facilities, and medical care were unsatisfactory by Western standards. Food was a particular difficulty, since it was deficient both in quantity and quality. The attitudes of the captors seemed capricious; the crew members began to associate improved food and a decrease in physical maltreatment with hopes for imminent release and the converse as a sign that negotiations were going poorly. Despite the "confessions," members of the crew attempted to communicate to the free world their lack of sincerity. These actions included references to dead relatives and awkward and uncharacteristic phrasing in letters. Propaganda photographs often showed smiling faces in association with obscene gestures.

The North Koreans attempted through lectures, field trips, and written material to convince crew members of the injustices of their "imperialist" government.

In the fall of 1968 there were signs of increased friendliness by the Koreans, which was interpreted by the crew as an indication of increasingly good prospects for release. However, this turn of attitude changed abruptly in the first week of December and was supplanted by "Hell Week." It became apparent that the Koreans had learned the meaning of the obscene gestures, as well as the meaning of some details in the letters. Physical maltreatment increased to its most intense level since their capture, and there were extensive efforts to force information from the crew as to which of its members had collaborated in the efforts to foil the propaganda plans of the North Koreans.

Suddenly and unexpectedly on December 19, the maltreatment was reversed. On December 23 the crew was returned to South Korea; they walked across a bridge connecting the Koreas. Each man received a screening physical examination, was provided with an American uniform, and had a holiday meal. The crew was then flown to San Diego Naval Hospital for a more intensive evaluation and intelligence debriefing. They arrived on Christmas Eve into an emotional scene provided by welcoming relatives.

METHOD

After arriving at the Naval Hospital in San Diego, each crew member was examined by an internist and by one of the six Navy psychiatrists designated as part of the medical evaluation team. Extensive laboratory examinations were performed and consultations by other specialists were requested when indicated. Crew members were not interviewed by Naval intelligence officers until they had first been given medical and psychiatric clearance, in order to rule out medical disease or psychiatric problems needing immediate attention.

The psychiatric examination consisted of a private interview lasting at least one hour. Brief notes were taken but the interviews were not tape-recorded. Some crew members were seen on more than one occasion, either at their own requests or because further evaluation was indicated. In addition to the clinical interviews, which varied according to the style of the individual psychiatrist, each crew member completed a standard form requesting demographic and background information. All also completed the Minnesota Multiphasic Personality Inventory and a sentence completion test.

The information obtained and the clinical observations were recorded on forms according to a prearranged protocol. This allowed a systematic comparison of the different psychiatrists' impressions, as well as a means of detailing symptoms, premorbid personality characteristics, and similar information.

FINDINGS

When the men of the *Pueblo* were first seen, they generally appeared subdued, voiced concern about wanting to see their families, and with rare exceptions had no evidence of any thought disorders, either organic or functional. However, there was a major contradiction between the verbalized affect and that which the examiners observed. Such comments as "This is the happiest day of my life" were *not* associated with euphoria. Quite the contrary, the majority of the crew seemed mildly depressed or anxious, although, almost to a man, they denied such feelings.

The men were characterized by a willingness to talk freely and to discuss their feelings toward their captors and fellow crew members. Expressions of overt hatred toward the North Koreans were common and there were many spontaneous comments denying that they had had any success in "brainwashing" crew

members. At times the protests were so insistent as to suggest a reaction formation. There were concerns of guilt about the capture and subsequent "confessions," but serious statements of this type were limited to a minority and most
men had ready rationalizations, such as "We really screwed them over" (in reference to widely publicized pictures in which the crew members were making
an obscene gesture).

There was in general a willingness to talk about other crew members and to
describe those who had done well and those who had tolerated the stress poorly.
The descriptions were fairly unanimous in singling out the ones who had done
poorly as well as those who had done well. In both instances the feelings were
strongly expressed but did not seem to us as intense as those reported following
the Korean conflict in similar circumstances. The consistency of the crew's reports permitted the examiners to believe that the physical and emotional experiences the men had endured during the preceding 11 months were accurately
reported.

The crew as a whole had experienced anxiety and apprehension during the
first days immediately following their capture. Feelings of guilt that followed the
"confessions" for the most part were transitory and were ameliorated by shared
rationalizations with other crew members. Many men, initially encouraged by
rumors of imminent release in October, ascribed their lowest period to the disappointment and despair that followed the failure of repatriation at that time.
Anxiety again increased in response to "Hell Week" prior to their actual release
in December.

However, there were some men who had more striking reactions than these
generalized ones. Forty men admitted experiencing significant anxiety, but all attributed this to the unpredictability of their treatment by the North Koreans.
Sixteen men gave a history of subjective depression, often accompanied by feelings of despair, hopelessness, and suicidal ideation. An attempt was made to
define the nature of the object loss(es) that were thought to be responsible for
the depressive reactions. For 11 men this was separation from family. Four men
stated that their failure to live up to personal ideals in regard to the "confessions"
was a significant factor. For seven men the anticipated loss of their Navy careers
was important. One man worked through a transient depression secondary to
traumatic castration by shrapnel. (Note that some men experienced more than
one significant loss.) Suicidal ideation was uncommon, but six admitted this and
two actually made three attempts; two of these, however, seemed to be more
manipulative than serious. As a result of profound depression during the period
of incarceration, one man did make an attempt to end his life, but the degree of
despair that he was suffering at that time probably precluded his effectively killing himself.

Despite the unusual and monotonous food, which was prepared with less
than Western standards of sanitation, the majority of the men ate what was
available. One member of the crew sincerely stated that he thought the food was

Figure 1. Range of Weight Losses During Confinement

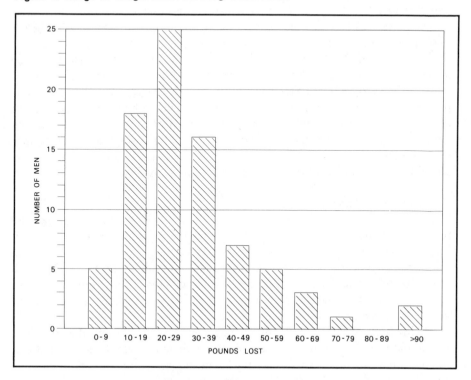

not bad. Many members of the crew stated that they felt that they were being fed as well as most of the soldiers in the Korean Army and better than much of the civilian populace of North Korea. Most members of the crew suffered significant weight loss; see Figure 1.

Complaints of pain in the lower extremities and hands were diagnosed as neuropathies in seven men. These neuropathies were thought to be secondary to the loss of fat supporting the peripheral nerve sheaths, rather than to a lack of vitamins. These symptoms responded well to conservative management and the return of the men's weight to normal levels for their height and age.

Twenty-seven men complained of anorexia during their confinement. Of these, nine were anorexic throughout their stay, 15 were anorexic early during internment but later regained their appetite, and three who originally ate well became anorexic at a later time. The anorexia that was reported was not clearly related to anxiety or depression. Discussion of food was a favorite subject during the internment. In fact, one of the ship's cooks kept himself busy by planning menus.

Disturbances of sleep patterns were more clearly related to feelings of depression or anxiety and most of those who complained of a change from their usual sleep pattern admitted to feelings of depression and/or anxiety. One man was hypersomnic, ten complained of difficulty falling asleep, and three men were bothered with early wakening. Dreams were common and a variety of themes were presented, including wish-fulfillment dreams—"fishing back home with Dad," dreams of abandonment—"I returned home and everyone had left," and sexual dreams. Dreams of retaliation were rare; however, many of the men fantasized that the United States would strike with vengeance at the *Pueblo* and possibly at the area where the crew was interned, killing them all for their breach of conduct and loss of ship. Several men reported dreaming about escape but, after successfully negotiating the barriers of the military compound, they decided to return and remain with their comrades, feeling marked guilt for having abandoned them. A number of the men denied recalling any dream material.

Methods of Coping

When crew members were asked about the methods they used to cope with their incarceration, they frequently stated that they maintained a faith in their commanding officer, religion, and country. It became apparent that there was considerable group support provided by the natural leaders who emerged in each of the rooms in which they had been randomly billeted. Crew members referred to their room as a "haven" and convinced themselves that it was not bugged since they talked freely about their captors.

Many of the rooms apparently operated like leaderless group therapy; there was ample time for extended review of their individual lives, hopes, aspirations, and accomplishments. Sexual topics for discussion were common and often quite candid, although some men denied any sexual libido during their captivity. There was no report of any homosexual activity. Many of the crew members commented that they felt that they had gained from the experience through increased understanding of themselves and others. Furthermore, there were statements reflecting a belief in increased maturity on the part of the individual, greater appreciation for the American way of life, and a firm resolve for meaningful application to a purposeful life in the future.

The crew's attitudes toward their North Korean captors were almost uniform and varied only in the degree of negative sentiment expressed. The crew was exposed to some of the techniques of "thought reform" that were previously described, but not to such a marked degree as that used during the Korean conflict. These efforts consisted primarily of requests for the life histories of the men and frequent lectures stressing the "oppression of the peoples" by the "imperialistic warmongering capitalists." One young man was ridiculed for lying when he stated that he owned a car. Although some became involved in the lectures to the extent of arguing points, many passed the time by daydreaming. The issues discussed during the lectures would then be rediscussed back in the room, where the logic of the lectures would be destroyed by the leaders of the group.

However, in fantasy or in fact, one idea that was widely accepted by the crew at the time of their release was that the North Korean people hated the Americans with such a passion that it was safer to remain in prison than to try to escape. To a man the crew denied that they had been "brainwashed" by their captors. They insisted instead that they had come to learn that they were superior to their captors who were just like "animals, barbarians, programmed robots," etc. However, non-Western phrases, such as "the Democratic Peoples Republic of North Korea," would occasionally creep into their speech.

Physical illnesses during confinement were common. Baldridge, the ship corpsman, reported [1] that at some time during detention virtually all crew members suffered at least one of the following illnesses: diarrhea, upper respiratory infections, fevers of unknown etiology, pedal edema, respiratory infections, urinary tract symptoms, stomatitis, and skin infections. One man had infectious hepatitis. One man was operated upon by North Korean physicians for shrapnel injuries and another had a tonsilectomy. Usually treatment was symptomatic and involved taking "white tablets." Acupuncture was used on occasion but was not regarded as beneficial. Many of the physical complaints seemed to be related to poor nutrition and vitamin deficiency. A list of the physical conditions diagnosed at the crew's return is presented in Table 1.

Table 1. Physical Illnesses Present at the Time of the Crew's Return

Diagnosis	Number of Men
Malnutrition	31
Dermatomycosis	26
Orthopedic problems (e.g., fractures)	17
Diminished visual acuity	16
Peripheral neuropathy	7
Shrapnel wounds	7
Recent trauma	6
Macular degeneration	5
Tooth infections	3
Hepatitis	2
Atrophy of optic nerves	1

It was apparent that some men did much better than others in their ability to tolerate the stress of incarceration. The men who did better would in general be characterized as bright and schizoid. These men had an ability to isolate their affect and entertain themselves with fantasy. Most of them comfortably rationalized that the Koreans would release them eventually; otherwise they would have been killed immediately. In general, the factors associated with poor adjustment were youth, immaturity, and personality characteristics of obsessive-compulsiveness, passive-dependency, or emotional instability.

The stresses on the crew members did not end with their release from North Korea. On their return to the continental United States many feared that they would be punished and ridiculed. The heroes' welcome was confusing and their anxiety about the upcoming Board of Inquiry continued to be high.

They initially remained a very well-integrated group and identified with the group. They maintained a daily military routine and after about three weeks most were found fit for limited duty for a period of three months. All were scheduled for reevaluation three months later, at which time it was anticipated that the Board of Inquiry would be completed.

At the time of the reevaluation, approximately 12 weeks later, the difference in the crew was striking. In contrast to the bland but cooperative demeanor initially seen, many men were openly antagonistic and hostile toward both the Navy and their fellow crew members, despite the fact that the findings of the Board of Inquiry were more benign than initially anticipated by the crew. There was evidence of considerable acting out with alcohol and drugs, minor traffic violations, and squandering of back pay.

At the time of reevaluation one man had developed paranoid ideation and another had incapacitating symptoms of an obsessive-compulsive neurosis. At least one man developed symptomatic alcoholism shortly after leaving the Navy and another is reported to have committed suicide following his discharge from the Navy.[11]

CASE REPORTS

Case 1. A 25-year-old junior officer who was regarded as exceptionally strong and an inspiration to many other crew members reported considerable mental anguish, which he had hidden. He admitted to anxiety nightmares, transient suicidal ideation, and feelings of guilt over not having lived up to the military Code of Conduct.

He was described as conscientious, hard-working, and sensitive; he had some difficulty with close interpersonal relationships. His background data indicated that he came from an intact, stable family and had entered the Navy after graduating from college, where he had done well.

On his return to the continental United States, he expressed some increased guilt secondary to the "heroes' welcome." He was serious, acted appropriately, and adjusted well over the next several months.

Case 2. A 20-year-old unmarried enlisted man, a high school graduate of average intelligence, did poorly during confinement. Since he was of dark complexion, he was accused of being a South Korean and was severely maltreated during the early part of the confinement. His feelings of guilt concerning confession were significantly diminished after a supportive talk with one of the officers, but disturbances of sleep and appetite continued. He had tension headaches, lower back pain, recurrent nightmares of an abandonment theme, and periodic suicidal ideation.

His personality traits were those of friendliness, extroversion, and a tendency to rely heavily on others (passive-dependency). Background information indicated that he was one of an extremely large, close-knit family. He had enlisted after high school graduation and had been on active duty one month when captured. On his return to the continental United States he quickly recompensated and within a few weeks had rationalized or repressed his previous distress.

Case 3. A 23-year-old unmarried, exceptionally bright high school graduate had tolerated his confinement remarkably well. When he was not killed during the first 24 hours, he decided that he was a "political prisoner" and would not be severely mistreated. He suffered little or no anxiety or depression and comfortably rationalized his "confession." He spent most of his confinement time in fantasy, and he rethought the electrical wiring of computers and other technical matters. He felt that he had "straightened himself out" and found new resolve in his plans for attending college.

His personality traits were those of isolation and introversion, with considerable fantasy (schizoid). His background data indicated that he was from a broken home and had not performed to his highest ability while in high school. His enlistment in the Navy had followed a period of nonskilled work after high school graduation.

On his return to the continental United States after the confinement he was confident, comfortable, and enthusiastic about his new plans. Several months later his enthusiasm had dissipated and his previous personality characteristics predominated.

DISCUSSION

In contrast to the case in the prisoner of war camps during World War II and the Korean conflict, there were no deaths during the detention of the *Pueblo* crew. There were several possible reasons for this. First, the period of imprisonment was comparatively short; second, the Koreans did not wish any deaths because of possible adverse propaganda; and third, the organization of the crew and group support offered protection against the "give-up-itis" previously described as a frequent cause of death. That the Koreans actually encouraged the development of leadership represents a marked departure from the practice during the Korean conflict of isolating the leaders. Their motives are not known, but the result seems to have clearly reduced the stress upon the more dependent and vulnerable members of the crew.

"Brainwashing" techniques were unsuccessful in converting any man to Communism or even in persuading him to reject any American principles. However, it appeared to be successful in convincing the crew that the Korean people felt persecuted by the United States and that such animosity existed that escape would mean being killed by the people. The idea was presented that they were not so much imprisoned as being protected from the North Korean people.

Many features of the demeanor and behavior of the *Pueblo* crew following their release from North Korea were very similar to previous descriptions[7,12,13] of American POWs following repatriation from Oriental countries. This repatriation syndrome consists of an initially bland affect, identification with the prisoner group, anxiety concerning return to the United States, and the development of outspoken anger and hostility after several days or several weeks.

Greenson,[5] writing on the psychology of apathy, stated that it develops when the ego restricts perception and response to internal and external stimuli as a defense against overwhelming feelings of annihilation. Recovery from apathy occurs bit by bit as the patient gives up the constant image of disappointments and as lost objects are reinstated into the world and are once again charged into libidinal cathexis. The recovery from apathy of repatriated POWs is often marked by anger, hostility, and recriminations against other prisoners. In our view, what occurs is the expression of bottled-up anger and guilt, which must be experienced and worked through. Anger toward the captors must be repressed for survival. Guilt because of surrender and subsequent cooperation, no matter how slight, has often been rationalized through group identification or on occasion projected onto other prisoners.

The above formulations suggest that psychotherapeutic procedures play an important role in the rehabilitation of prisoners of war, since failure to recognize the need to work through the intrapsychic conflicts can result in unhappiness for both the man and his family. This seems to be demonstrated by our findings with the *Pueblo* crew. Certainly the most striking observation was the change between the initial evaluation and the findings three months later. There was great public pressure to return the men by Christmas for sentimental reasons; there was a quick reunion with their families and then a rush to complete their medical evaluations in order that they could be interviewed for intelligence purposes. In retrospect there is a question whether the needs of either the men themselves or the U.S. Navy were best met by hastening their return to the United States.

Since there are now at least several hundred American POWs in Southeast Asia, consideration of repatriation procedures is far more than an academic question. Our experience and review of the literature leads us to believe that the lengthy sea voyage back to the continental United States described by Lifton[7] provides a good way of dealing with some of the issues mentioned above.

A hospital ship staffed with medical specialists, including psychiatrists, would allow for careful medical evaluation and the use of psychotherapy aimed at uncovering and working through the conflicts associated with imprisonment and the return to the United States. For those men imprisoned for many years, there would be an opportunity to acquaint them in a psychotherapeutic setting with facts associated with their personal lives, such as deaths and divorce. Providing American newspapers, magazines, and recent films could assist their return to America.

To put such a recommendation into action would be difficult in this jet age since there is certain to be great public pressure for the men's immediate return to the United States and reunion with their families. We believe that rushing the process of repatriation is contraindicated and would complicate the rehabilitation of these unfortunate men.

ACKNOWLEDGMENTS

Lt. Cdr. R. E. Bullock, Lt. Cdr. D. B. Martindill, Lt. Cdr. J. M. Clemente, and Cdr. C. W. Erwin were the other members of the Psychiatric Evaluation Team. Lt. Cdr. H. B. McIntyre provided neurological consultation and assisted in the preparation of the paper. Lt. J. F. House, MSC, USN, Staff Clinical Psychologist, provided the psychological testing.

REFERENCES

1. BALDRIDGE, H. P. Experiences of the U.S.S. *Pueblo* (AGER-2) hospital corpsman. *U.S. Navy Medical Newsletter*, 1969, 54(6), 29–32.
2. BRANDT, E. *The last voyage of U.S.S.* Pueblo. New York: W. W. Norton and Co., 1969.
3. BRILL, N. Q. Neuropsychiatric examination of military personnel recovered from Japanese prison camps. *Bulletin U.S. Army Dept.*, 1946, 5, 429–38.
4. BUCHER, L. M. *My story.* Garden City, N.Y.: Doubleday and Co., 1970.
5. GREENSON, R. R. The psychology of apathy. *Psychoanalytic Quarterly*, 1949, 18, 290–302.
6. HARRIS, S. R. *My anchor held.* Old Tappan, N.J.: Fleming H. Revell Co., 1970.
7. LIFTON, R. J. Home by ship: reaction patterns of American prisoners of war repatriated from North Korea. *American Journal of Psychiatry*, 1954, 110, 732–39.
8. MAYER, W. E. Why did so many GI captives cave in? *U.S. News and World Report*, Feb. 24, 1956, 56–72.
9. MURPHY, E. R. *Second in command.* New York: Holt, Rinehart and Winston, 1971.
10. NARDINI, J. E. Survival factors in American prisoners of war of the Japanese. *American Journal of Psychiatry*, 1952, 109, 241–48.
11. SCHUMACHER, F. C. *Bridge of no return.* New York: Harcourt Brace Jovanovich, 1971.
12. SEGAL, H. A. Initial psychiatric findings of recently repatriated prisoners of war. *American Journal of Psychiatry*, 1954, 111, 358–63.
13. WOLF, S., AND RIPLEY, H. S. Reactions among Allied prisoners of war subjected to three years of imprisonment and torture by the Japanese. *American Journal of Psychiatry*, 1947, 104, 180–93.

Article 26 The Prisoner of War

Coping with the Stress of Isolation

EDNA J. HUNTER

Social isolation produces stress, and although persistence of effects in short experimental isolations may last only a matter of hours or days, it has been suggested that the psychological disabilities resulting from prolonged extreme stress may be permanent. Numerous studies over the past thirty years have presented evidence which supports the hypothesis that there are long-term effects resulting from the stresses of wartime captivity.[1-5,11,12,15,16] In a recent long-term follow-up of World War II and Korean prisoners of war, Beebe[1] pointed out that the multidimensional character of the stresses of captivity "severely limits inferences about the etiologic role of specific components; e.g., malnutrition, social isolation, sensory deprivation, physical punishment, compulsory education, and the like."

The present study concerned itself primarily with one specific component, *social isolation*. Suedfeld [17] has pointed out that the amount of research on the effects of social isolation *per se* is very small, and that most studies of "social isolation" deal with psychological rather than physical isolation. Another obvious problem with the experimental literature is that isolation and confinement are of relatively short durations.

Most investigators would agree that complete isolation, especially when prolonged, is to some degree stress-provoking for most individuals, and, further, that any particular stressor affects different individuals to varying degrees. Certainly work with infra-human organisms has demonstrated that social isolation

Edna J. Hunter, Ph.D., is a clinical research psychologist, Assistant Director for Administration, Center for Prisoner of War Studies, Naval Research Center, San Diego, Calif.

Source: This research was supported by the Department of the Navy, Bureau of Medicine and Surgery, under work order request number 18–75–WR–0000 dated July 1974 and by the Office of the Surgeon General, Department of the Army under military interdepartmental purchase request number 7401 dated July 1974. The opinions and assertions contained herein are the private ones of the author and are not to be construed as official or as reflecting the views of the Department of the Navy or the Department of the Army. Based on a paper presented at the International Conference on Psychological Stress and Adjustment in time of War and Peace, Tel Aviv, Israel, January 1975.

during early stages of life can have severe adverse effects. One of the recently released prisoners from Southeast Asia commented that he felt "periods of isolation had a definite effect on personality traits," and that while in isolation he could see these changes and realized he had to do something about them.[13] For the human organism, there is considerable evidence that "isolation for an extended period leads to effects which would normally be considered pathological symptoms; hallucinations, delusions, the development of obsessive rituals, periods of severe anxiety, feelings of unreality, and the like." [17]

Although individuals may vary in their ability to adjust to differing degrees of stress, Hocking[7] has contended that "subjection to prolonged, extreme stress results in the development of 'neurotic' symptoms in virtually every person exposed to it." In a study by one of the American prisoners of war recently released from North Vietnam,[18] an attempt was made to measure the relative amount of stress produced by the various stress situations which the returnees encountered during their captivity. Data for that study were derived from the responses of the investigator's fellow prisoners of war (POWs) who were asked to rank-order the various captivity stresses as to the degree that each was perceived as stress-provoking. Social isolation ranked as the third most important stress of captivity, exceeded only by the stresses produced by: (1) the event of the capture itself, and (2) physical torture during captivity.

Although there seem to be numerous similarities between the treatment of the American POWs in Southeast Asia and the treatment of the POWs of the Korean War and those involved in the *Pueblo* incident, the experience of this more recent group of POWs was, in a number of ways, highly unique. The four major aspects of the Vietnam POW experience which distinguished it from that encountered by other American prisoners of past wars were: (1) the longer duration of the captivity period, with the majority confined for more than five years (some men were captive in excess of eight years); (2) the more highly select population with respect to age, rank, educational level, and training (the majority were officers and flyers); (3) the greater psychological stresses used by the captor; and (4) the extended periods of social isolation for the majority of the POWs (for some up to as much as five years spent alone in a cell).

PURPOSE OF THE STUDY

The Navy and Marine prisoners of war who were returned to the United States from Southeast Asia early in 1973 had experienced varying periods of social isolation during captivity; these periods of solitary confinement ranged from less than one month to almost five years. Thus, these men presented a unique opportunity to examine the effects of prolonged social isolation. The present study was a preliminary pilot-type project designed to gain some general insights into psychiatric residuals of prolonged social isolation through looking only at the group extremes with respect to both time spent in captivity and time in solitary confinement. The major hypothesis set forth was that those prisoners of war who had been subjected to prolonged periods of social isolation during captivity would present more abnormal psychiatric symptoms after release than those with limited solitary confinement.

Three factors should perhaps be noted, however, that might cause one to make a prediction that there would be only minimal abnormal psychiatric symptomatology at the time of repatriation: (1) the euphoria among the prisoners noted at the time of homecoming probably had a "masking effect," and even where problems were observed by the psychiatrist, he may have tended to write them off as merely situational; (2) these men were a highly select group as to age, intelligence, and training even before capture and even more select at release, because had they not been better than average in physical constitution and emotional stability they would never have survived; and (c) studies of prisoners of past wars indicate a belated onset of psychiatric symptomatology.

THE SAMPLE

There were 164 Navy and Marine returnees,* but because some men had experienced relatively brief captivity periods and because postrepatriation reports by the prisoners indicated that captor treatment improved substantially subsequent to 1969, only those men who had been held captive for at least five years were selected. The sample for this particular study was comprised of *all* Navy and Marine Corps returned prisoners of war who had spent sixty months or more as captives in Southeast Asia (N = 100).

METHOD

Each of the prisoners of war released from Southeast Asia in 1973 was given a thorough medical evaluation according to a carefully planned computer-analyzable schedule, *The Initial Medical Evaluation Form* (IMEF).† Evaluations of the releasees were begun at Clark Air Force Base in the Philippine Islands and were completed during the weeks immediately subsequent to release at hospitals within the continental United States that were located near the men's homes.

Data for this study were derived primarily from two sections of the IMEF: (a) Form VI: The Psychiatric Questionnaire, and (b) Form VII: the Psychiatric Examination. Both sections were completed by the particular psychiatrist assigned to evaluate the mental status of the man at the time of return. The Psychiatric Examination was divided into three major subsections that consisted of (1) a mental status examination; (2) a narrative summary of the captivity his-

---

* In addition to the 164 Navy and Marine returnees, there were 77 Army, 325 Air Force, and 25 civilian POWs released in February and March 1973.

† The IMEF was drawn up through the coordinated efforts of all branches of the military service, in liaison with the medical staff of the Center for Prisoner of War Studies in San Diego, California.

tory of the man, personal and family history, and his current level of functioning; and (3) a subsection that included a problem list, treatment plans, diagnoses, and prognosis. The Psychiatric Questionnaire covered aspects of the captivity experience that could perhaps, in some way, contribute to a fuller understanding of the range and magnitude of the stresses of each man's individual captivity experience and the various mechanisms, such as denial, rationalization, intellectualization, etc., utilized by the man in order to cope with those stresses.

Dividing the total sample in half, it was found that fifty of the men had experienced social isolation lasting seven months or less, and this group was designated as the Lo Solo Group. The remaining fifty men who had been subjected to more than seven months' social isolation were in the Hi Solo Group. Using the 2 x 2 chi square test or the *t* test for independent means, comparisons were made between the two groups for all items contained in the Psychiatric Examination and Psychiatric Questionnaire.* The degree of social isolation, that is, the amount of time spent in solitary confinement, was based upon the psychiatrist's response to the question: "How much of the period of captivity was spent alone, by himself?" It should be noted that spending time "alone" did not necessarily mean that the man was totally isolated from other prisoners, as he was often still able to communicate through tap codes or other means, even though in a cell by himself. It has been shown by other investigators[18] that there is a high correlation between stress and condition of communications while in solitary. Therefore, it is unfortunate that information on the actual amount of communication possible while in isolation for each man was not available from data sources used for this study.

RESULTS

Mean Captivity and Solo Time

On the average, the 100 Navy and Marine POWs in the total sample had experienced exceptionally long periods of captivity, with a mean captivity of 73.9 months for the total sample, 72.5 months for the Lo Solo Group, and 75.3 for the Hi Solo Group. During the captivity period, without exception, all men experienced some period of time alone, but the amount of time varied considerably from man to man. Mean duration of solo time (not necessarily endured in one increment) for the Lo Solo Group was 3.3 months, for the Hi Solo Group, 21.2 months.

Demographic Differences

No statistically significant between-group differences were found with respect to race, branch of service, or years of formal education. (Both groups showed an average of slightly more than fifteen years of formal schooling.) There were, how-

---

* Data analyses were carried out by Gary Lester of the Data Analysis Branch of the Center for POW Studies.

ever, statistically significant between-group differences for certain other demographic variables. The Hi Solo men were older and of higher rank at time of
capture. Mean pay grade was 3.6 for Hi Solos and 2.7 for Lo Solos. Mean age
at time of casualty for men with limited solo time was 28.9 years; for those with
prolonged solo time, 31.9 years. As might be expected, along with higher mean
age, a higher percentage of the Hi Solo Group was married (76.0 percent to
68.0 percent for the Lo Solo Group) and had more children. These findings
would seem to corroborate returnees' reports that the older, higher-ranking men
were often separated from the other prisoners.

Social Isolation Effects as Reflected in Psychiatric Ratings

Between-group comparisons were made on the basis of the presence or absence of an abnormal rating on items contained in the thirteen categories of the
Psychiatric Examination. Many differences were found between the Hi Solo and
Lo Solo Groups, but only those between-group differences which were statistically
significant (p ≤ .05) will be presented and discussed here. For all differences discussed, the direction was that the Hi Solo Group was found to have more
abnormal ratings than the Lo Solo Group.

For example, those with prolonged solitary demonstrated significantly more
guilt than those in the Lo Solo Group. The Hi Solo Group also received more
frequent abnormal ratings on suggestibility and rate of speech. Moreover, the Hi
Solo Group received significantly more abnormal ratings on overdeveloped superego, unrealistically high need for achievement, and the need for achievement not
commensurate with the man's abilities (see Table 1).

Table 1. PSYCHIATRIC RATINGS: Between-group Differences for Returned Navy and Marine POWs Who Experienced Prolonged Captivity and Solitary Confinement[1] and Those with Limited Social Isolation[2] during Captivity

Psychiatric Rating	Lo Solo Group (N = 50) %	Hi Solo Group (N = 50) %	χ^2 (df = 1)
Presence of guilt feelings	16.3	41.7	7.58†
Abnormal suggestibility	16.0	40.8	7.51†
Abnormal rate of speech	16.0	34.0	4.32*
Superego overdeveloped	10.0	34.7	8.73†
Unrealistically high need/achievement	2.0	17.0	4.83*
Need/achievement not commensurate with abilities	4.0	21.3	6.67†

[1] All Navy and Marine returnees who were captive 60 months or more and who spent more than seven months in solitary confinement.

[2] All Navy and Marine returnees who were captive 60 months or more and who spent seven months or less in solitary.

* p < .05
† p < .01

Age or Social Isolation Effects?

Since findings showed that the men in the Hi Solo Group were, on the average, older and of higher rank, we must question whether the statistically significant between-group differences with respect to postrepatriation psychiatric ratings are merely a function of age, rather than related to time spent in solitary confinement. Therefore, another comparison was made using only those returnees who were thirty years of age or older at the time of capture. Within the original sample of a hundred men who had been held captive for five years or longer, thirty-three men in the Hi Solo Group had attained that age at time of capture, and twenty-one men in the Lo Solo Group were that age or older. When between-group comparisons of these older subsamples were made for all postrepatriation psychiatric ratings, statistically significant differences were very much evident, indicating that perhaps social isolation, and not mere age alone, is at least, in part, related to the psychiatric differences noted at the time of release (see Table 2).

Table 2. PSYCHIATRIC RATINGS: Between-Group Differences for Older[1] Returned POWs Who Experienced Prolonged Solitary Confinement and Older Men with Limited Solo Time

Psychiatric Rating	Older Lo Solo Group (N = 21) %	Older Hi Solo Group (N = 33) %	χ^2 (df = 1)
Does not appear stated age	4.8	39.4	8.01†
Feelings of guilt	10.0	43.8	6.58*
Abnormal suggestibility	19.1	46.9	4.30*
Ambivalence	5.0	28.1	4.24*
Overdeveloped superego	14.3	40.6	4.20*
Unrealistically high need for achievement level	0.0	19.4	4.60*
Apprehensive during psychiatric interview	81.0	54.6	3.92*

[1] All Navy and Marine Returnees who were held captive in excess of 60 months and who were 30 years of age or older at time of capture (N = 54).
* p < .05
† p < .01

In comparing the older Hi Solo Group (N = 33) with the older Lo Solo Group (N = 21), with only one exception the Hi Solo Group received more frequent abnormal psychiatric ratings. Again, the men who experienced prolonged solitary confinements were rated as having higher levels of guilt and more frequent ratings of abnormal suggestibility (usually significantly *less* suggestible than the Lo Solos) and unrealistically high need for achievement. In addition, the Hi Solo Group showed significantly more ambivalence and were more often judged as not appearing their stated ages, usually in the direction of appearing *older* than actual chronological age.

The one area where the older Lo Solo Group received more abnormal ratings than those men who experienced prolonged solitary confinement was in appearing overly apprehensive during the psychiatric interview.

Severity of Treatment by the Captor

We still must ask one further question: Can these differences be accounted for wholly by time spent in social isolation, or were there also between-group differences with respect to the severity of treatment by the captor which could perhaps account for the abnormal psychiatric ratings at time of release? Were the men who experienced excessive solo time also the ones who received the harshest treatment during captivity? In an attempt to answer that question, the older Hi Solo (N = 33) and Lo Solo Groups (N = 21) were compared on all aspects of their treatment by their captor; and statistically significant between-group differences were found for three items (see Table 3). The men who were

Table 3. SEVERITY OF TREATMENT BY CAPTOR: Between-Group Differences for Older[1] Returned POWs Who Experienced Prolonged Solitary Confinement[2] and Older Men with Limited Solo Time[3]

Captor Treatment as Indicated on Psychiatric Questionnaire	Older Lo Solo Group (N = 21) %	Older Hi Solo Group (N = 33) %	χ^2 (df = 1)
Use of torture devices or procedures (such as "the ropes," etc.)	28.6	60.6	5.28*
Lack of adequate shelter or clothing	47.6	75.0	4.13*
Withdrawal or diminishing food or water	9.5	35.5	4.50*

[1] All Navy and Marine POWs who were held captive in excess of 60 months and who were 30 years of age or older at the time of capture (N = 54).
[2] Solitary confinement in excess of seven months.
[3] Solitary confinement for seven months or less.
* $p < .05$.

subjected to the longest periods of social isolation were also the ones who appeared to have received the harshest treatment from the captor. Seventy-five percent of the Hi Solo Group reported lack of adequate shelter or clothing as compared to 47.6 percent of the Lo Solo Group. Withdrawal or diminishing the ration of food or water as a means of punishment or coercion was also reported more frequently by the men who experienced prolonged periods of solitary. Additionally, actual physical punishment in the form of torture devices or procedures such as "the ropes" were used on the Hi Solo group more often than on the Lo Solo Group.

DISCUSSION

The major hypothesis that these prisoners of war who had experienced prolonged periods of social isolation during captivity would present more abnormal psychiatric symptoms after release than those with limited solitary confinement appears to be confirmed by the findings of this study. Between-group comparisons of the Hi Solo and Lo Solo Groups showed the Hi Solo Group members receiving significantly more abnormal ratings for presence of feelings of guilt and ambivalence. This group was also more apt to have other than average ratings on suggestibility, superego development, and need for achievement. Physiologically, the men who spent extensive periods in solitary tended to appear older than actual chronological age. However, it was the older man who experienced *limited* solo time who appeared more apprehensive during the psychiatric interview. What do these differences really indicate?

We have already mentioned the more severe treatment that appeared concomitant with high solitary time, which points up again the difficulty mentioned by Beebe[3] of making inferences about the etiologic role of specific components of the captivity experience.

Lazurus[9] mentioned four factors within the psychological structure that influence one's coping with stress: (1) patterns of motivation, (2) ego resources, (3) defensive dispositions, and (4) general beliefs about the environment and one's resources. He further pointed out that "stress cannot be defined exclusively by situations because the capacity of any situation to produce stress reactions depends on characteristics of the individual."

Reactions of the prisoner of war run the entire gamut of human coping mechanisms. Ford and Spaulding[6] enumerated the wide variety of ego-defense mechanisms used in coping, such as faith, reality testing, denial, rationalization, and humor, and pointed out that those prisoners who handled stress poorly were those who appeared to be passive-dependent personality types and were more limited in the number of ego-defense mechanisms utilized. Suedfeld[17] has stated that "the optimal adjustment to isolation and to sensory deprivation is apparently made by those subjects who are able to relax and enjoy the flow of fantasy," and that "the less information is available, the more the individual attends to and elaborates residual stimuli, whether internal or external."

As one former POW put it: "Beyond their bodies, they kept their minds occupied by similar mental gymnastics. They built houses, roads, and bridges; they drilled oil wells; they dreamed and fantasized; they thought of families and classmates and friends; they did physics and math problems to several decimal places in their heads; they calculated money saved; and they looked inward at self." [14]

Prisoners often spend hours on end contemplating what they have done with their lives and what they intend to do with them if they are lucky enough to survive their imprisonment. This type of contemplation often results in very positive philosophical changes in a man's attitude, value system, and philosophy of life.[10]

A physician who survived his Southeast Asian captivity experience expressed the belief that in order to survive, *mental* strength was infinitely more important than physical strength, and found that accepting isolation year after year required ultimate mental discipline. The lesson he learned was that one can rise above the isolation environment; that "man is an adaptive animal . . . and though he is a product of his environment, and may be subjugated and destroyed by a strange one, he must constantly attempt to work within it, to mold it, and to rise above it. *It is the key to survival.*" [8]

In attempting to interpret the findings of this study, we are forced to pose the question as to whether *personality factors* were at least partially responsible for particular individuals' being subjected to extended periods of solitary, as well as asking whether prolonged social isolation resulted in the types of between-group differences which were found. It may be that these differences existed between these particular men even *prior* to captivity. The more rigid men with overdeveloped superegos and unrealistically high need for achievement might be less pliant and more resistive in their interactions with the captor, and thereby more apt to antagonize the captor into subjecting them to harsher treatment, including longer periods of solitary confinement. Rutledge[14] pointed out that "often (but not always) POW conduct in the face of captor demands was directly responsible for his solitary confinement."

Note further that it was the men with only limited amounts of solitary confinement who appeared more apprehensive at the time of release. Apprehensiveness as a typical response mode could perhaps also explain those individuals' shorter periods of social isolation.

CONCLUSION

Although a number of statistically significant between-group differences on postrepatriation psychiatric ratings were found for prisoners of war who experienced prolonged social isolation and those who experienced limited solitary confinements during captivity, no definitive statement can be made on the basis of these data as to the specific effects of social isolation *per se*.

REFERENCES

1. BEEBE, G. W. Follow-up studies of World War II and Korean War prisoners: II. Morbidity, disability, and maladjustments, 1974 (in press).
2. BRILL, N. Q. Neuropsychiatric examination of military personnel recovered from Japanese prison camps. *The Bulletin of the U.S. Army Medical Department*, 1946, 5, 429–38.
3. COHEN, B. M., AND COOPER, M. Z. A follow-up study of World War II Prisoners of War. V. A. *Medical Monograph*, 1954.

4. EITINGER, L. Psychosomatic problems in concentration camp survivors. *Journal of Psychosomatic Research*, 1969, *13*, 183–89.
5. EITINGER, L. Schizophrenia among concentration camp survivors. *Int. J. Psychiat.*, 1967, 3, 403–6.
6. FORD, C. V., AND SPAULDING, R. C. The *Pueblo* incident: A comparison of factors related to coping with extreme stress. *Arch. Gen. Psychiatry*, 1973, *29*, 340–43.
7. HOCKING, F. Psychiatric aspects of extreme environmental stress. *Dis. Nerv. Syst.*, 1970, *31*, 542–45.
8. KUSHNER, F. H. To live or to die. *AMEDD Spectrum*: U.S. Army Med. Dept., Vol. I, No. 1, 16–21, 1974.
9. LAZARUS, R. S. *Psychological stress and the coping process.* New York: McGraw-Hill, 1966.
10. MILLER, W. N. Dilemmas and conflicts specific to the military returned prisoner of war. In H. I. McCubbin; B. Dahl; P. Metres; E. J. Hunter, and J. Plag, eds., *Separation and reunion: Families of prisoners of war and servicemen missing in action.* Washington, D.C.: U.S. Government Printing Office, 1975.
11. NARDINI, J. E. Psychiatric concepts of prisoners of war confinement (The William C. Porter Lecture—1961). *Military Medicine*, 1962, *127*, 299–307.
12. NEFZGER, M. D. Follow-up studies of World War II and Korean prisoners of war. *Am. J. Epidemiol.*, 1970, *91*, 123–38.
13. ROWAN, S. A. *They wouldn't let us die.* Middle Village, New York: Jonathan David Publishers, 1973.
14. RUTLEDGE, H. E. A comparison study of human values of the Vietnam prisoners of war experiencing long-term deprivation. Unpublished Master's thesis: United States International University, 1974.
15. SEGAL, J. Long-term psychological and physical effects of the POW experience: A review of the literature, *NPRU Pub. No. 74-2*, 1974.
16. STROM, A., ed. Norwegian concentration camp survivors. Oslo: Universitetsforlaget, 1968.
17. SUEDFELD, P. The individual in experimental isolation. Paper read at a symposium, "Man in Isolation," at the annual meeting of the American Psychological Association, Montreal, August 1973.
18. VOHDEN, RAYMOND A. *Stress and the Vietnam POW.* Industrial College of the Armed Forces, Washington, D.C., Student Research Report No. 091, 49 pgs., March 4, 1974.

Coping with Unusual Stress: Concentration Camps

X Internment camps have played a role in government policies sporadically through history. These camps differ from those for prisoners-of-war in that they are designed for a country's own citizens. They are not penal institutions since the inmates usually have broken no laws. The basis for the internment is race, religion, ethnicity, or political philosophy. The most famous example is the Nazi concentration camps of World War II. While nothing can match the ruthless destruction or the scale of the Nazi camps, the internment of Americans of Japanese descent in the Western United States and the countless "preventive detention" camps for political prisoners in authoritarian regimes in South America, South Africa, and the Soviet Union all present their victims with similar psychological stresses.

In the spring of 1942, 117,000 Japanese-Americans were given notice that they had to leave their homes and enter relocation camps within a few days. Two-thirds of these people were American citizens, who were almost totally unprepared for the wholesale loss of their rights. They had to sell many of their belongings hurriedly (and often at great loss), store the rest with friends or in government warehouses, and find people to take over their homes and businesses. With armed soldiers presiding they were herded into camps surrounded by barbed wire and lighted guard towers. Unlike the inmates of the Nazi camps they were not physically abused; they had sufficient food, clothing, and shelter. There was a semblance of normalcy as schools, jobs, and churches were established inside the camps; but the realities of confinement, regimentation, monotony, and overcrowding could not be ignored.

The assault on personal and ethnic self-esteem was the most severe problem. Children interpreted internment as a punishment for being Japanese and, accepting the values of the Caucasian majority, became ashamed and apologetic for their ethnic background.[6] Young adult Nisei ("second generation") found some solace in keeping busy organizing and staffing camp activities, hoping for a change in their circumstances, and redefining their cooperation in the evacuation and internment as their part in the war effort. Older Issei ("first generation") men, robbed of their roles of responsibility for and authority over their families, responded with depression and withdrawal.[4] Aside from the severe economic setbacks suffered by many families, and the premature end of the education of some Nisei (resulting in permanent occupational handicaps), most people recovered well after their release. By 1960 Americans of Japanese descent ranked very high in education, professional status, and income levels, and one observer has concluded that internment toughened Japanese-Americans while reinforcing their determination to do well within the American system.[6]

The Nazi concentration camps, functioning as combination relocation and extermination centers, represent an entirely different magnitude of physical and mental trauma. Although European Jews had experienced many years of persecution in various ghettos and in flight from the Nazi advance, "expectation failed to fulfill its protective functions; for what actually happened, was 'beyond expectation.'"[2]

In the first article Paul Chodoff examines the stressful conditions in the German concentration camps and how the prisoners dealt with them. Physical stresses included malnutrition, exposure, forced labor, crowding, sleeplessness, beatings and other torture, exhaustion, epidemic diseases, and "death marches." The brutal physical conditions helped make the prisoners more vulnerable to the massive psychological stresses, such as the ever-present threat of death, the separation from family members, the daily humiliation, the general uncertainty, and the constant assault on individual self-esteem and identity. Nazi guards took sadistic pleasure in ordering inmates to engage in degrading acts that evoked extreme disgust and shame, thus provoking tremendous conflicts between inmates' self-images and their desire to survive.[2,5]

The initial response to these conditions was shock and terror, but eventually a certain degree of apathy took over. This denial of feeling served a temporary protective function, unless it became an overwhelming and usually fatal lassitude (the "musselmann" syndrome). Docility and submissiveness were found to be adaptive in the circumstances, and often led to feelings of dependence on the Nazi guards and an ambivalent rather than overtly hostile attitude toward them. Identification with their Nazi oppressors relieved some of the psychic and interpersonal tension, as did quarrels with other inmates and daydreams of revenge. Companionship, usually in pairs, provided some human warmth although the high rate of inmate death and the need for emotional self-preservation dictated that relationships not be too deep.

In the second article Joel Dimsdale focuses on the effectiveness of various coping strategies used by survivors of concentration camps. These include a selective focus on the good, a general will to live or survival for some special purpose (like helping a relative or seeking revenge), various forms of psychological removal (for example, shifting time focus to past or future), cultivating some sense of mastery (for example, by working in the underground resistance), the mobilization of hope, group affiliations, fatalism, and identification with the oppressor. Dimsdale points out the importance of interaction and overlapping of these strategies and that too heavy reliance on any one strategy was unhealthy and could prove devastating if it should collapse.

The problems did not end with liberation. In the third article Peter Ostwald and Egon Bittner discuss the postwar adjustment of concentration camp survivors. After the war survivors had three alternatives: to return to their homes (usually to find them destroyed and their families gone), to enter Displaced Persons camps (where idleness and dependence became the way of life), or to emigrate. The nature of these experiences had a significant effect on the ultimate adjustment a person was able to make. Within a few years after the war doctors began to recognize among some survivors a common pattern of psychological symptoms, which has been called the "concentration camp syndrome." Anxiety (expressed in irritability, nightmares, psychosomatic illnesses), depression, and obsessive preoccupation with recollections of persecution have been noted. Ostwald and Bittner discuss one strategy of long-term adjustment in which survivors allow the attainment of material success to become the focus of all their energies. They cannot relax or enjoy their prosperity, though, because the threat of doom seems ever-present.

As the effects did not end with liberation, nor did they end with the first generation. In recent years the children of the survivors of Nazi persecution have begun to receive some attention, as they have been seeking psychiatric counseling in disproportionate numbers. Concentration-camp survivor parents tend to be overprotective, to overvalue their children (expecting them to make up for all their lost relatives), and to be so preoccupied with the past that they view their children's activity as an interference and a burden.[7,8] These children experience feelings of guilt, depression, or alienation. Even with the extreme trauma that the Nazi concentration camps represented, some people were able to make the adjustments necessary to survive, and to deal with the effects that continue to be a part of their lives more than thirty years later.[1,3]

REFERENCES

1. BETTELHEIM, B. *The informed heart.* New York: Free Press of Glencoe, 1960.
2. BLUHM, H. O. How did they survive? Mechanisms of defense in Nazi concentration camps. *American Journal of Psychotherapy,* 1948, 2, 3–32.
3. FRANKL, V. E. *Man's search for meaning.* Boston: Beacon Press, 1959.
4. KITAGAWA, D. *Issei and Nisei: The internment years.* New York: Seabury Press, 1967.
5. NIEDERLAND, W. G. An interpretation of the psychological stresses and defenses in concentration-camp life and the late aftereffects. Pp. 60–70 in H. Krystal, ed., *Massive psychic trauma.* New York: International Universities Press, 1968.

6. OKIMOTO, D. I. *Americans in disguise.* New York: Walker/Weatherhill, 1971.

7. SIGAL, J. J., AND RAKOFF, V. Concentration camp survival: A pilot study of effects on the second generation. *Canadian Psychiatric Association Journal,* 1971, *16,* 393–97.

8. SIGAL, J. J.; SILVER, D.; RAKOFF, V.; AND ELLIN, B. Some second-generation effects of survival of the Nazi persecution. *American Journal of Orthopsychiatry,* 1973, *43,* 320–27.

Article 27 The German Concentration Camp as a Psychological Stress

PAUL CHODOFF

Homo homini lupus (Man is a wolf to man).

The bloody chronicles of recorded history have, time after time, demonstrated the truth of this bitter adage, but never more clearly than in the treatment of the Jewish minority under their control by the German Nazis of the hitlerian reich. Therefore, in a symposium on the psychological aspects of stress, the concentration camp experience can serve as a paradigm of how the human organism reacts to stressful conditions which approach the outermost limits of human adaptability.

All of the concentration camps set up by the *Geheime Staats Polizei* (SS) in Germany and occupied Europe were not alike and the differences between extermination camps and labor camps were certainly significant. However, the conditions faced by the inmates of all concentration camps can only be described, in the words of A.P.J. Taylor, as "loathsome beyond belief." In addition to the out-and-out extermination measures, the physical stresses endured by the prisoners included malnutrition, crowding, sleeplessness, exposure, inadequate clothing, forced labor, beatings, injury, torture, exhaustion, medical experimentation, diarrhea, various epidemic diseases, and the effects of the long "death marches" from the camps in the closing days of the war. The physically depleted state of the prisoners, the brutal and primitive conditions in which they lived, and the entirely inadequate medical facilities were responsible for an extremely high death rate, and also had the effect of increasing the susceptibility of the inmates to the nonphysical stresses which they had to face. Chief among these latter was the danger to life, everpresent and unavoidable. It is possible, to some degree, for a healthy personality to defend itself against a peril which, though very grave, is predictable, and is at least potentially limited in time, as in the case of soldiers

PAUL CHODOFF, M.D., is Clinical Professor of Psychiatry, George Washington University School of Medicine, Washington, D.C.

Source: Archives of General Psychiatry, Vol. 22 (January 1970), 78–87. Copyright 1970, American Medical Association.

in combat who can at least hope for relief and rotation out of the fighting zone, but for the concentration camp inmate, as has been described by Viktor Frankl from his own experiences in Auschwitz,[5] the absolute uncertainty of his condition was a barrier to the erection of adequate psychological adaptive measures. In addition to the threat to life, the prisoners had to face the catastrophic trauma of separation from their families and, consequently, to the agonizing uncertainty about their own future there were added equally agonizing doubts as to whether they would ever see their relatives again. The very least price one had to pay to survive in the camps was to suffer the grossest kind of daily humiliation at the hands of an organization superbly qualified to mete out humiliation. Massive frustration of their basic drives had the apparent dual effect of driving the sexual life of the prisoners underground and of rendering the insistent demands of hunger all-dominating so that fantasies about food occupied much of their conscious awareness. If not himself the victim of casual violence or deliberate cruelty, the prisoner at least frequently witnessed such exhibitions. Since it was impossible to retaliate effectively he had to smother his aggressive feelings, for even the appearance of resentment against the torturers could lead to his own destruction. Regimented, imprisoned, without a moment of privacy during the 24 hours, the prisoner's human worth, and even his sense of an individual human identity, was under constant and savage assault. His entire environment was designed to impress upon him his utter protoplasmic worthlessness, a worthlessness which had no relationship to what he did, only to what he was. Reduced from individual human status to the status of a debased being, identified not by name but by a number and a badge of a particular color, a conviction of his ineluctable inferiority was hammered into the prisoner by the SS jailers who needed to justify their own behavior by convincing themselves of the inferior, subhuman status of the Jews in their charge. The concentration camp inmates lived in a kafkaesque world in which the rules governing their existence were senseless, capricious, and often mutually contradictory, as, for instance, when impossible standards of cleanliness were demanded, while, at the same time, the inadequate opportunities and senseless rules about toileting made even an elementary decency impossible.

At this point, I am going to interpose exerpts from a description—recorded in her own words at my request—of some aspects of life at Auschwitz by a former inmate who has also been a patient of mine. I believe that this account will convey to you something of the physical and psychological stresses which confronted the concentration camp inmate, as well as a suggestion of the adaptive measures called forth by these stresses.

LIFE IN A CONCENTRATION CAMP

"We arrived to Auschwitz in the early part of April. That's a story by itself— the arrival and the happenings. We lived in blocks. It's called a block in German but it's a barracks and it has 1,500 people and the barracks had rows and rows of three-tier bunk beds—they were very poorly built, out of just boards and lumber and each bunk bed had 12 people—in our case 12 women, and we happened to choose the upper bunk because we thought it would be the best. It was OK for us but it caused an awful lot of heartache afterwards because, as I said, the bunk beds were not built very well and the boards broke under the weight of us in the beginning before we started losing weight and there were many broken or fractured skulls or bones below us.

"Each person was supposed to have a blanket. It so happened that the blankets were stolen so we never ended up with enough blankets for all of us. Maybe by the end of the evening we would have two or three blankets left for 12 people. I think the biggest crisis after the day's hardship was finding a blanket to sleep on and the fights we had keeping the blanket. I must say I was one of the lucky (I say lucky) or exceptional persons who was an assistant to the Capo. It came very handy at the time. I suffered an awful lot because of it later, but it gave me protection. Maybe I had the blanket . . . I don't even remember because all this did not matter.

"Anyway, we got up at two o'clock in the morning. We were awakened by the girls who brought the coffee in big barrels. It was black and it had charcoal in it and maybe chicory. It was rather lukewarm and they thought it was enough to hold us through the day. And no one wanted the coffee. You would rather have an hour sleep or so, but you had to have it. The barracks had to be emptied, and we had to get back and line up for the *appel*—roll call—and that was my job, getting people up. I was assigned to three or four bunks and that means I was assigned to wake-up approximately a 150 to 200 people who did not want to face the day's reality. I mean, because we knew what was coming. So I was up, pulling blankets off of people—'get up,' screaming, carrying on, even hitting. Once I hit someone and she looked at me and it was one of my mother's friends. I apologized. I felt terribly bad. I kept on doing it. I got them up. I think the whole process took nearly half an hour to make 1,500 people get out of the block. We were up but there was no question of dressing because we got one dress when we entered Auschwitz and I got a very, very long petticoat. It came all the way down to my ankles. It came in very handy because I was able to use it little by little as toilet tissue which was nonexistent, 'till there was nothing left. So I just had a dress to wear and I think most people, that's all they had, and shoes and a toothbrush that hung around my neck. It was my most prized thing . . . a toothbrush.

"I know I got out of my bunk a half an hour earlier than anyone else because we were determined to keep clean and we went to the washroom that was at least, oh, two or three bunkers away from us and we stripped ourself completely naked and scrubbed ourselves with cold water because we didn't want any lice on us. We checked our clothes for lice, because typhus was spreading already at that time and, anyway, we had to line up for *appel*—and the 1,500 people were divided in three groups so that means I had 500 people to line up and in each line I think there were maybe 10, 15, I'm not sure. But somehow, I don't know how the rumor started, but someone gave the word that they were going to pick so many people to be cremated—that the first four or five rows are going to be taken to be cremated. Well, I tried to get anyone to get in the first five rows. It was impossible. I had to make them do it. I was shoving and pushing. But we were lined up around two-thirty in the morning and the Germans did not come out till five or six o'clock I think it was for the *appel* and there were 30,000 people in the C lager and it took them a while. They always miscounted because, after the first counting was over we were being punished for someone was missing. We had to get down on our knees with our hands up and wait till they counted over and over and over and over till they decided that they had a number. But meantime, we didn't have the big selections as when Dr. Mengele came around. Just the little officers did their jobs and decided a girl with a pimple on her face, someone who was a little bit more run down than should be, or someone with a little bandage were selected. Naturally those people were taken out automatically to the crematorium. So we had to be very careful that you shouldn't have any scars showing or you should look fresh and not unwell. Well, by 11 or 12 we were allowed to come in to our bunks and then lunch-time came.

"It is very hard for me to recall food—what we ate or not because I decided that all those things are unimportant. That I am going to have control over my body, that food is not important, so I can't remember, but I think it was one slice of bread and a slice of margarine on it, and some kind of jelly made out of red beets.

"I think in all my eight months that I have been there, maybe twice we were able to take a walk—because we were being punished for one reason or the other and we had to stay in our bunks, or a new transport came in and they would not allow us out to watch the transport coming in so we were locked in our block or barracks. So the days were spent in the barracks.

Then came the evening roll call. That started about two o'clock in the afternoon and then again the same circus—lining up, waiting, miscounting and counting, and counting, and counting, and it was seven to eight o'clock in the evening before the evening meal came and before we were allowed to come in— sometimes nine-thirty or ten o'clock. If someone was missing we were punished by kneeling. Otherwise we stood—and we had to stand sometimes in hot sun and sometimes in rain.

"Anyway, evening came around and then we had our meal. The meal con-sisted of mush. It was some kind of a liquid, thickened with something. It had a few potatoes in the bottom of the barrel and then it had some kind of a chemical taste to it so that the women decided it was some kind of a tranquilizer, but I doubted the Germans spent money on putting tranquilizers in. But another thing, we stopped menstruating and they thought that, whatever chemical taste it had in it, this is the thing that stopped us from menstruating so we didn't menstruate at all.

"Again, I was one of the lucky ones who distributed the food. I think I was fair when it came to distributing food—I did not, I would not allow myself to bring any more up to my bunk than we were allowed to except for my mother who was quite ill—she had diarrhea and she could not digest the food. She was getting weaker and weaker day by day. You know she was at that time 43 years old and it is hard to think that I have reached that age. How would I have been able, how would I have reacted under the circumstances that she did at this age. I feel very young, by the way. But she seemed to me very old at that time and well anyway, evening meal over, we were allowed to take a walk. But again, when transports were coming in they locked the doors and there was complete curfew and we were not allowed to go out and we were just well, listening to the screams or the silence or the smell of the burning flesh and wondering what tomorrow was going to bring to us. Some were wondering. I knew I was going to make it. The human torture, getting up in the morning and things, smelling the burning flesh and the flying soot. The air was oily, greasy. There were no birds."

*   *   *

"I think they were very worried about our physical well-being and cleanliness so we had to be shaved. That's right, several times while I stayed in Auschwitz, we were shaved of hair—head, under arms and intimate parts of our body by Jewish inmates—standing. We were standing on a stool so that the Jewish inmates would be able to reach us easier and we were surrounded by SS men who seemed to enjoy it very much. Well it didn't bother me either. I had no feelings whatsoever. I couldn't care less at this point of the game. It really didn't matter."

*   *   *

"As I said, we had selections every single day—some just slight—just picking people out as I mentioned before because of scars, because of pimples, because of being run down, because of looking tired or because of having a crooked smile, or because someone just didn't like you. But then they were beginning to liquidate Auschwitz and we had major selections where you were selected either to go to work or to the crematorium and in this case Dr. Mengele was involved in it. He was quite an imposing figure and his presence—I don't think everybody was scared because, rather I wasn't. I was hypnotized by his looks, by his actions.

"The barracks had two massive doors and we were inside. They did not let us out. It was in October and it was rainy and we were holding pots of water—we didn't sleep because of the rain because we had to keep ourselves dry . . . and then the doors swing open quite dramatically . . . great entrance with Mengele in the center accompanied by two SS women and a couple of soldiers . . . he stands with his whip on one side and his legs apart. It's unbelievable. It looked like Otto Preminger arranged the theme for the whole thing. It seems to me now that it was like a movie.

"Anyway the Capo came out and she gave us orders to undress and line up in front of the barracks. It had two rooms. One was a storage room and one was the Capo's room—and Mengele stood in between and he had one leg lifted on a stool, his right leg, and he was leaning on his knee and he had a switch in the same hand and while we were lining up I was able to observe what he was doing. Till I had to face him I really had no feelings. I couldn't describe how I felt but I saw the switch go. It was a horsewhip—left, right, and I noticed that those who were motioned left were in a better condition, physical condition than the ones who were motioned to go right. Anyway, this went on and on and not a sound, even if it meant life or death. I don't know how other people felt about it, but I was quite well informed, I was accused of being able to face the truth, of being able to know because I had my mother with me. The others said it was easy for me to believe all this because I had my mother with me. I had no great loss. They loved their mothers, they cared, and they wanted to believe their mothers were safe somewhere in another bunk, somewhere in another camp. It made them stronger, knowing their parents were not cremated, but I had my mother and I knew my father was OK at that time in another bunk at camp.

"Anyway, my turn came. I had a choice to make. Not only Mengele had a choice to make, I had. I had to make up my mind. Am I going to follow my mother or is this it? Am I going to separate from her? The only way I was able to work out the problem was that I was not going to give myself a chance to decide. He will have to decide. I will go ahead in front of my mother—that was unusual, she being my mother, out of courtesy I followed her all the time in any other circumstances—but in this case I was going first and my mother followed me and I went. I think my heart was beating quite fast, not because I was afraid—I knew I would come through, but because I was doing something wrong. I was doing something terribly wrong. Anyway I passed Mengele. I didn't see him. I just passed and I was sent into the room where I would be kept alive and I turned around and my mother was with me, so this was a very happy ending."

"As I said in the evenings if I had a chance I went over to talk to my friend . . . to the fence, to the electric fence, and at each end of the fence they had the watch-towers where the Nazis were able to observe us and . . . just for the fun of it . . . the girl who was right next to me—I think they just wanted to see if they can aim well because I don't know why—they shot her right in the eye, and she lost her eye. Another time another girl was at the same place. Her friend threw a package of food over to her and she ran towards the fence to catch it and she touched the wire and there she hung. She looked like Jesus Christ spread out with her two arms stuck to the wire."

*   *   *

"In the end, people were losing weight, and they were getting skinnier and skinnier and some of them were just skeletons but I really did not see them. I just wiped the pictures out of my mind. I was able to step over them and when I came out from the concentration camp I said I did not see a dead body—I mean, I feel that I didn't see them. Even if I can see them. This is what's killing me now that I have never felt the strain, the brutality, the physical brutality of the concentration camp. I mean like my aunt, my young aunt. She was 13 or 14 when she was exposed to brutality and death and she talks with a passion of what they did to her. Then when I meet another woman who is in her 60's and she will tell me her sufferings, I can't stand them. I broke all the friendship up with them. I don't want them. I can't stand them, because they bore me. I just can't stand listening to them, and I have nothing to do with them."

COMMENT

What enabled a man or woman to survive such a hell? We have no real answers to this question and must resort to such generalizations as the almost miraculous and infinite adaptability of the human species. It needs to be emphasized that while particular varieties of individual defensive and coping behavior played a role in whether a prisoner would live or die, such behavior was far less important than were such chance factors as where the prisoner happened to be when a "selection" for the gas chamber was taking place or the mood of the selector at the time.

However, accounts by survivors do agree in describing a fairly consistent sequence of reactions to concentration camp life. This sequence was inaugurated by a universal response of shock and terror on arrival at the camp, since the SS made it their business to impress the new prisoner with their malevolent power, while many of the old prisoners, displaying "the camp mentality," manifested by irritability, egotistical behavior, and envy, were often cold and unsympathetic to new arrivals. This fright reaction was generally followed by a period of apathy, and, in most cases, by a longer period of mourning and depression. The period of apathy was often psychologically protective, in that it served to provide a kind of transitional emotional hibernation, but, in some cases, apathy took over to such an extent that the prisoner became a "mussulman," who gave up the struggle to live and did not survive very long.

Among those prisoners who continued to struggle for life, certain adaptive measures gradually gained ascendancy and came to be characteristic of the long-term adjustment. Regressive behavior, of greater or lesser degree, was almost universal, being induced in the prisoners by the overwhelming infantilizing pressures to which they were subjected and by the need to stifle aggressive impulses. As a consequence of the complete reversal of values in the camps, it is likely that such behavior served an adaptive function, since regressive prisoners immersed in fantasies were likely to be docile and submissive toward the SS and thus to have a better chance of escaping retaliatory measures. A consequence of such regression was that many prisoners, like children, became quite dependent on their savage masters, so that attitudes toward the SS were marked more by ambivalence than by conscious overt hostility. Some prisoners went so far as to employ the well-known mechanism of identification with the aggressor, imitating the behavior and taking on the values of the SS. In the dreams of more than one female survivor whom I have examined, SS troopers were always tall, handsome, and god-like figures, and I believe you will be able to detect elements of such a reaction in the description of my patient's encounter with Dr. Mengele. Those who have seen or read the play, *The Deputy*, will remember the satanic figure of "The Doctor" who was modeled on the same Dr. Mengele, "The Angel of Death of Auschwitz," whom I have heard described in such awed terms by survivors of Auschwitz that I—who have never seen him—can visualize his tall, radiant, immaculately dressed figure sitting nonchalantly astride a chair as—like Osiris, the judge of the Kingdom of the Dead—he gestures with his riding whip, selecting the prisoners lined up in front of him for either death or life.

It appears that the most important personality defenses among concentration camp inmates were denial and isolation of affect. As might be expected in the case of this probably most ubiquitous of all defenses, there was widespread employment of denial as manifested by my patient who would not see the corpses she was stepping over, and by the poignant picture of her fellow inmates who refused to believe that the smoke arising from the crematorium chimneys came from the burning corpses of their mothers. Isolation of affect, which could be so extreme as to involve a kind of emotional anesthesia, seemed to have functioned particularly to protect the ego against the dangers associated with feelings of hostility toward an external object which treats the self as if it were an inanimate thing and not a person. My patient who says, "It didn't bother me. I had no feelings whatsoever" when being shaved while naked in front of SS troopers, was certainly isolating her affect from her cognition. When combined with an ability to observe themselves and their surroundings, this kind of tamping down of affect along with sublimatory processes helped such gifted individuals as Frankl,[5] Cohen,[4] and Bettelheim[1] to produce some rather remarkably objective reports of life in the camps. Some form of companionship with others was indispensable, since a completely isolated individual could not have survived in the camps, but the depth of such companionship was usually limited by the overpowering egotistical demands of self-preservation, except in certain political and religious groups. Daydreams of revenge served the purpose of swallowing up some of the submerged aggression, while aggression could also be discharged through quarrelsome and irritable behavior toward the other prisoners, as illustrated by the description of the fighting over food and blankets, and by projection on the SS who were then seen as even more formidable, endowed as they thus were with the unexpressed hostility of the prisoners. Since the existence of mental illness of any degree of severity would have been incompatible with survival, an interesting adaptive consequence of imprisonment was that new psychosomatic or psychoneurotic disorders rarely developed while already existing ones often markedly improved, and suicide—except under conditions as to be almost indistinguishable from murder—was also an infrequent phenomenon.

As soon after the end of the war and liberation as the recovery of some physical health permitted, most survivors made their way back to their homes. More often than not, their worst fears were confirmed, and they usually found that not only had their relatives and friends perished in the holocaust but they also found their homes and communities destroyed or uninhabitable. Such frustrating and disappointing aspects of postliberation reality were responsible for the inevitable disruption of the rosy fantasies of postwar life which had proliferated during the imprisonment and which have been called "the post-disaster Utopia." [10] In the regressed state of most of the exprisoners at this stage, such a narcissistic blow, as well as real disappointment of their idealistic hopes that a better world would now arise from the ruins of Europe, resulted in much bitterness, resentment, and depression, even, in some cases, in temporary flurries of antisocial or paranoid behavior. A large number of the liberated prisoners, homeless, alone, bewildered, and without resources, took refuge in the displaced person (DP) camps which were set up in various parts of Europe, and, in some cases, remained in them for years with the result that their neurotic symptoms became fixated because of the monotony of DP camp life and its fostering of passivity and hypochondriacal preoccupation.

As the immediate postwar epoch drew to a close, the surviving remnant of concentration camp prisoners gradually settled themselves in more or less permanent abodes in their own countries of origin, or in other lands, especially in the United States and Israel. For this latter group, to the multiple traumata they had already endured there was now added the need to adjust to a new environment, to new customs, and to a different language.

At this point one might expect the grisly story to come to an end for most of the survivors, with the passage of time allowing the gradual envelopment of their fears and memories in psychic scar tissue. This is not what happened, however, for as I have described in another publication,[2] in the late 1950's and early 1960's articles began to appear in the medical literature of many countries describing features of personality disorder and psychiatric illness still present many years later in a large number of these survivors of the holocaust, in some cases cropping up after latent periods of months or years after the persecution.

Although figures are not available for the overall incidence of psychiatric sequelae among the survivors of the Nazi persecution, it is clear, judging from reports from many countries, including Germany, the United States, Israel, Poland, and Norway,[2] that they are of high frequency.

Long-term, unfavorable personality alterations in survivors have taken two widely overlapping directions. Some individuals develop tendencies toward seclusiveness, social isolation, helplessness, and apathy. They are passive, fatalistic, and dependent, wanting only to be taken care of, and to be let alone by a world whose requirements they are no longer interested in trying to fulfill. Other survivors regard their environment with suspicion, hostility, and mistrust, reflected in their attitudes toward other people which range from quiet, envious bitterness to cynicism and quarrelsome belligerence.

The most distinctive long-term consequence of the Nazi persecution, however, is the concentration camp syndrome. Invariably present in this syndrome is some degree of felt anxiety along with irritability, restlessness, apprehensiveness, and a startle reaction to such ordinary stimuli as an unexpected phone call or a knock at the door. These anxiety symptoms are almost always worse at night, and are usually accompanied by insomnia and by nightmares, which are either simple or only slightly disguised repetitions of the traumatic experiences. Psychosomatic involvement of almost all the organ systems has been reported, the most common being weakness, fatigue, and gastrointestinal troubles. A very characteristic feature is an obsessive ruminative state in which the individual is preoccupied with recollections of his own persecutory experiences and with the often idealized life with his family before the persecution began. Interviews with concentration camps survivors often leave the interviewer with the uncanny sensation that he has been transported in time back to, say, the gray inferno of Auschwitz, so vivid and compelling is the wealth of detail with which they describe the events which befell them and which they witnessed, so that the interviewer gets the impression that nothing of real significance in their lives has happened since the liberation. This feeling, that life was permanently interrupted by the concentration camp period, may be why my patient still feels "young" at her present age. Most individuals find their memories unwelcome and obtrusive, but there are a few who actually appear to derive pleasure from them. Depression, at times associated with "survival guilt," is also a very common manifestation of the concentration camp syndrome.

Studies dealing with the very few surviving individuals who were infants or very young children in the concentration camps are particularly interesting. Anna Freud and Sophie Dann have made a fascinating study of a group of six children who had all been in the concentration camp at Theresienstadt before the age of 3 years.[6] When seen in England after the liberation, these children showed severe emotional disturbances and were hypersensitive, restless, aggressive, and difficult to handle, but they had also evolved a remarkably stable sibling group and a group identity during their internment which seemed to protect them against the worst pathogenic effects of the absence of a maternal figure. A follow-up of the later fates of these children indicates that, though they all had stormy experiences during adolescence, most of them achieved some degree of stability by early adulthood. Sterba has reported on a group of 25 displaced children and adolescents[9] who had lost both parents and had been in concentration camps or in hiding for three to five years. She describes how attempts to place these children with foster parents were greatly hampered by the disappointment and dissatisfaction, expressed by the children toward whatever was done for them. They, too, displayed a desperate need to cling together, apparently deriving more security from these peer relationships than from even the most benevolent relationship with the adults on whom they were displacing all the angry fear engendered in them by the loss of their parents. Among my own cases, there were six who were 5 years of age or younger in 1939, and who therefore underwent the experiences of the persecution at an extremely early period of their lives. In addition to various degrees of overt psychoneurotic symptomatology, I found them all to be emotionally volatile people whose moods fluctuate markedly and who react to a mild degree of stress, such as an unexpected event, with exacerbations of anxiety. Their personalities are marked by various admixtures of a bitter, cynical, pessimistic attitude toward life and a child-like and total kind of emotional dependency. Although intimacy and closeness are of the greatest importance to them, they tend to show self-defeating patterns of excessive expectation and bitter or despairing withdrawal when these expectations are disappointed. They are extremely sensitive to actual or threatened separation from those on whom they have become dependent. It seems clear that the most damaging consequences to the personality maturation of these individuals resulted from the absence of a reliable and secure interpersonal environment, particularly the lack of adequate mothering in the early years. This applies not only to those children who lost both of their natural parents but also to those whose mothers were forced to appear and disappear actually by force of necessity but certainly, to the perception of the children, in a cruel and capricious manner.

To return to the concentration camp syndrome, its nature has been the subject of some controversy but, on the whole, it appears reasonable to regard at least the anxiety core of the concentration camp syndrome as a special variety of traumatic neurosis resembling in some respects the ordinary combat stress reaction, which is the paradigm of the traumatic neurosis. A more interesting analogy is the Japanese A-bomb disease or neurosis described by Lifton in survivors of the Hiroshima bombing,[8] who for many years afterwards have suffered from vague and chronic complaints of fatigue, nervousness, weakness, and various physical ailments, along with characterological changes and feelings of what Lifton calls "existential" guilt. Also, although the analogy is not as direct, it is possible, without stretching the bounds of imagination beyond credibility, to draw significant parallels between the concentration camp experience and the poverty ghettos of our great cities.[3]

Although there can be no doubt that the primary cause of the concentration camp syndrome is the multiple, massive emotional and physical stress to which the prisoners were exposed during internment, these stresses cannot alone account for the origins and development of the later added features of the syndrome or of the persistence of these symptoms in many cases in spite of the passage of time and infinitely more favorable circumstances. For such secondary developments, which differ in character and intensity in different patients, other, additional explanations must be invoked. One theory seeks to ascribe not only the immediate postwar, but also the chronic persistent symptoms of the concentration camp syndrome to organic brain disease resulting from injury, illness, and malnutrition incurred during the internment. However, my own experience is in agreement with that of a majority of observers who have reported that brain injury was relatively unimportant in their cases, and that the development and persistence of certain features of the concentration camp syndrome are far more dependent on the immediate and later postwar experiences of the survivor than on organic factors. It is during this postwar period, too, that basic personality strengths and weaknesses would have their effects on the ability of the survivor to cope with what fate might bring his way. Postwar experiences having a significant influence on the ultimate adjustment of the individual include loss of immediate members of family, relatives, homes, and livelihoods, a dashing of the inflated hopes for a postwar Utopia, a long and debilitating residence in DP camps, downward change in socioeconomic status, and emigration to strange lands with different language, tempo of life, and customs.

I will conclude with a consideration of depression and guilt among concentration camp survivors, because these symptoms seem to me to strike to the core of the psychopathology they display, and to constitute a kind of existential signature of the persecution. Depression is one of the most characteristic symptoms of concentration camp survivors, and, along with the traumatogenic anxiety state and the obsessive rumination over the past, is a hallmark of the concentration camp syndrome. More likely to occur is an unvarying feeling of emptiness, despair, and hopelessness in older people; most of the survivors become depressed episodically, particularly at holidays, anniversaries, and in connection with events which remind them of the past, like the Eichmann trial. It has been suggested that depression may represent a delayed mourning reaction, particularly insistent in its demands because the concentration camp prisoners had been unable to engage in ceremonial mourning for their dead.[8] Certainly the brutal destruction of so many family members, relatives, and friends needs no explanation as a cause for depression, but it is not so obvious why the depression of the concentration camp survivors is so often tinged with feelings of guilt. In some cases, such feelings are attributed to specific episodes, such as when a prisoner had taken an action which led to the saving of his own life at increased risk to another and in such instances it may be possible to detect the predisposing influence of insuf-. ficiently resolved earlier conflictual experiences. This was true in the case of the patient whose account of life at Auschwitz you have heard and certainly in general it would be difficult to devise a more diabolical environment than the concentration camp for fostering guilt-producing conflict over aggressive drives or between self-preservative and superego impulses. However, there remains a significant number of exconcentration camp inmates whose guilt and depression are attached to nothing more specific than the very fact that they survived when so many were lost. My encounter with a large number of survivors of the persecution has left me not completely satisfied that the depression and guilt they feel can be explained entirely by the usual combination of precipitating stress and morbid predisposition which psychiatrists invoke to explain such symptoms. Perhaps this is because the immense scale of the tragedy in which these survivors were immeshed seems to render the ordinary language of psychopathology inadequate and invites a grander, more transcendent dimension of explication. The stubborn, even prideful, refusal to forget displayed by certain survivors seems to involve something more than masochistic personality mechanisms or a revival of past, incompletely resolved emotional conflicts, and, instead, suggests a desperate attempt to rescue their dead from the limbo of insignificance to which they have been consigned by bestowing upon their destruction a benison of meaning.

Viktor Frankl himself emerged from Auschwitz with the conviction that the need to find some meaning in an otherwise incomprehensible universe constituted an elemental human hunger equal in importance to other such putative primary motivational forces as the pleasure principle of Freud.[5] In a similar vein, Lifton found [7] in his study of A-bomb survivors in Hiroshima that for them the order of the universe had somehow been violated and that they continued to identify themselves with the dead and be unable to resolve their guilt through the usual mourning experiences because they could find no spiritual or ideological explanation, no "reason" to explain the disaster and thus to release them from this identification with the dead.

There is an ancient Talmudic legend of the Lamed-Vov, the 36 just men who take upon themselves the sufferings of the world. Perhaps those concentration camp survivors who ceaselessly lament the past are performing a similar function and their sufferings can be thought of both as a memorial to their dead and as an act of existential expiation for a species capable of such an outrage upon a common humanness.

REFERENCES

1. BETTELHEIM, B.: *The Informed Heart*, New York: Free Press of Glencoe, 1960.
2. CHODOFF, P.: "Effects of Extreme Coercive and Oppressive Forces," in Arieti S. (ed.), *American Handbook Psychiatry*, New York: Basic Books. Inc., Publishers, 1966, vol. 3.
3. CHODOFF, P.: "The Nazi Concentration Camp and the American Poverty Ghetto: A Comparison," in the *Proceedings Psychiatric Outpatients Centers of America*, 1967.
4. COHEN, E. A.: *Human Behavior in the Concentration Camp*, New York· W. W. Norton Company, 1953.
5. FRANKL, V. E.: *Man's Search for Meaning*, Boston: Beacon Press, Inc.. 1959.
6. FREUD, A., AND DANN, S.: An Experiment in Group Upbringing, *Psychoanal. Stud. Child.* 6:122, 1951.
7. LIFTON, R.: Psychological Effects of the Atomic Bomb in Hiroshima. *Daedalus* (Summer) 1963.
8. MEERLOO, J.: Delayed Mourning in Victims of Extermination Camps, *J. Hillside Hosp.* 12:96, 1963.
9. STERBA, E.: Some Problems of Children and Adolescents Surviving From Concentration Camps, read before the Second Wayne State University Conference on the Late Sequelae of Massive Psychic Traumatization, Detroit, Feb. 21, 1964.
10. WOLFENSTEIN, M.: *Disaster*, New York: Free Press of Glencoe, 1957.

Article 28 The Coping Behavior of Nazi Concentration Camp Survivors

JOEL E. DIMSDALE

The Nazi concentration camps killed 72 percent of eastern and central European Jewry.[6] The inmates faced total rule by a hostile bureaucracy; extensive overcrowding; starvation; destruction of values, status, society, and family; and the threat of impending death.

Such genocide efforts are hardly new; other examples from this century range from the Turkish attack on the Armenians in 1912 to the Bangladesh tragedy of 1971. What appears to be unique about the Nazi effort was the bureaucratization of the stress and death immersion experience. It was only through a careful study of the machinery of destruction, transportation lines, and propaganda for internal and external dissemination that the SS was able to achieve its aims so successfully. This success in bureaucratizing which characterized the SS genocide troubles the citizen of this century about possible future applications.

Although the concentration camps may not teach us much about coping with everyday stress, they do teach us much about coping with severe stress— stress not unlike that faced in prisons or disaster situations. The effects of the concentration camps on the survivors are well documented.[2] But very little has been done to specifically study coping in the camps, asking the question, "How did anyone keep going and survive such stress?" This is not surprising since the concept of coping, which essentially focuses on how a person responds to stress, is relatively new.[5]

Coping has connotations similar to the concept of defense mechanisms but includes a broader range of actions the person can use to help vitiate the impact of stress. Coping includes such processes as individual flexibility and habits of

JOEL E. DIMSDALE, M.D., is with the Department of Psychiatry, Harvard Medical School, Massachusetts General Hospital, Boston.

Source: The American Journal of Psychiatry, Vol. 131 (1974), 792–97. Copyright the American Psychiatric Association.

information gathering as well as a variety of defense mechanisms. We have descriptions of a number of coping strategies from a variety of study settings— e.g., burn victims,[4] graduate students taking comprehensive examinations,[9] and normal adolescents.[10]

There has been some difficulty defining coping more precisely because it has been held that almost *any* response of the person to stress is coping behavior. A more specific definition of coping is possible if one analyzes behavioral responses in terms of short-term and long-term consequences and also in terms of internal versus external changes. Truly functional coping behavior not only lessens the immediate impact of stress but also allows the person to maintain some sense of self-worth and unity with his past and anticipated future.[11] This paper examines further the distinction between functional and nonfunctional coping behavior in the network of coping strategies used by concentration camp survivors.

Although very little work has been done concerning coping among concentration camp survivors, there are strongly held beliefs about the phenomenon. In some cases these beliefs are based on personal experience and in others on so-called popular opinion. Bruno Bettelheim was one of the first psychologists to write on this phenomenon; his writings are based on his experiences in Gestapo (internal intelligence) political camps rather than the concentration camps that developed later for the explicit purpose of extermination. In his work he described the forms and stages of intellectual withdrawal he observed in himself and others.[1] Viktor Frankl wrote about his experience in the concentration camps, emphasizing the importance of the way in which the person interpreted the stress and the importance of finding some meaning to his suffering.[3] While both Bettelheim and Frankl provided much insight, it is important to ascertain the generality of their experience and the incidence of other coping strategies.

There are two other widely held beliefs with regard to coping in the camps. One is that the individual was completely powerless to influence his fate; the other is that only the barbaric inmates survived. Both of these beliefs are based more on popular opinion than a study of the concentration camp survivors.

METHOD

Survivors were located through rabbis, word of mouth, and the assistance of an organization of concentration camp survivors. In every case the survivors contacted were regarded as relatively emotionally stable, although the criteria for assessing their mental health were not rigorous. "Health" was defined simply as ability to function in society—through work or family. The second characteristic of those sought for interviews was that they be willing to talk about their experience. Semistructured interviews were held in the survivors' own homes. Interviews were conducted in each survivor's native tongue, and interpretation was used as needed. The interview sessions lasted two to eight hours and were tape-recorded; those portions of the tapes directly pertaining to coping were later transcribed.

RESULTS

The following two case vignettes are noteworthy because of their general strong coping tone and because of the differences between the two means of coping.

Sara was 20 years old in 1939. Before the Germans invaded Poland her family had talked at great length about emigrating to Palestine, but the parents were frightened of leaving their own country. On the first day they were taken into custody Sara lost her parents and was separated from her husband and infant son. She said that from the first day

> I had a belief, a religious belief. I was convinced that all the wrong things had to change and we would be free. My mother put in me the belief that if someone is doing right, he will not always suffer. In Auschwitz I saw my husband at a distance, and I shouted to him that we would meet again, never to give up. I picked the good from the bad situation; they took us to the plaza and killed half of us, but I survived. They took us to a labor camp and told us we would all die there, but I lived. Then to Auschwitz, but always I kept remembering that I got through these things in the past. In Auschwitz I was beaten many times, the shooting, the smell, the electric fence, but I knew to survive I had to believe, believe that such a bad thing cannot win.

After the war Sara and her husband moved to America; she continues to believe that her small child survived the war and is living somewhere. The after-effects of the camp, she said, have been "to make me nervous. I try to control myself, but I have ulcers. I try to control myself, to talk myself into feeling better. It is just like in the Middle East today—you must have hope; I hope for a solution, a happy solution, that they can live together side by side." Her overriding coping strategy has been the mobilization of hope. It is an active sort of hope, a conviction that things will improve. She was evidently sustained by a preexisting mode of belief, firmly established, that ultimate outcomes must be benign, that suffering can be endured.

Jacob was 30 years old when he, along with his wife and child, was taken from Germany to Dachau. He is the sole survivor of his family.

> On the first day I learned that the SS would kill me eventually. I knew that I would die, that there was no possibility a man could go from the camps. The uncertainty was removed and also the fear; I became active in the camp underground. Each day we had many tasks—to save this person, watch the new transports, etc. We were so active in the camp underground that we did not pay any attention to the danger. I worked as a carpenter in the camps and always kept a cyanide pill if the SS should suspect me, catch me, and interrogate me."

After the war Jacob moved to an eastern European country and got into conflict at the government agency where he worked. "I spoke out against them and the director came. We argued and he fired me and put me in a mental hospital. The psychiatrist pointed out to me that I was not perceiving danger properly, that I was insulting powerful men in the government who could do me great harm. In a sense it is true; I don't feel the right proportion between the danger and the consequences." He later escaped to Israel. In talking about his current life, he states "Now I don't feel fear but I feel very tired."

The feeling one gets from Jacob is that he has coped through defying and actively resisting danger. This mechanism characterized his life in the concentration camps as well as afterward.

The differences between these two cases are striking. The apparent constancy of each person's coping strategy is also striking. However, when each interview is examined statement by statement, it is apparent that many coping strategies are used by a given person—let alone by a variety of people.

CLASSIFICATION OF COPING

The rest of this paper will describe various coping strategies and their implications. This description is based on an analysis of each coping statement made by the survivors. These coping strategies are summarized in the following list:

1. Differential focus on the good.
2. Survival for some purpose.
3. Psychological removal, including: intellectualization, belief in immortality, time focus, humor, and "musselmann."
4. Mastery—environmental and attitudinal.
5. Will to live.
6. Hope—active and passive.
7. Group affiliation.
8. Regressive behavior.
9. Null coping—fatalism.
10. Anticoping—surrender to stress.

A prominent coping strategy is the *differential focus on the good*. This is essentially a "figure-ground" problem; a person at all times has a choice as to what to focus on—foreground or background, good or bad. In most instances camp inmates adjusted their demands for pleasure so that these demands were consistent with the environment. Thus many focused on the small gratifications of getting through the food line without a beating and ignored the larger tragedies of the camps. Also involved here is the simple appreciation of beauty, even in the midst of ugliness. One survivor commented on the sunset against the distant fields.

Sarah: "In the camps I saw many prisoners beaten to death when they could work no more. However, that was a good period for me; we had no work on Saturday and were in camps where we could walk in the fields and sometimes had the luck to catch a potato or a carrot."

Survival for some purpose was an extremely powerful motivating strategy. The person who had to survive to help a relative, to bear witness and show the world what had happened, or to seek revenge—this person was using a strong coping style. It is a style described at some length by Frankl.[3]

Sarah: "I wanted to find my child, I wanted to be back with my husband. The feeling that somebody needs you is more important for survival than just the feeling that you have to survive. There were times that I did not feel like waiting for my food, thinking that I could get it tomorrow, but then I remembered my sister needed me and my husband needed me."

Naomi: "Once I saw a woman beaten with a whip by an SS man because she stole a carrot in the field where she worked. To see that young girl stretched out naked on the ground with the SS man beating her, I vowed I would never forget it."

The third major coping strategy is that of psychological removal. It essentially revolves around insulating oneself from the outside stress, developing ways of not feeling so that "I'm not here," and "This is not happening to me." Such strategies were discussed by Bettelheim.[1] These general denial strategies were omnipresent and were often effective in shielding the person from a complete realization of the shock.

Josef: "We were not very clever, we did not want to believe that anything was going on. There were rumors going around but we preferred to believe the version that said that the Jews are collected in camps where they are treated all right."

John: "We talked together as friends about concrete things, not about feelings. I think all of the feelings were blocked; if you felt too much, you felt bad. To feel was to feel unpleasant, better not to feel at all, don't think about it."

The strategies of removal include certain subcategories. Intellectualization is a very successful way of withdrawing from the impact of the situation. The "feeling I" no longer has to bear it and instead is replaced by a "photographic dispassionate I."

Many people were able to decrease the impact of the threat of death by withdrawing into considerations of immortality—personal immortality, an immortality of the Jewish people, or the immortality of a political movement such as the Communist party.

Laura: "I believed in my personal survival. You felt that it could happen to everybody around you, but death could not happen to you."

Tanya: "We Jews have always been great sufferers, and I knew that we would grow up and survive this too."

By possessing some religious belief the person was able to remove himself from the camp world to another world of different values and meanings.

Henry: "Many times I would say 'Shema Yisroel,' * and it would give me extra strength. To say the 'Shema' was to say that somewhere, even in this period of great loneliness and antihumanity, there is someone to whom you can address yourself."

Time distortion was another way of not feeling the stress. Focus on the peaceful past or a presumably peaceful and happy future was helpful; likewise, focus only on the present moment was helpful in removing considerations of what had happened already and the probabilities of continuing imprisonment.

Henry: "Life was carried on from day to day, one day to another, we lived from one meal to another. Any thinking that I did was no longer than a couple hours or the next day into the future. I did not think any longer than the immediate period and its needs."

The role of humor in mediating stress is well known; it is a potent agent for psychological removal from the stress situation.

Chaim: "I remember a kapot joking with us. 'Behave properly because yesterday a man was *even killed* for not behaving.' We thought it was very funny, the irony of such a warning in Auschwitz."

The final and most complete stage of withdrawal was withdrawal into a "musselmann" ‡ stage, a stage characterized by profound apathy, complete indifference to the surroundings, and a lack of any response to the environment, both physical and interpersonal. Almost all who reached this stage and who were not immediately rescued by their comrades never returned from it. They were immediately selected out and killed. Almost all inmates had brief episodes of semi-musselmann behavior in which they completely ceased to feel, ceased to struggle to survive, and became psychologically "not here" in the camp. These brief episodes were usually brought on by sudden massive onslaughts such as those that occurred when the inmates arrived at the camp.

Henry: "In the beginning I was comfortably blind; it was as if I had reached the moon or another world. It was like I was in a state of shock, not feeling, nor realizing what was going on."

It is this musselmann stage that especially draws our attention to the question of the efficacy of various coping strategies. Apparently some strategies may have

---

* Jewish prayer beginning "Hear O Israel, the Lord our God, the Lord is One."

† Guard chosen from the ranks of prisoners.

‡ Camp slang for the severely apathetic and fatalistic inmate.

effectively mediated between the person and the impact of stress but may at the same time have virtually guaranteed the person's death. More difficult to assay are the long-term effects of certain habitual coping strategies. One wonders whether the same coping habits are functional in other circumstances or are easily changed.

Another major category of coping revolves about the concept of *mastery*. If the person is still able to express some autonomy through mastery of a portion of the universe, external or internal, then he does not feel defeated by the stress. The concentration camps were unique in the degree of assault on individual autonomy. Survival was enhanced by counteraction—ways of actively manipulating or resisting the stress system. Survival was also enhanced by counterthought—resisting the internal impact of the system, not allowing oneself to be crushed or dehumanized by it. The concentration camp provided very little room for autonomy through counteraction, and yet the opportunities were used to the fullest. Once it was realized that the purpose of the camps was to kill, anything the inmate did consciously to stay alive was an expression of his autonomy. Gathering information, helping fellow prisoners, resisting the camp machinery even in the most trivial ways—all these activities bolstered the person.

> David: "The thing that kept me alive was to focus on where I could find a blanket, something to chew, to eat, to repair, a torn shoe, an additional glove."

No matter how rigidly the SS controlled mastery through action, there was still attitudinal mastery. This, too, faced a massive onslaught; it was difficult to maintain any self-esteem under camp conditions, and yet many inmates were able to continue thinking of themselves as human instead of "ungeziefer" (vermin), as they were called by the SS.

> Tanya: "On Yom Kippur my mother always wanted me to fast and my father always told me that I was always bad and would break the fast too soon so why not spare the little stomach the hunger pains and eat. Every time I would start with good intentions, and father would bribe me with delicacies. When I was in Auschwitz, for the first time I fasted. Mother was dead so I could not please her; I did not believe in God at that point; I just wanted to show the Germans that I could be Jewish if I wanted to be, even in Auschwitz."

The most basic coping strategy was simply the *will to live*. It may seem trivial but in many respects it was one of the most important aspects of coping. For some, this almost protoplasmic unthinking reaching out for life was the only "coping" mediating between the individual and surrender to stress. In the concentration camps there was no question but that to prolong life meant to prolong suffering. Many people performed the mental calculations, decided it would be better to die, and yet continued to try for survival semiconsciously.

David: "The next night after I moved blocks, my entire old cellblock was eliminated. What I remember from that night of death—the terrible cries from my old block and my old comrades[sobs]—I covered my ears with my blanket, but I could not overcome the shouts and cries and my own fears. I only wanted to live through that night and see again the daylight."

The *mobilization of hope* was basic to survival; it acted as a kind of kindling to action. Without it coping was limited to actions that would be easily fatigued by the repeated stress of the camps. There were two forms of hope: one was an active hope, a belief that the camps were abberrant and would not last—a conviction that "this cannot go on forever." The second hope was more passive, conveying the attitude "Where there is life, there is hope."

Sarah: "All during the camps I believed that somehow somebody would come and save us. I always believed that I must have hope; I just knew that I had to believe in my survival, that such a bad thing cannot win."

Henry: "The ray of life, the sparkle of hope kept me going. When I was in one of my worst moments, my friend clapped me on my shoulder and said 'Everything will be all right, cheer up.' That was like a prescription, it helped me. It was as if I was in a desert of blackness and one fine ray of light was slipping through, and it was enough. A breath of hope allows a person to live."

Group affiliation was important in providing information, advice, and protection. It was also important in reinforcing the person's sense of individuality and worth. The group was one place where one was not a number but a comrade. Groups ranged from the political (e.g., the Communist party) to collections of people from the same country or to small family groups. The size of the group was not of key importance; the two-person friendship group, which was very common, was extremely effective in mediating between the person and stress. If an inmate was unsuccessful in affiliating with a group within the first few days of internment, his chances of survival were very limited.

Sarah: "We had a group in Auschwitz singing operas and shows. We were not always in the mood for it, but sometimes in a quiet moment we would talk about the past and everybody would play a part from the show he had seen. In Auschwitz all the time the ovens were going, day and night, and they told us that they would burn us. But my sister and I consoled each other."

Occasionally the camp inmate was able to receive help from other inmates or from camp guards by extremely *regressive behavior.* The motivation to help a crying child or adolescent was very strong. Such crying seemed to elicit a helping response from the older inmates, even in Auschwitz.

Ruth: "When I was in jail I cried so much that the whole jail came over to see what happened. I cried, I cried, I cried. I cried that I had been in jail, that I was alone, I shall never see my brother, my father, nobody, I'm gone forever, and the other prisoners helped me."

Tanya: "We kids did not struggle to survive, we just treated the camp like a hostile, dangerous playground, and sometimes people would help us."

All of the coping strategies described above lessened the impact of stress. Two other major strategies are "coping" only in a very special sense of the word. One is a kind of *null coping* where the person does nothing, internally or externally, to mediate the stress but instead relies on fate or others. Although such behavior may not lessen the impact of stress, it does alter the perception of it. To experience severe stress passively was not as bad as to experience the stress and to blame oneself for not being able to avoid it. In this sense, a fatalistic attitude was helpful.

We have talked about coping and null coping. There is one last major stance, which is a kind of *anticoping*. Here, the person completely surrenders to the stress and acknowledges that "it is right and the Self is wrong." Even this strategy is slightly functional in that it removes the dissonance of the situation. To the extent that the inmate was able to identify himself as vermin belonging in the camps, he would feel his internment to be just. This was not a common strategy, but there are occasional hints of it in the greater-than-life magnification of the SS.

Ruth: "That horrible man Mengele is still living. He lives, he lives in Argentina in the jungle on an island. He is surrounded by the army. He was the evilest man I ever met, he was the handsomest man I ever met, he was dark with blue eyes and tall and looked like a god. The first time I saw him I remember saying to myself, 'He is so beautiful, he cannot possibly be bad.' "

The classification of coping strategies given earlier is a loose one because there is much common ground among the different strategies. Coping behavior is complicated; more often than not a brief coping statement may be found to contain a number of different strategies. The following statement illustrates this complexity.

Tanya: "I thought, I tried, I did all kinds of things to survive and to help others survive, but after a while it was only God that allowed me to survive" (strategies: counteraction, survival, group affiliation, religion, and fatalism).

The very difficulty of connecting these statements with any one strategy in the list illustrates the dynamic and changing process that coping represents. In the example from Tanya just cited there appears a certain trend, a turning from one strategy to another and then to another. This is probably the most accurate description of an individual's coping behavior.

One feels that various coping strategies may be more or less effective, that there may be a hierarchy of functional coping strategies. To answer this question convincingly requires a far larger sample than this study provides and also requires more precise definition of coping strategies. But some inferences may be drawn from our sample.

It may be a sign meriting concern to see a person overwhelmingly favor one strategy, such as a strategy of withdrawal. This strategy is essentially a process of building a wall between the experiencing self and an oppressive reality. Over time it becomes extremely difficult to coax such a person into dealing with reality again.

> Anna was five years old when she witnessed the shooting of her family. She lay hidden in a cave alone for the next two years of her life, coming out only to scavenge for food. At age eight she was finally caught and placed in Bergen-Belsen. Anna learned very early that the world was hostile and venomous, and she learned to withdraw her concerns from the world. In this way she tolerated and survived the war years. Today she lives an outwardly normal life but states "I must keep frantically busy, so that I don't stop and remember what the world is like; that is my way of going on."

The person who copes by focusing on survival for some concrete purpose is in danger when this reason for survival is destroyed. It is tragic to see a husband who struggled to survive and be with his wife again only to discover later that the wife had been killed in the war.

> Saul: "Only when I returned home did I become aware of what I had lost. In Auschwitz I only suffered physically, but spiritually hope maintained me in life. When I saw that all hope was lost—no family, no wife, no child, no anybody anymore alive, then came the shock."

No less tragic is the person who theoretically attains his goal but finds that its taste is bitter. Those who coped by anticipating future revenge against the Germans were not satisfied. The enormity of what they had suffered made any single act of revenge seem trivial.

Those who coped by compromising their integrity too much suffered severe guilt. The survivor guilt phenomenon is universal.[8] When combined with guilt about actions taken in the camp in cooperation with the SS or to harm a fellow inmate, the guilt became overwhelming.

Those who survived by focusing intently on later being a witness to the world and telling what had happened entered a world that was not interested in what had happened to them.

> Albert is such a man, a man who described himself as a camera in the camps, constantly recording so that he could obtain justice later. After the war no one ever asked him about the camps, and he felt very guilty, that he had cheated in some fashion by surviving without bearing witness. His interview lasted eight hours at one sitting; it was his first opportunity to "bear witness." At the end he said "I have failed. Nothing I can say can conjure up to you what we went through; I have failed."

CONCLUSIONS

A study of coping among concentration camp survivors is important not only because of the significance of the event itself but also because of the similarity between the Nazi concentration camps and continuing events.

It is of further interest because in some circles the concentration camp had acquired a general allegorical meaning. There is something about the camps that stirs us profoundly and resonates familiarly. Perhaps every century has certain overriding symbols that characterize it. A brief survey of 20th-century literature shows an absorption with the image of man trapped.[7] Dostoevski, Kafka, and Ibsen had all worked on this theme long before the Nazi concentration camps. Since World War II writers have dealt with concentration camps, sometimes explicitly but very often implicitly: Camus, Sartre, Solzhenitsen; the list goes on and on. What has become apparent is that almost every major modern author has been dealing with concentration camps in some fashion as a symbol that conveys more than their historical existence. For many, the image of man in the camps is quite simply the image of man and his fate, thrown into a world that is hostile, with no chance for escape and no real change. The way that man deals with his feelings about such a condition, the way he goes on to create a world within these confines, may not be dissimilar to the way that millions coped with their immersion in concentration camps.

REFERENCES

1. BETTELHEIM, B.: Individual and mass behavior in extreme situations. J. Abnorm. Soc. Psychol. 38:417–452, 1943.
2. EITINGER, L.: Concentration Camp Survivors in Norway and Israel. London, Allen & Unwin, 1964.
3. FRANKL, V.: Man's Search for Meaning. New York, Washington Square Press, 1963.
4. HAMBURG, D.: Adaptive problems and mechanisms in severely burned patients. Psychiatry 16:1–20, 1953.
5. HAMBURG, D., ADAMS, J.: A perspective on coping behavior. Arch. Gen. Psychiatry 17:277–284, 1967.
6. HILBERG, R.: The Destruction of the European Jews. Chicago, Quadrangle Books, 1961.
7. KITCHIN, L.: Compressionism, in Drama in the Sixties. London, Faber & Faber, 1966, pp. 43–74.
8. LIFTON, R.: Death in Life. New York, Random House, 1967.
9. MECHANIC, D.: Students Under Stress. Glencoe, Ill., Free Press, 1962.
10. SILBER, E., HAMBURG, D., COELHO, G., et al: Adaptive behavior in competent adolescents. Arch. Gen. Psychiatry 5:354–365, 1961.
11. VISOTSKY, H., HAMBURG, D., GOSS, M., et al: Coping behavior under extreme stress. Arch. Gen. Psychiatry 5:423–448, 1961.

Article 29 Life Adjustment After Severe Persecution

EGON BITTNER

Persecution of European Jews between 1933 and 1945 represents an ugly chapter in history, and it is easy to understand why investigations into details of this tragedy are fraught with anguish. Millions of people succumbed to the gradually intensifying stress of Nazi anti-Semitism, and those few who survived have been too busy or too upset to want to talk about the terror. When some of the brutal facts came to light at the end of World War II, observers were at first shocked and repelled:

> Incredible savagery . . . a studied policy of neglect, starvation, torture, and extermination . . . so shocking it overtaxes imagination to grasp the reality . . . 40,000 emaciated, apathetic scarecrows . . . defecate and urinate where they sit or lie. . . . Some 8,000 to 10,000 naked decomposing corpses lay all over the camp (Belsen).[7]

After the living were sorted from the dead there was yet optimism because some of the survivors were noted to make unusually rapid recoveries from severe illness.[2] Early reports described how certain concentration camp inmates had survived by depersonalization and withdrawal that seemed almost to render immunity against "normal diseases."[6]

Another defense mechanism—unconscious identification with the oppressor[3] —was noted among the victims, whose survival gave "startling demonstration of mankind's capacity for endurance."[8] But as the postwar years posed new problems of resettlement and acculturation to this already severely traumatized population, psychiatric sequelae became more troublesome. Depression, anxiety, resentfulness, a tendency to relive the traumatic experiences obsessively in dreams

PETER OSTWALD, M.D., and EGON BITTNER, PH.D., are with the Department of Psychiatry, University of California School of Medicine, and Langley Porter Neuropsychiatric Institute, San Francisco.

Source: The American Journal of Psychiatry, Vol. 124, No. 10 (April 1968), 87–94. Copyright the American Psychiatric Association.

or daytime fantasies, and numerous somatic complaints were reported in psychiatric clinics throughout the world.[1,9] The persecution victims' eligibility for financial compensation brought with it new questions about differential diagnosis and management.[10]

Today the awful problem of extreme coercion and oppression in terms of its effects on human behavior is an important topic for psychiatry.[5] Of particular interest to psychiatrists called upon for evaluation and treatment of persecution victims is the quality of their postwar life adjustment. While clinical study invariably reveals severe symptomatic residuals or "late effects" of persecution,[4] the overt socioeconomic adjustment of many former concentration camp inmates is often paradoxically good. Their small businesses, restaurants, and delicatessen stores seem to be flourishing by dint of hard work and enterprising management. To the casual observer—and even to many examining physicians—the ex-persecutee now appears, "well-adjusted," affluent, and comfortable. The question that concerns us is in what relation, if any, does this seeming success stand to the past experiences and the psychiatric problems of the victims of persecution.

To clarify the matter, we reviewed case records of 60 subjects and tried to assess their current level of socioeconomic adjustment. We also attempted to obtain interviews with some persecution victims from nonpsychiatric sources. Invitations were sent to "successful" victims of Nazi persecution who had never requested psychiatric assistance, but this elicited the response of only one man, and he turned out to be suffering from a type of chronic reactive depression which has often been described in the European psychiatric literature.[1] Thus, without being able to base our observations on matched samples of "successful" and "unsuccessful" victims we decided to look carefully and critically at the clinical data already available for the purpose of trying to find out what kind of life adjustment these people had found possible and whether this adjustment had a bearing on the pattern of symptoms and disabilities.

DESCRIPTION OF THE STUDY POPULATION

Our information is derived mainly from clinical interviews containing data about present complaints, life adjustment, and past history. Virtually all patients were seen more than once and in many cases relatives and other interested persons were interviewed about matters connected with the patient. In addition, medical records, neurologic examinations, psychologic tests, and other relevant data were available in a significant number of cases.

Roughly 39 percent of our patients were referred by attorneys helping them to pursue their claims under the West German restitution laws. Forty-seven percent were sent directly by the German consulate or consular physicians for psychiatric evaluations. Fourteen percent came on their own, seeking psychiatric examination or treatment. Exactly the same number of patients were male as female.

Twenty-five percent of the patients were born before 1919, 60 percent between 1920 and 1929, and the remaining 15 percent between 1930 and 1938. Approximately 46 percent of the population came from the territories encompassed by post-World War I Poland, Czechoslovakia, and Hungary; 54 percent were born and reared in Germany and Austria. All but five percent of the patients were the offspring of parents who were both Jewish, but all had at least one Jewish parent.

Considering the over-all socioeconomic status of the parental home, we estimate that 25 percent of our subjects came from what might be deemed upper-class circumstances. These were people who reported that their parents were executives, professionals, and proprietors of large businesses, and who gave descriptions of past affluence, large homes, private education, book and art collecting, travel, etc. Sixty-seven percent had a middle-class background; their parents tended to be white-collar workers, retail merchants, and independent craftsmen; their life circumstances were described as modest but comfortable. Only seven percent of the patients gave information leading us to believe that they came from lower-class situations.

Fifty-four percent of the patients married or remarried after the war, in almost every instance a partner who also had experienced the Nazi terror at first hand. Thirty-five percent were married before the war and were still with their spouses or were widowed. Eleven percent never married. Though the number of never-married persons is above expectations by United States standards, our "method" of sampling scarcely permits us to draw inferences from this fact.

The same can be said for the finding that seven of the 32 postwar marriages were childless. We can state, however, that all single persons saw their failure to marry as resulting from wartime experiences under Nazi persecution. Similarly, childless couples maintained uniformly that they avoided offspring out of a feeling of pessimism and resignation: "Thank God we didn't have any children. What kind of a world is this to bring children into? Life is enough of a struggle for just the two of us." On the other hand, those who lost children as a result of the persecution were among the saddest and most guilt-ridden patients in our sample:

> *Case 1.* Mrs. S. escaped from Germany after agreeing to leave her children with relatives who assured her they would also emigrate to the United States so that the family could be reunited after the war. The news of their extermination in gas chambers along with every other member of her family precipitated a severe depressive reaction which Mrs. S. has never been able to resolve, in spite of remarriage and the attainment of an externally comfortable way of life. A quarter-century later she still berates herself for "abandoning" her children who "if they were alive today could give me a sense that life is worth living." She is markedly hypochondriacal, and has made one suicide attempt.

The war-time experiences of our subjects were varied. Twenty-five percent were forced into exile, for the most part during the last couple of years preceding the outbreak of hostilities on the European continent. While these persons were spared the immediate threat of genocide, they all report experiences of abject suffering.

> *Case* 2. Mrs. J. was ten years old when her parents fled to Shanghai where both had to work to make a marginal living. Without daytime supervision the child roamed the streets of the Japanese-occupied city. A favorite "game" of her teen-age gang was "body-counting," i.e., discovering corpses floating in the river or abandoned in disease-infested slums. Because of pulmonary tuberculosis her parents were not allowed to accompany the teenager's resettlement in the United States, and on the boat she became involved with a soldier returning from overseas duty. Their long and unsuccessful marriage was but one of a series of personal disasters during the years of her attempted readjustment.

Sixty percent of our patients were incarcerated in concentration camps. Many of these had also spent some time in hiding and in ghettoes. For them the threat of genocide was inescapable.

> *Case* 3. Mr. P. was 14 when storm troopers piled him and his parents on cattle-cars filled with German Jews bound for Russia. Temporarily housed in filthy, ice-cold barracks, exposed to beatings and insults, the boy knew that these quarters had only just been vacated by the bullet-riddled bodies piled up in a nearby ditch. His parents were soon to join the victims but the boy was put to work in an airplane factory. Slave-laborers "died like flies" from disease, malnutrition, and brutal punishment. When finally transferred by truck to an extermination camp, the patient was strafed by Allied aircraft. In the camp he pretended to be a non-Jew, avoided sick call, stole food, and played on the compassion of others to avoid the gas chambers.

Finally, 15 percent of our population survived in ghettoes and/or in hiding. From the accounts available to us, we can say that the fugitive existence of those who escaped the Nazi dragnet and subsisted in forests and bunkers was as suffused with terror and martyrdom as was the life of concentration camp inmates.

> *Case* 4. Mr. K. was six years old when soldiers occupied his home, forcing the boy to witness public executions, beatings, and humiliation of Jews. After living in a ghetto, his family was able to flee into a large forest, thus avoiding transportation to an extermination camp. Lacking adequate food and shelter, the refugees readily fell victim to troops who periodically hunted them down, shooting their prey on sight. This is how he lost his father.
>
> When their little hideout was finally overrun and his mother was shot, the boy assumed it was the end for him too. He shivered all night among the dozen or so corpses lying about, but in the morning realized it was still possible to go on. For two years he managed to survive by hiding in foxholes covered with branches, eating grass and leaves, and speaking to no one until the Russian troops arrived.

SURVIVAL PROBLEMS AFTER THE WAR

Liberation from Nazi oppression did not by any means result in the hoped-for alleviation of suffering or danger; thousands of concentration camp victims survived only long enough to die in the ensuing weeks. Those who did remain alive for the most part either tried to return to what was left of their homes (36 percent of our sample) or settled in the displaced persons camps awaiting an uncertain future and hoping ultimately to emigrate to a non-European country (31 percent). The remainder (33 percent) made immediate efforts to emigrate, stayed in exile for several more years before coming to the United States, or drifted from one place to another with an uncertain and often helpless expectation of the future resembling that of the inmates of displaced persons camps.

"Going home" invariably had the most profoundly distressing effect on an ex-persecutee because it forced him to be confronted with the stark realities of the genocide as a truly effective policy of extermination of villages and families. The fantasy that "I can now be reunited with my loved ones and return to a normal life" had to be abandoned when it became clear that often not a single member of the patient's family had survived.

> *Case 5.* Mr. G. was 16 years old when liberated by Russian troops and, since it had been agreed that the dispersed family members should try to find each other, he immediately went back to his home town. The emaciated, starved, and terrified teenager discovered that every member of his large family except a distant cousin had been killed.
>
> His overwhelming grief led to renunciation of previously high moral standards, while his intense anger found a ready outlet in illegal black market activities. He did extremely well as a smuggler and did not hesitate to kill border guards or others who got in his way. Several years of successful thievery enabled him later to live comfortably in a large city, obtain an education, and emigrate to the United States.

The inertia of existence in displaced persons camps forced many into idleness or into part-time jobs for which they were not trained and in which they had little interest, with a resulting pattern of frequent job-changing that often persisted for some time.

> *Case 6.* Mr. D.'s experience in dentistry was made use of by the occupation troops before they sent him to a series of forced labor gangs and concentration camps. There he worked in stone quarries, railroad repair, and other unskilled jobs. Hospitalized after the liberation, he recovered quickly and became a baker's helper in a displaced persons camp.
>
> After migration to the United States, he held work briefly in an iron foundry, doing carpentry on the side. He purchased a chicken ranch that went broke. He now owns and operates a small restaurant that is doing well because both he and his wife work long hours there and the children help on weekends.

In 37 percent of our cases resettlement in the United States was facilitated by the presence of a distant relative or personal sponsor.

Case 7. Mrs. W., wife of a diplomat murdered by the Nazis, sent her laboratory equipment to a brother in the United States, hoping to continue her career as a biochemist. Trapped in Berlin, she lived the entire war underground, hiding in abandoned houses or living a clandestine underground existence.

She was one of the first postwar refugees to arrive in the United States and newspaper clippings showed her carrying an urn containing her husband's ashes. Immediately offered jobs, she felt too sick and exhausted to return to work. Instead she began to manufacture handbags, relying on a sewing skill that had come in handy during the war when secret documents, money, and contraband had to be smuggled in carefully designed secret pockets and hand-sewn belt-linings. She lived with her brother for a while, but then remarried and now has a successful private business with customers throughout the country.

Such success stories must be counterbalanced with some of the examples of failure. Our records show that during the early phases of existence in the United States, only five percent of our subjects were able to establish independent businesses, enter professions, or do managerial work. Thirty-five percent were unemployed, 23 percent did unskilled labor, and 33 percent had salaried employment or white-collar positions.

Case 8. Mr. N., who fought in the underground of an occupied country after escaping from a concentration camp, has been unable to settle into regular employment. Occasional jobs as a waiter are given up because of intense nervousness. A doctor advised farming but lack of experience led to the collapse of this venture.

His wife went to work while the patient was only occasionally occupied as a handyman. He is being retrained as a television repairman, but finds this difficult because "my hands shake all the time and I keep dropping the small parts."

Failure in postwar adjustment provoked suicidal thoughts in many of our disabled victims of persecution, and there were several suicide attempts, one of them successful.

Case 9. After an energetic struggle to obtain the education that Nazi occupancy of his homeland had made impossible, Mr. H. received a master's degree and a predoctoral fellowship to study industrial problems in Germany. Confronted there by the postwar affluence and certain signs of neo-Nazi sentiment, he recalled the terror of the war years, became depressed, and was unable to complete his work. He returned to the United States feeling guilty, inadequate, and filled with despair.

In frankly psychotic phases of his depression he heard accusing voices and felt himself again to be persecuted. Intensive treatment in several private and state hospitals did not suffice to reverse the suicidal trend, and the patient died from a massive self-inflicted gunshot wound.

PRESENT LIFE ADJUSTMENT

An analysis of employment patterns shows considerable advancement, especially among the male population. During the initial period of settlement in the United States almost one half of the men were either unemployed or working in menial, unskilled occupations. At the time of the study only four elderly men (13 percent) were unemployed and only one (three percent) worked as an unskilled laborer. Ten of our total population (17 percent) achieved a substantial degree of material security and independence. Several hold excellent positions as professionals, executives, and independent managers of businesses.

It is of special interest that the great majority of the most successful ex-persecutees were inmates of concentration camps, and that they tend to come from the territories encompassed by prewar Poland, Czechoslovakia, and Hungary. Among German-born ex-persecutees, only the concentration camp victims have had the same proportion of successes.

That these are successful life adjustments in an *external* sense only can best be recognized from the fact that the amount of complaining and unhappiness among the successful victims is just as great as among the unemployed or semi-skilled patients. To establish this point we rated the intensities of all symptomatic complaints by the 60 patients on a 15-point scale according to three categories. These categories comprise complaints expressed in terms of (1) bodily symptoms, (2) internally perceived mental difficulties with concentration, memory, mood etc., and (3) interpersonal and social maladjustment. The small size of our population forced us to condense the 15-point estimates to a three-point rating; the lowest point indicates denial of symptoms or only minimal self-expressed dissatisfaction; the intermediate point defines definite symptomatic interference of moderate to moderately severe intensity; and the highest rating indicates that patients emphasized complaints very strongly or saw themselves as totally disabled with respect to complaints falling in this category (see Table 1).

Particularly striking is the finding that persons whose success in business or professional life is readily apparent voiced as many complaints about discomfort in the area of interpersonal and social maladjustment as did the unemployed or marginally adjusted patients. One heard the following kind of comment from persons in the successful group:

> We've all been damaged, doctor, and I think we're really a bunch of rotten apples. We may look okay on the outside, but when you get to know us you'll see that we're different and sick inside and no matter what happens our lives will never be normal again. Life has no real meaning any more.
>
> We can't be together with Americans because nobody understands what we've been through. Our only friends are the other refugees, and whenever we see each other the conversation gets back to what happened in the concentration camp. We try to avoid going to the movies or watching TV because it only brings back bad memories. We haven't been away on a vacation for 15 years because there's so much work to do at the store. Now that we've at least got a little income it should be possible to enjoy life, but we

Table 1. Severity of Complaints Related to Degree of Outward Success (Present Socio-economic Adjustment) of 60 Persecution Victims

| Complaint Type | Severity* | Degree of Outward Success ‡ | | | | | |
| | | 1 | | 2 | | 3 | |
		N	Percent †	N	Percent †	N	Percent †
Bodily symptoms	1	1	10	3	15	12	40
	2	5	50	4	20	5	17
	3	4	40	13	65	13	43
Internally perceived mental difficulties	1	2	20	4	20	2	7
	2	5	50	4	20	7	23
	3	3	30	12	60	21	70
Interpersonal and social maladjustment	1	4	40	8	40	9	30
	2	2	20	7	35	5	17
	3	4	40	5	25	16	53

* Severity of complaint: 1 = none to mild

† Percent within each grouping   2 = moderate to moderately severe

‡ Degree of outward success: 3 = severe to disabling

1 = persons who have achieved a substantial degree of material security and independence

2 = persons who are gainfully employed in white-collar or skilled occupations and meet the demands of life adequately

3 = persons who lead sheltered lives, are dependent, and meet the demands of life with marginal adequacy

can't. Our children are still going to school. My heart isn't too good. At my age you can't even get a good insurance policy, so who's going to take care of the family when I'm gone?

Complaints of social alienation and interpersonal discomfort among those who are most successful were regularly associated with a high degree of somatic complaining, whereas psychologic symptoms were much more easily expressed by the middle- and low-income groups. A similar pattern of social alienation, coupled with an intense compulsion to work, was detectable among the *women* whose life adjustment appeared outwardly successful. These ladies, who now live in pleasant homes and can afford certain luxuries, spend most of their energies cleaning up, doing menial chores, and reinforcing the external appearance of success.

I've always got to find something to keep me busy. My kitchen is spotless, and each room has to look just perfect. I can't have anyone come to the house unless I've had a day or two to get things prepared. I stay up until three or four o'clock in the morning sewing the children's clothes or fixing curtains.

Psychiatric diagnoses among successful persecution victims are similar to those of the unemployed and marginally adjusted, but their ratings of disability tend to be lower. In the total group of 60 patients there was a very high proportion (77 percent) of stress-induced syndromes, including "late effects of the concentration camp syndrome" and chronic reactive depression (uprooting depression). Eighteen percent of the patients suffered from multiply-determined chronic illnesses in which the stress of persecution served to intensify or activate preexisting psychoneurotic, psychophysiologic, or personality disorders. In the remaining five percent, a diagnosis of psychotic reaction was applicable, schizophrenia in two instances and a terminal suicidal psychotic depression in one case.

THE SYNECDOCHE OF SUCCESS

One-fourth of our subjects described their present activities and circumstances in ways indicating an assiduous interest in success and often a remarkable degree of accomplishment. In most instances this tendency is expressed in business pursuits of men who by the sheer dint of wits and scrupulous husbanding of resources managed to acquire small business establishments. In other cases, involving women, we saw a consuming concern for housekeeping at a level of exquisite pedantry. With respect to external appearances, the most impressive feature of these lives is their tidiness. There is a single-minded dedication to the attainment of financial gain or to the management of orderly and well-appointed households. Moreover, these activities seem to absorb all interest and energy.

It is almost as if these patients have found something they could do well and have allocated to it all the powers of their lives, making it the substance of their existence. Thus, each seemed to have constructed around himself a small world within which he could live with a sense of relative autonomy and security precisely because of its narrow boundaries. That this world has been filled with activities that ordinarily are thought to have merely instrumental value (i.e., the making of money or housekeeping) on behalf of some other interest is dimly but painfully perceived. Thus, we have seen men who freely dedicate themselves to making money because they feel that their life chances are reduced to this and nothing else! Not only are all other avenues to fulfillment, growth, and gratification closed to them, but even the money they earn does not serve them. If they think of the ulterior purposes of their businesses at all, they view them as promoting the interests of their children.

We have conceived of this arrest of interest among persecution survivors as the synecdoche of success. A part of success—acquisition—takes the place of the whole, and maximal energy is directed towards attaining only the outward manifestations of success—affluence and good appearance. This takes its toll on the patient's family life and personal health; there continue to be severe symptoms as the synecdoche of success precludes opportunities for relaxation, spontaneity, interpersonal pleasure, and communal comforts.

In our treatment of these patients and their families we shift entirely away from the theme of persecution that, once elaborated for the purpose of the history, can only become an obsessional reliving of nightmares. Instead there is a painstaking working through of present reality and its opportunities for simple pleasure, which are often overlooked by the successful persecutee. The chronically victimized person is afraid of recognizing and expressing the positive emotions that were so long unavailable to him. In many instances he needs the therapist to replace benevolent parents or grandparents who perished in the genocide.

It is important that the children of persecutees be helped. Growing up in the United States, they may be troubled by the scotomatized thinking of parents who work day and night without finding gratification. Their environment looks tense, dreary, and emotionally impoverished. Money may be spent largely for display—a teasing expression of affluence that embarrasses onlookers because it so clumsily tries to hide the spender's disability.

In broader terms, the predicament of the successful persecutee could be interpreted as follows. Men who have once lost the basis of their existence and who have regained it through a combination of good luck and hard work can no longer take it for granted. In a manner of speaking, they cannot lift their gaze above the level of the wherewithals lest these disappear. Thus, they must find rewards in the unrewarding and seek to enrich the unenrichable. Profoundly distrusting all signs of security, relying on no one but themselves, they are forever engaged in shoring up the boundaries of their existence against unpredictable disaster. The threat of doom is not a memory for them but an ever-present reality. It is almost as if the phantom arm of persecution victimizes them even now. While they have seemingly regained a place in the world, they cannot join it.

Nor is this modality of "adjustment" wholly new with the present generation of Nazi persecution victims. This is a rather important point, for it could be said that the worldly success of the small shopkeeper is no innovation in the Jewish tradition and that our more successful subjects, whether in making money or in housekeeping, merely perpetuate an established style of life. Our response to this is that this pattern existed wherever Jews lived under persecution. Never before, however, was the Jewish community shattered and never before were the survivors of pogroms doomed to live solitary lives. Thus, earlier, interest in worldly success was always embedded in the texture of religious culture that augmented and widened horizons. None of these resources is available to our subjects.

SUMMARY

This is a study of 60 victims of concentration camps and other devasting aspects of Nazi persecution. We found that what some of the subjects described as a successful life adjustment after persecution was often based on an aggressive and single-minded pursuit of financial and only externally demonstrable signs of success.

In many instances a tremendous residual of post-traumatic depression, anxiety, and hostility was thus obscured, at least to the view of outsiders, including physicians called upon for medicolegal evaluations. We call this condition among persecution survivors the "synecdoche of success" and propose that it results from a massive threat to existence which for these persons is not merely a memory but an ever-present reality.

REFERENCES

1. BAEYER, W. R. VON, HAEFNER, H., AND KISKER, K. P.: Psychiatrie der Verfolgten. Berlin: Springer, 1964.
2. BERBLINGER, K.: personal communication about the recovery of persecution-connected tuberculosis cases.
3. BETTELHEIM, B.: The Informed Heart. Glencoe, Ill.: The Free Press, 1960.
4. CHODOFF, P.: Late Effects of the Concentration Camp Syndrome, Arch. Gen. Psychiat. 8:323–333, 1963.
5. CHODOFF, P.: "Effects of Extreme Coercive and Oppressive Forces—Brainwashing and Concentration Camps," in Arieti, S., ed.: American Handbook of Psychiatry, vol. 3. New York: Basic Books, 1966, pp. 384–405.
6. COHEN, E. A.: Human Behavior in the Concentration Camp. New York: W. W. Norton & Co., 1963.
7. COLLIS, W. R. F.: quoted in Nazi Mentality (editorial comment), Amer. J. Psychiat. 102:131–132, 1945.
8. FRIEDMAN, P.: Some Aspects of Concentration Camp Psychology, Amer. J. Psychiat. 105:601–605, 1949.
9. KRYSTAL, H.: presentation at the second Wayne State University workshop on the Late Sequelae of Massive Psychic Traumatization, Detroit, Mich., February 21, 1964.
10. LEDERER, W.: Persecution and Compensation: Theoretical and Practical Implications of the "Persecution Syndrome," Arch. Gen. Psychiat. 12:464–474, 1965.

Coping with Unusual Stress: Disasters

XI Throughout history mankind has been confronted with, and has survived, seemingly overwhelming disasters, both from natural causes such as floods, earthquakes, and tornadoes and from his own handiwork such as atomic bombs and airplane crashes. The minimization of physical and psychological damage, and sometimes even survival itself, are dependent on the ability of people to adapt to new and unfavorable conditions with little preparation. There is general agreement on the overall pattern of reactions. The immediate reaction is shock. Panic is not common and usually occurs only when people perceive themselves to be in great danger and feel their escape routes are limited and closing, not closed. People who have direct responsibility for someone else (children or elderly parents for example) almost never panic (See chapter 7 of Baker and Chapman.)[1]

After the initial shock people are in a dazed, apathetic state, either relatively immobile or wandering aimlessly. This "disaster syndrome," also characterized by passivity and suggestibility, can last several hours. Gradually the individual begins to function more effectively, usually with his first thoughts and efforts aimed at determining the safety of his family. Afterward people may be troubled by "survivor guilt," depression over their losses, or insecurity about the future; but most are able to accept their experiences, to adjust to their changed circumstances, and to rebuild their lives. (A number of informative studies of the social and psychological consequences of natural disasters were carried out in the last twenty years, Disaster Studies, National Academy of Sciences.)

In the first article Irving Janis discusses the experiences of the survivors of the atomic bomb explosion in Hiroshima. In their case there was a "sense of sudden and absolute shift from normal existence to an overwhelming encounter with death" (see Lifton,[5] p. 465). Added to the grief for the death of loved ones and worry about the missing and injured, was the uncertainty of their own future (due to the unpredictable effects of radiation exposure) and that of their community, whose social and economic structures were destroyed. As frightening rumors were circulated and the gruesome reality of radiation sickness became apparent, people had to deal with severe anxiety feelings and also with the sense of guilt that survivors experience. This guilt is based on two thoughts: "Why did I survive when so many others died?" (similar to the feelings of concentration camp survivors (see Part X), and "I did not offer enough help to the injured and dying." Janis concludes that while temporary acute reactions were common, in general there was no long-lasting apathy or depression; people who survived the physical impact also successfully weathered the psychological impact.

In the second article Don Schanche examines the effects of a disastrous tornado on the residents of a small American town. Within the space of a few minutes thirty-two people were killed, 2500 injured, and half the town destroyed. Victims felt overwhelmed, anxious, depressed, and angry. Some felt compelled to be active, even frenetically active, while others found that talking about the storm offered some release for their feelings. In his fascinating account of the survivors of a plane crash in the Andes, P.P. Read (1974) also describes the need of the participants to recount their experiences over and over again after their rescue. Schanche reports the interesting finding of a local psychiatrist that his severely neurotic patients coped surprisingly well with the crisis; one patient attributed his success to the experience he had gained in dealing with his personal crises every day.

The role of community caregivers in helping victims with practical and emotional problems is also examined. Schanche warns of the tendency of various relief workers, including doctors, clergy, teachers, and police, to become too involved. They often try to do everything, offering victims so much help that there is little room for self-sufficiency, thus encouraging dependence. The positive attitude of the townspeople, seen in their enthusiastic plans for rebuilding an improved town, is evidence of the effectiveness of their coping abilities.

In the third article Freda Birnbaum, Jennifer Coplon, and Ira Scharff discuss the aftermath of the 1972 flood in Pennsylvania, with special attention to the role of social workers in preventive intervention. Although there was little loss of life, many lost their homes, most of their belongings, and their businesses or jobs. The elderly were especially hard-hit, as they had neither the time nor energy to rebuild what they lost. Interim housing away from familiar schools and neighborhoods, crowded conditions, and heavy work schedules for those involved in clean-up operations all contributed to the tension and stress.

Social workers can be very helpful in dealing with both practical issues (filing applications for loans and temporary housing) and emotional problems (offering encouragement and support to those who are feeling overwhelmed or depressed). "Outreach" group discussions can be set up to bring people together at a time when the usual social networks are disrupted. The intervention of trained workers can be an effective force in helping the members of a disabled community to deal with their problems and to recover their normal level of functioning. The particular goals that the authors outline for professionals involved in relief operations are helping community leaders resume active roles despite their personal losses, acting as information sources in evacuation centers, and encouraging the organization of self-help programs to combat apathy and depression.

Birnbaum, Coplon, and Scharff suggest that individual reactions to the flood were determined largely by prior patterns of handling problems. Studies of the survivors of a marine explosion[4] and of the Andes plane crash,[6] on the other hand, showed more uniform reactions among the victims, despite differences in predisaster personality and background characteristics. Hocking[3] argues that the explanation for these conflicting findings is that the more severe the stress, the less significant the background factors are in people's reactions.

Another concept whose validity may depend on the extent of the disaster is that of the postdisaster "therapeutic community," in which people alter their usual behavior and give aid and support to victims outside their own families, even to relative strangers. Among the Andes survivors this alteration of the climate of the community was apparent in the effort to make decisions by consensus, and by the rigorously equitable sharing of limited resources.[6] If the whole community is destroyed, though, as in Hiroshima, there are not likely to be sufficient material and emotional resources left for community support to develop. Perhaps the most important conclusion is that most people are surprisingly resilient in the face of disaster. They cope with massive stresses, pick themselves up after unexpected blows, and effectively respond to the challenge of rebuilding their lives.

REFERENCES

1. BAKER, G. W., AND CHAPMAN, D. W., eds. *Man and society in disaster.* New York: Basic Books, 1962.
2. Disaster Studies, Disaster Research Group. Washington: National Academy of Sciences—National Research Council, 1952–present.
3. HOCKING, F. Extreme environmental stress and its significance for psychopathology. *American Journal of Psychotherapy*, 1970, 24, 4–26.
4. LEOPOLD, R. L., AND DILLON, H. Psycho-anatomy of a disaster: A long term study of post-traumatic neuroses in survivors of a marine explosion. *American Journal of Psychiatry*, 1963, *119*, 913–921.
5. LIFTON, R. J. Psychological effects of the atomic bomb in Hiroshima: The theme of death. *Proceedings of the American Academy of Arts and Sciences*, 1963, 92 (3), 462–497.
6. READ, P. P. *Alive: The story of the Andes survivors.* Philadelphia: J. B. Lippincott, 1974.

Article 30 Aftermath of the Atomic Disasters

IRVING L. JANIS

Does an atomic disaster give rise to delayed psychological effects that are qualitatively different from those caused by other types of wartime disasters? Are there any unusual syndromes—comparable to the delayed biological effects— that characterize the psychological state of the survivors after an atomic explosion? The material on postdisaster reactions to be presented in this chapter provides an empirical basis for formulating some tentative answers to these questions.

In the last chapter we have seen that there was relatively little that was unique about the immediate reactions of the A-bombed survivors. It was noted that there was an exceptionally high incidence of narrow-escape experiences and of disturbing perceptions of the casualties; nevertheless, the emotional effects of such exposures do not appear to differ from those seen in persons exposed to heavy bombardment or incendiary attacks. The symptoms of acute emotional shock observed in a small proportion of the A-bombed survivors apparently were the same as those seen in other types of disasters. The widespread feelings of fear and apprehensiveness seem to have been typical "objective" anxiety reactions of the sort to be expected whenever people are exposed to sudden danger. From the fragmentary evidence, it seems that overt panic was not of frequent occurrence and was probably evoked when survivors were trapped in the presence of rapidly approaching fires or were caught in other special circumstances where they were helpless in the face of imminent danger.

The possibility remains, however, that there may have been some unique postdisaster reactions. Insidious, delayed effects might have shown up in the form of unusually persistent anxiety reactions, prolonged apathy, or other sustained symptoms that are indicative of a failure to re-establish normal emotional equilibrium. Conceivably, the exceptionally intense stress of an atomic disaster might even have had the effect of weakening psychological stamina to the point

Irving L. Janis, Ph.D., is Professor of Psychology, Yale University, New Haven, Conn.

Source: Chapter 3 in Irving L. Janis, Air War and Emotional Stress. New York: McGraw-Hill, 1951.

where acute psychoses, traumatic neuroses, or other forms of chronic mental disorder would be prevalent. Or perhaps the atomic disasters had a profoundly demoralizing effect, giving rise to extreme changes in the social and political attitudes of the survivors.

Such possibilities will be examined in our survey of the observations on postdisaster reactions. As will be seen, the evidence points to some fairly severe psychological sequelae; but, again, none of the effects appears to differ from those which have been noted among the English, German, and Japanese people who were exposed to "conventional" air attacks.

SUSTAINED FEAR REACTIONS

After the acute danger phases of the atomic disasters had come to an end, the sources of emotional stress had by no means subsided. The A-bomb shattered the normal pattern of community life and left the survivors in an extremely deprived state. For many days there was practically no medical aid for the tens of thousands suffering from acute burns, lacerations, and other severe injuries. Injured and uninjured alike were homeless, without adequate clothing or shelter. Food was in such scarce supply that starvation and malnutrition became widely prevalent.[1]

In addition to the extreme physical deprivations, there were many other sources of emotional stress. With the economic and social life of their community so completely disrupted, the survivors faced a bleak and insecure future. Moreover, during the postdisaster period most survivors experienced grief over the death of relatives or close friends and many were continually worried about those who were missing, seriously injured, or unexpectedly afflicted with radiation sickness. Under such conditions, emotional recovery from the traumatic events of the disaster could hardly be expected to proceed rapidly.

Various sources of information indicate that severe anxiety persisted among some of the survivors for many days and possibly weeks after the bombings. One of the most frequent types of sustained emotional disturbances appears to have been a phobic-like fear of exposure to another traumatic disaster. This reaction consisted in strong feelings of apprehensiveness accompanied by exaggerated efforts to ward off new threats.

A vivid description of anxiety states evoked by minimal signs of potential danger has been given by Dr. T. Hagashi, a physician in Hiroshima, who was one of the special informants on postdisaster reactions interviewed by USSBS investigators:

> "Whenever a plane was seen after that, people would rush into their shelters. They went in and out so much that they did not have time to eat. They were so nervous they could not work. . .

> ". . . Most of the people were very, very uneasy and afraid that another bomb would be dropped. They lived in that condition for days and days."[2]

Hersey describes a few illustrative incidents, such as the following:

> It began to rain. . . . The drops grew abnormally large, and someone [in the evacuation area] shouted, "The Americans are dropping gasoline. They're going to set fire to us!" [3]

That sustained fear reactions occurred at Nagasaki as well as Hiroshima is indicated by some of the statements in the USSBS morale interviews. For example:

> ". . . after that atomic bomb I was constantly afraid." [Domestic worker in Nagasaki]

> "There were no words that can describe the terror it caused. . . . We were so scared that another would fall that we stayed in the woods for two days wondering what to do next." [Housewife in Nagasaki]

> "I later heard it was an atomic bomb, but didn't venture out of the house for a week or so because we were told it was dangerous." [Housewife in Nagasaki]

> "[I left because] I had the fear of another atomic bomb at Nagasaki." [Housewife in Nagasaki]

Further indications of sustained apprehensiveness among the populace comes from the anxiety-laden rumors which are reported to have been widely circulated during the postdisaster period. Both Siemes[4] and Hersey[5] state that there were rumors that American parachutists had landed in the vicinity of Hiroshima shortly after the A-bomb attack. The latter author also reports that several weeks after the disaster stories were circulating to the effect that "the atomic bomb had deposited some sort of poison on Hiroshima which would give off deadly emanations for seven years; nobody could go there all that time." Brues[6] reports similar exaggerated fears of lingering danger at Nagasaki. He states that there was a widely circulated rumor that Nagasaki would remain uninhabitable for years to come and that this rumor was still creating concern when his party of investigators visited the city several months after the disaster.

In several of the eyewitness accounts from Nagasaki there are allusions to such rumors. For example, one woman reports:

> I heard that people who had not been wounded and seemed to be all right would begin feeling out of sorts and all of a sudden drop dead. It made me panicky. Here I was bustling around now, but I might go off myself. . . .

> The story was going around that the ruins of Urakami ran for two miles from north to south and that if you walked through them you would get diarrhea and if you tried to take care of many of the dead you would come down with some terrible disease, and sometimes you would start coughing up blood. Was this going to happen to me too? I wondered. [Fujie Urata Matsumoto's story in *We of Nagasaki*.][7]

Dr. Nagai claims that directly after the explosion the damaged area actually was so powerfully radioactive that people who merely walked around in it developed acute enteritis with diarrhea and those who worked in the ruins came down with incapacitating or fatal attacks of blood disease.

To some extent, fear rumors may have been touched off or reinforced by the unexpected appearance of many cases of radiation sickness. During the weeks following the atomic explosion numerous unusual signs of organic pathology began to appear among survivors: loss of hair, high fever, excessive fatigue, hemorrhagic spots under the skin, and other severe symptoms of radiation sickness.[8] A number of the morale interviews contain references to the surprising occurrence of severe illness and sudden death among the ranks of seemingly intact survivors. For example:

> "Next evening, my son, who was burned—his face, hands, and legs—came home on foot. . . . At first he seemed all right and I never thought he was going to die as he used to eat three times a day. But after two weeks his teeth began to loosen and his hair started falling out and three weeks later he died." [Housewife in Hiroshima]

> "Six more of my men died a month later. They were well at first, but their hair started coming off about twenty days later and their teeth; their gums started bleeding, and another two or three days later they finally died." [Mechanic in Nagasaki]

> "This friend of mine was well when we worked together helping the other people, but after a few days he said he lost his appetite. Then his hair started falling out and the next day he just fell over dead. There were many people who just dropped dead as the days went on. I suppose it was due to the concussion." [Electrician in Nagasaki]

> "Some of the folks when they came seemed normal. But about one month later their hair all dropped off and they died. Death was caused by gas. The people that were [?] over—their faces were beyond description. If you haven't seen it for yourself, it couldn't be understood. The children, two or three years of age, even if they were living at our place, were dead with the hair on their heads all falling off. . . . The people even after they have recovered—their faces are all disfigured so it is really a pitiful sight." [Female high school student in Hiroshima]

From descriptions such as these, it is apparent that over a long period of time the survivors were likely to see the human damage caused by the violent release of nuclear energy; such experiences probably augmented the sustained emotional disturbances created by the disaster.

With respect to overt avoidance behavior, there is one well-established fact from which some inferences can be made. Within twenty-four hours after the mass flight from Hiroshima, thousands of refugees came streaming back into the destroyed city. According to one of the USSBS reports, road blocks had to be set up along all routes leading into the city because there were so many people who wanted to search for missing relatives or to inspect the damage.

The strong motivation to return to the destroyed city is illustrated in several of Hersey's case studies. For example:

... Mrs. Nakamura, although she was too ill to walk much, returned to Hiroshima alone. . . . All week, at the Novitiate, she had worried about her mother, brother, and older sister, who lived in the part of town called Fukuso, and besides, she felt drawn by some fascination, just as Father Kleinsorge had been.[9]

Although both Hiroshima and Nagasaki required almost complete rebuilding and lacked an adequate food supply, the inhabitants gradually returned to live in improvised shacks. Within three months the population in each city was back to about 140,000.[10]

The fairly prompt return of large numbers of survivors to the target cities is itself a noteworthy postdisaster reaction. This behavior points up the obvious fact that despite whatever potential radiation hazards might persist after an atomic explosion, there are no immediate, impressive signs of lingering danger that impel people to stay away. From what happened at Hiroshima, it is apparent that special problems of disaster control are likely to arise in connection with keeping unauthorized persons out of stricken or contaminated areas (unless avoidance tendencies have been built up by public information about the dangers of radio-activity). Apparently there were strong "approach" motives among the survivors: to search for the missing, to salvage possessions, or to satisfy curiosity. Of central importance to our present inquiry is the inference that such motives were capable of overriding reluctance to return to the scene of the disaster. From the material presented earlier, we know that apprehensiveness about another attack may have been prevalent immediately after the disaster and, later on, fear of contamination may have developed; but evidently such fears were generally not so intense as to prevent resettlement in the target cities. In any case, the fact that such large numbers of survivors returned to the target cities during the days and weeks following the disasters implies that the A-bomb did not produce a unique mass avoidance of the disaster locale.

DEPRESSION AND APATHY

Among some of the survivors, severe reactions of guilt and depression are known to have occurred during the postdisaster period. Dr. Nagai gives a vivid description of his own guilt feelings.[11] Despite being injured, he had worked assiduously during the disaster rescuing people and rendering medical aid until he collapsed from loss of blood and utter fatigue. Nevertheless, he blamed himself for numerous shortcomings: by remaining at the hospital with the members of his first-aid squad, he was neglecting his own wife and children, as well as his injured neighbors who were expecting him to care for them; while devoting himself to directing the rescue work of patients, he was aware of the "selfish" motive of wanting to achieve social recognition for his heroism; several nurses who subsequently succumbed to radiation sickness had complained to him of feeling weak, but not recognizing the early symptoms, he had forced them to keep going; later on, while lying ill and exhausted, he experienced intense fear of another bomb attack and could not get up the nerve to cut across the shelterless wastes to the ruins of his neighborhood, where his wife lay dead.

In the context of reporting his personal reactions, Nagai develops the general thesis that practically all survivors were affected in the same way:

> We of Nagasaki, who survive, cannot escape the heart-rending remorseful memories. . . .

> We carry deep in our hearts, every one of us, stubborn, unhealing wounds. When we are alone we brood upon them, and when we see our neighbors we are again reminded of them; theirs as well as ours.

Nagai believes that persistent "survivor-guilt" is an inevitable consequence of atomic bombing, because most survivors could not avoid behaving negligently in one way or another: people who were in the heart of the city were able to survive only by running away from the fires without stopping to rescue others; people who were in a position to give aid could not simultaneously perform all the duties and obligations of rescuing the wounded, rushing to their own families, assisting neighbors, carrying out their civil defense assignment, saving valuable materials at the office or factory where they worked, preserving treasured household articles, etc.

Although there are independent observations which indicate that some survivors experienced temporary guilt reactions following the A-bombings, there is no satisfactory evidence to support the claim that such reactions persisted in large numbers of survivors or that, four years after the war, the "rents in the ties of friendship and love . . . seem to be getting wider and deeper." Nagai is able to cite a few examples of persistent guilt feelings in the eyewitness accounts he collected from his neighbors. The translators of his book, however, inform us that: "In his editing, Nagai has preserved entirely the plain, unsophisticated character of the narratives, while focusing each one in such a way as to point up the theme that a spiritual wreckage, more vast than the material, must result from atomic war." Moreover, it is doubtful that Nagai had the opportunity to observe the postwar behavior of very many of his fellow survivors inasmuch as he had been continuously confined to his home, bedridden due to chronic leukemia.

Other sources of information provide no substantial basis for concluding that persistent guilt or depressive reactions were an inordinately frequent consequence of the atomic bombings.

Some of the evidence cited in the preceding chapter indicates that at least a small percentage of the survivors felt depressed during or immediately after the disaster. But in the entire sample of USSBS morale interviews, there were found only a few cases who made comments suggesting that they had experienced feelings of guilt, sadness, hopelessness, or apathy during the postdisaster period.

At the time of the interviews, three months after the bombings, a very small percentage expressed attitudes of pessimism or gloom. In discussing their future, most of the survivors described fairly concete plans for increasing their economic security. Although practically all of them were deeply concerned about the food shortage and other economic difficulties, very few voiced feelings of resignation or despondency. In response to the question, "Do you feel you are better or worse off than you were during the war?" only 20 per cent of the Hiroshima cases stated "worse off now." The comparable figure cited by the USSBS morale report for Japan as a whole is almost the same: 17 per cent.[12] (Only the Hiroshima sample was used in the present analysis of pessimistic responses; the impression received from reading the Nagasaki interviews was that negative responses, e.g., "worse off now," occurred even less frequently in the Nagasaki interviews than in those from Hiroshima.)

From a detailed examination of answers to all relevant questions, it was found that about one-half of the Hiroshima cases expressed some degree of concern about the future. Only 11 per cent, however, expressed clear-cut pessimism. Directly comparable percentages are not available for other urban areas in Japan, but the USSBS morale report gives the following information:

> Three months after the surrender, fifty-three per cent of the Japanese people gave pessimistic answers to the question: "Now that the war is over, how do you think you and your family will fare in the next two or three years?" Only twenty-five per cent reported fair satisfaction with their prospects. It is apparent, again, that the majority of the Japanese people were exceedingly depressed in the post-surrender period. Typical responses were: "We have no plan." "We are living from day to day." [13]

From this statement, it would appear to be highly improbable that there was a significant difference between Hiroshima and other Japanese cities with respect to the relative incidence of interview responses expressing pessimistic attitudes about the future.

Although the interview data provide little evidence of widespread gloom, despondency, or hopelessness among the A-bombed survivors, there are some independent observations which have been interpreted as indicating a high degree of overt lethargy. USSBS investigators in the Medical Division visited Hiroshima three months after the bombing and noted that the city still had not recovered to the point where adequate shelter and essential utilities were available: Only a few shacks had been constructed for homeless people; there was no garbage or sewage collection; leaking water pipes all over the city remained unrepaired; etc. In the Medical Division's report the slow and haphazard restoration of Hiroshima is interpreted as indicating an absence of initiative among the populace. The same sort of apathy is reported at Nagasaki.

> At the time the Allied Military Government entered Nagasaki, about 1 October, the population was found to be apathetic and profoundly lethargic. Even at this time the collection of garbage and night soil had not been reestablished, restoration of other public utilities was lacking and the hospital facilities were inadequate.[14]

The claim that there was widespread apathy or lethargy among the A-bombed survivors is evidently based solely on the fact that the restoration of housing, public utilities, and hospital facilities had proceeded at a very slow rate. However, when a city has been almost totally destroyed, with over half its population killed or injured, the rate of restoration probably is not an adequate indicator of the motivational state of the remainder of the city's population. Restoration would undoubtedly depend to a large extent on the amount of aid received from the rest of the country.

It should also be borne in mind that apathy and absence of cooperative activity have been reported by the USSBS Morale Division as characteristic of the entire Japanese nation after the war was terminated by the unexpected surrender, which came shortly after the A-bomb attacks.

> The war left Japan with its cities laid waste, its industrial system disorganized, and its merchant fleet almost obliterated. Millions of Japanese were unemployed, underfed, homeless. Countless others were casualties from bombing or had been displaced in evacuation. The nation as a whole had suffered the extreme hardships of the war and tasted the bitterness of defeat. It had been disillusioned about its leaders and left uncertain about its own future.

> . . . recognized common goals and accredited common leadership were lacking. The cement that held the nation together during the war lost its grip, and the people, in many places, became a disorganized mass, split among themselves, seeking individual solutions to their desperate personal problems and conscious only of the immediate day-to-day task of staying alive.[15]

When the factors mentioned in the above excerpts are taken into account, together with the other findings from the morale surveys, it appears unwarranted to conclude that the A-bombs produced an exceptionally high degree of apathy or depression among those who survived at Hiroshima and Nagasaki.

REFERENCES AND NOTES

1. USSBS Report, *The Effects of Atomic Bombs on Health and Medical Services in Hiroshima and Nagasaki,* U.S. Government Printing Office, Washington, D.C., 1947.
2. These quotations and similar ones are taken from the original protocols of the USSBS interviews in Hiroshima, Nagasaki, and the towns surrounding those two cities.
3. JOHN HERSEY, *Hiroshima,* Alfred A. Knopf, New York, 1946.
4. FATHER SIEMES, "Hiroshima—August 6, 1945," *Bull. Atomic Scientists,* Vol. 1, May, 1946, pp. 2–6.
5. HERSEY, *op. cit.*
6. A. M. BRUES, "With the Atomic Bomb Casualty Commission in Japan," *Bull. Atomic Scientists,* Vol. 3, June, 1947, pp. 143–144.
7. T. NAGAI, *We of Nagasaki: The Story of Survivors in an Atomic Wasteland,* Duell, Sloan and Pearce, Inc., New York, 1951.
8. Los Alamos Scientific Laboratory, *The Effects of Atomic Weapons,* U.S. Government Printing Office, Washington, D.C., 1950; USSBS, *The Effects*

of *Atomic Bombs on Health and Medical Services in Hiroshima and Nagasaki, op. cit.*

9. HERSEY, *op. cit.*
10. Report of British Mission to Japan, *The Effects of the Atomic Bombs at Hiroshima and Nagasaki,* His Majesty's Stationery Office, London, 1946.
11. NAGAI, *op. cit.*
12. USSBS Report, *The Effects of Strategic Bombing on Japanese Morale,* U.S. Government Printing Office, Washington, D. C., 1947.
13. *Ibid.*
14. USSBS, *The Effects of Atomic Bombs on Health and Medical Services in Hiroshima and Nagasaki, op. cit.*
15. USSBS, *The Effects of Strategic Bombing on Japanese Morale, op. cit.*

Article 31 The Emotional Aftermath of "The Largest Tornado Ever"

DON A. SCHANCHE

"And the wind got mad and the trees got sad." Sabina DeJarnette 12; killed in Xenia, 1974.

A routine atmospheric storm passed over Xenia, Ohio on a Friday evening a few weeks ago. It was not alarming—just a little thunder and lightning accompanied by wind gusts and heavy rain. Yet some of the 27,000 residents of the southern Ohio town panicked. A number of men, women and children dropped what they were doing and ran for shelter as if they had heard an air-raid siren. One mother screamed in fright and hurried her two children into a bathroom to hover protectively over them in the tub. Her husband, a recently reformed alcoholic, cracked under the stress and walked resignedly into the gusty rain and back to the bottle. Many Xenia children cried. One asked his father to get a gun and kill the storm. Telephone circuits were jammed with callers seeking reassurance that it was only a passing squall.

In most American cities, such reactions to a thunderstorm would be considered bizarre, to say the least. In Xenia they not only were understandable, but appropriate. Since April 3, this year, every dark cloud that has passed over the Ohio town has seemed like some kind of cosmic bad joke. It was on that spring Wednesday, between 4:38 and 5 P.M., that 32 people were killed and almost half of the seven-square-mile town destroyed in a tornado that may have been the largest ever observed on earth.

Many saw it coming. At first it looked like a huge, mushroom-shaped black cloud with three narrow stems. Then the stems merged into one devastating funnel, six-tenths of a mile wide. Winds in the funnel reached 318 miles an hour, four times the force of a hurricane. It roared as deafeningly as a thousand trucks on a highway, and swirled with what looked like millions of dead birds and other broken artifacts of nature and man, swept up from the ground. Traveling at 48 miles an hour, the tornado took only 22 minutes to cut a horrifying, 16-mile path

Source: Today's Health, August 1974, published by the American Medical Association. Reprinted by permission of the author.

from the outskirts of Xenia, directly through the main intersection of town and straight up U.S. Highway 42, through nearby Wilberforce.

It was as if a great crushing machine had rolled across the landscape, knocking over a building here, pulverizing another there, and witlessly hurling their contents throughout the town. Even today, with many of the most severely damaged buildings leveled and the rubble carted away, the pathway of the storm is clearly visible, marked throughout its length and breadth by bleak empty lots, cratered with debris-filled basements. If you have been to Vietnam and have seen what a flight of B-52s can do to a peaceful, bucolic place, then you have witnessed a scene very much like the one that the tornado created in Xenia.

The power of the storm was incredible. Canceled checks that were sucked out of file cabinets in Xenia that afternoon fluttered to the ground that night in Chagrin Falls, Ohio, 200 miles away. Automobiles were wrapped like untidy Band-Aids around shattered trees. A 57-car Penn-Central freight train, carrying autos from Detroit, was scattered along the right-of-way in the center of town, as if spilled from a toy chest. One house, occupants and all, was lifted up and deposited across the street on the slab of another house that had just blown away. Of 2,757 homes damaged by the storm, 1,095 were totally destroyed. The three-story high school and both junior high schools were demolished. Most of Central State University, in Wilberforce, was leveled. Virtually every flower, bush, and tree in the path of the tornado was uprooted, shredded, or defoliated. Twenty-five people died quickly; seven others lingered a day or two in hospitals before they expired. About 2,500 were injured, some very seriously.

All public buildings and offices were hit, but police, fire, and city officials organized quickly in response to the sudden, overwhelming tragedy. The walking wounded helped the fallen. The unhurt took in the homeless. More than a thousand National Guardsmen patrolled the rubbled streets and helped to arrest 35 looters, including an out-of-towner who said he was a minister. Two sergeants from the National Guard died in a fire that was the first aftermath tragedy of the tornado. Doctors, nurses, off-duty policemen, firemen, and other rescue workers from as far away as New Jersey came to lend what help they could.

During the week after the storm, Xenia and Wilberforce were hit with snow, hail, ice, and freezing rain. Both towns were without telephones, electricity, and gas. Public drinking water was unsafe. Bumper-to-bumper traffic, mostly sightseers, jammed the few open streets. "We had no windows—snow, ice, sleet, and the coldest rain I ever felt in my life. It was one week of hell," says Xenia Police Chief Raymond B. Jordan.

There are perhaps 25,000 intensely personal and acutely painful stories of what happened in Xenia and Wilberforce that day and in the days that followed. The overwhelming majority of the residents survived unharmed, most with even their property intact. Yet no one completely escaped the effects of the tornado. Some of these, like the incidents of panic during the recent Friday night thunderstorm, are becoming apparent weeks later. Such natural disasters seem invariably to suck an emotional storm in their wake that frequently cuts a wider swath than the disaster itself. In this psychological aftermath, the nonvictims often suffer as much, if not more, than the direct victims of the tragedy.

"It has become increasingly clear that in a natural disaster no one escapes completely," explains Ann S. Kliman, psychologist and disaster expert who was called to Xenia after the storm to consult with local mental health authorities. "All are either direct or indirect victims." She says the reactions of the indirect victims are likely to be similar to those of survivors of such desolating catastrophes as the atomic bomb attacks on Hiroshima and Nagasaki, even though the scale of destruction is much smaller.

"Direct victims usually respond to the disaster by an increase in anxiety reactions such as fear of another disaster, feelings of being overwhelmed, feelings of depression and rage, and inability to cope or frenetic coping activity," says Mrs. Kliman. "Indirect victims often respond by feeling conscious guilt—guilt that they escaped while others didn't—and by an increase in guilt-related and stress-induced physical symptoms, accidents, and explosive arguments at home, in school, and on the job."

"The ones I feel sorry for are the ones who did not become direct victims," sighs Richard A. Falls, M.D., who lost his own rambling, modern home in the storm. "In subtle ways, they express guilt for being spared: crying, yelling uncharacteristically at their husbands or wives, losing sleep. They tell themselves, 'I can't understand it; I have nothing to be upset about; the tornado missed my home.' I try to help these people understand why they feel as they do. Their town was destroyed. Their lives have been altered."

The tall, reflective, and weary general practitioner recognizes some of the psychological symptoms of direct victimization in his own behavior and that of his wife, Barbara, first woman member of the Xenia school board and nurse at the Greene County Vocational School.

"We dig out and try to interfere with the emotional crises that are ahead for people," Barbara Falls recently wrote in a letter to friends worried about the couple's fate. "The after sequences of such a total disaster are grisly—Dick and I will need our own counseling to remain healthy care-givers. We must remain whole—for ourselves and those depending on us for example and leadership."

"At first, I was anxious and overwhelmed with concern for my family and with the amount of work to do," says Dr. Falls. "Now I'm overwhelmed with feelings associated with depression and anxiety. I'm not the least bit ashamed that I was depressed as hell and that I still am."

Like many of his patients, Dr. Falls talks about the storm at every opportunity, aware that by ventilating his anger and frustration, the losses are somehow made easier to bear. Because he, too, was a direct victim, many of his patients find it easier to "let go" in his office. "They are almost hostile to some of the doctors who were not hurt by the tornado," he explains.

"I've been encouraging them to express exactly what they feel. Some of them have lost everything. They've got to go through the trials of rebuilding. They're angry. But they cover over their real feelings, perhaps unconsciously. I've been encouraging them to get out the real feelings. When they come in the office and want to swear or be angry, I help them express it."

Leonard W. Cobbs, M.D., the only psychiatrist who practices in Xenia, moved almost instantly after the tornado to perform what preventive psychiatry he could among the direct victims of the storm. While Dr. Falls and other staff members and volunteers sutured and treated the wounded, by flashlight, in powerless Greene County Memorial, Dr. Cobbs walked the corridors with his own flashlight and offered quiet, conversational solace.

"The fact that someone who didn't know them took the time and energy in the midst of chaos to talk and listen helped them," he says. "I found one man whose wife was killed. He couldn't find his three children and was verging on collapse. We talked. He got a hold on himself and went out to search for his children."

According to the psychiatrist, there were few immediate psychological problems after the storm. "The first reaction was shock," he explains. "Then functional necessity took over—looking for people, looking for things, exchanging information about one another's welfare. By Friday, two days after the tornado, with the cold and rain and with authorities moving people out of unsafe houses, you saw some depression."

Dr. Cobbs was surprised by the reactions of some of his own patients who had been under treatment for severe neuroses before the storm. They rallied and coped as well as or better than other, less troubled, victims. As one of his patients explained it: "I'm living in crisis every day. I'm used to crisis."

So far the storm has not resulted in the psychiatrist's being deluged with disturbed people seeking help. But he has seen indications that they will come. On Easter Sunday after the tornado, a severe wind- and rainstorm struck Xenia and triggered many panicky reactions. One of the most significant, he believes, was that 50 residents of nearby neighborhoods that had been spared damage in the tornado rushed to Greene Memorial for shelter.

"But most people are still feeling lucky," says Dr. Cobbs. "They're not feeling bad about it yet; they're feeling good about it. But they'll come in time. When we see them, they'll come as depressions. When the tornado watches start again in September, we'll see some panic reactions. And on April 3, 1975, I think we will see a lot of anxiety and depression—anniversary reactions."

While it may be years before the full extent of post-tornado reactions unfolds in Xenia, there were significant indications when I visited the town two months after the storm that problems were developing. The clearest signs were among the very young. Literally hundreds had simply refused to return to school. According to James DeFeo, one of two psychologists in the Xenia school system, many children were suffering from a tornado-induced neurosis appropriately called "school phobia." They were afraid to leave home. Perhaps more important, their parents often were afraid to let them leave.

[School phobia also showed up as an aftereffect of the 1971 Los Angeles area earthquake, but it is not caused only by natural disasters. Any serious traumatic event—such as a divorce, riots or violence in the school, even the death of a President—can trigger school phobia. "Like any other fear, it is usually a temporary thing, if it is handled correctly," explains Norma Gordon, a therapist at the San Fernando Valley Child Guidance Clinic, where many adults and children were counseled after the California disaster. The best thing, she says, is to get the child back to school as quickly as possible.]

"Some parents seem to have an increased dependency on their children since the tornado," DeFeo says. "They won't let them go. One woman insisted to a principal that her son was too upset to attend school, even though the child was sitting right there protesting that he wanted to go back to class." Even when they did return, some of the children seemed almost irrational. One kindergarten girl, for example, came back for the first time seven weeks after the storm and immediately panicked over a distraught fear that she would miss lunch. Her teacher calmed her with tenderness and a sandwich, but only after the incident caused a major disruption of the class.

Hal Bussey, a psychological social worker at the Greene County Guidance Center, in Xenia, says each passing storm exacerbated the problem. "The children got nonspecific headaches, stomachaches, and side pains; and school attendance dropped off even more." Bussey is particularly worried about one teenage boy he has been counseling. The youngster rode the bus to and from school each day, because his parents forced him to go, but he refused to attend school. He is not a malingerer—he is afraid, Bussey says, and unless both parents and child face the situation honestly, the young man may become suicide prone.

School problems added almost overwhelming emotional strains to many families. All but one of the city's public schools were either severely damaged or demolished. Elementary students were on double session through the end of the school year last month. Junior high and high school students were bussed to neighboring districts, an hour away, to occupy strange classrooms after the normal school day had already ended. They were in school or on the bus from 2:30 to 9:30 P.M. Bus drivers worked on an almost nonstop schedule from 6:30 A.M. to 9:30 P.M. There were fears that a school bus accident could add further tragedy to the stricken town.

Perhaps not typical, but not unusual, is one school principal who worked both double sessions, from morning to night. His wife, also in the school system, quit work at 2 P.M. and arrived home just as her children left to catch the bus for their late session. The family got together only on weekends. "For a family already bearing the stress of the tornado, it was a new and perhaps unbearable strain," says DeFeo.

Other incidents in the schools bore out the pattern of emotional aftershock. An eighth-grade class had a near riot, arguing the pros and cons of whether the destruction done to a mostly black neighborhood on the east side of Xenia was "God's judgment." About half of the children thought that it was, which was surprising in a town in which the races have lived in integrated harmony for years. The president of the city commission (tantamount to mayor) is black, as are Dr. Cobbs and many of the town's professional people, including the director of the local mental health board.

In some of the makeshift schools, sadistic practical jokers who themselves were showing neurotic storm reactions, panicked whole classes by screaming "Tornado coming!" in the hallways. "We've seen the more susceptible kids suffer the most," says DeFeo. For example, a first-grader who was slightly hyperactive before the storm became almost totally distracted and did odd things, such as stirring mustard into his milk. He was obsessed with guilt over the death of his three-year-old hamster, which was neglected during the confusion following the storm.

Some of the children and adults had multiple tragedies to deal with. Many divorced and separated parents, for example, were so shaken by the devastation that they reverted to old patterns of dependency, according to the Reverend Donald Young, chaplain of Miami Valley Hospital, in nearby Dayton, where many of the injured and dying were cared for. "They came back together in the aftermath, only to find that the problems that had driven them apart were still there. So they suffered the pain of breaking apart again."

DeFeo recalls a separated couple with two children, a boy and girl who were home alone when the tornado hit. The house was wrecked but the children survived. Father and mother rushed to the house from opposite sides of town and tearfully reunited in the wreckage. For a few days it looked like wedded bliss had returned, and the children believed the storm was "lucky." But old conflicts soon reappeared and the parents split for the second time. The children suffered a triple tragedy, in effect, with two separations and a tornado that destroyed their house. One of the youngsters now has severe emotional problems in which his social fears are irrationally intertwined with the storm. He views the tornado as both the cause and the effect of his parents' impending divorce.

Another man, whose wife walked out two years ago on him and their three children, lost his house, his job, and his children's school on April 3. His employer was wiped out; the house and school were demolished. A mental health worker, hearing of the case, called on him in the temporary apartment where he was relocated. "He's an angry man, furious at the whole world," she reports. "Yet he won't admit that he needs help. We've tried to motivate him toward going to the guidance center. He needs to vent his anger in a creative way. But he's not the type to say, 'I'm mentally ill.' It will be hard to help him."

Perhaps the most encouraging thing about Xenia, if anything can be encouraging in a field of rubble, is the way the community as a whole has girded itself for the problems to come. Businessmen and community leaders established a "Spirit of '74" committee to take advantage of an unparalleled social opportunity—the reconstruction, by rational, modern planning, of an entire town in such a way that its pre-tornado failings will be corrected and its future needs accounted for. Surprisingly, the town did not organize in the classic mold, where the rich and powerful impose their own views upon the poor. Every residential area in Xenia is represented on the committee and has its views heard in the planning sessions.

At the same time, the city has engaged in a massive program of preventive psychiatry, to at least minimize the psychological aftereffects that seem certain to become more acute. The mental health board, under its sociologist director, Robert Neal, drew together the local ministerial association, guidance center, school authorities, and other professionals to form teams of volunteer helpers who expect to be working for many months. Mrs. Donna Hart, a social worker who coordinates the teams under the clergymen's interfaith council, calls her volunteers "advocates," instead of caseworkers.

"The term 'caseworker' connotes mental illness," she explains. "Many people here are disturbed and will become more disturbed, but their behavior is normal for a disaster. We want our volunteers to be their advocates, to help relieve the bitterness and frustration, and to help the victims deal with the complexities of welfare agencies and federal aid.

"This is largely a middle-class town, and it is the proud middle-class people who often need the most help. The very rich can afford to take care of themselves, and the poor already know how the welfare system works. But most middle-class people do not know the workings of their own social welfare agencies. They fall through the cracks in the system. In most cases the needs are still physical ones—furniture, a place to live, clothes, money to get a new start, jobs," says Mrs. Hart. "They haven't felt the full emotional needs yet, because every day is still one of survival. It's all going to hit when things get back to 'normal,' when the Department of Housing and Urban Development leaves, when the insurance policies are all paid up. We're doing preventive work now."

One of the many cases that an advocate reported while I was there was that of Hazel and Glenn, an elderly retired couple, both of whom are disabled. Their house was demolished over their heads during the tornado. For three weeks they remained inside the wreckage, crawling on all fours to get food from their thawing freezer and refrigerator. Helpers found them still determinedly independent and reluctant to ask for aid. Yet they needed $200 to buy furniture for the temporary apartment to which they were taken, and $10 a month for gasoline so Glenn could go shopping in their car that miraculously survived destruction.

One of Mrs. Hart's co-workers wanted to give them more money than had been requested from the interfaith council's funds. "No," she said, "they're proud and self-sufficient, and we shouldn't hurt that self-sufficiency. We must give them only what they need."

Her firm but wise approach is not shared by all of Xenia's care-givers. Some have been too helpful. "Sometimes I get as concerned for our advocates as for the people they're trying to help," says Mrs. Hart. "The advocates become so deeply involved that they want to do too much."

According to Mrs. Kliman, the psychological consultant who came to Xenia from the Situational Crisis Service of the Center for Preventive Psychiatry, in

White Plains, New York, care-givers face almost as many problems as the victims of a disaster. "I call them hidden victims," she says. "Some of them tend to treat the direct victims like infants, because the disaster has been so terrible that they become overwhelmed and want to do everything for them. Then the care-givers face enormous burdens of overwork and underappreciation. You need to deal with *their* feelings, too."

That was one of the reasons Mrs. Kliman and a few other outside experts were brought to Xenia. One of her first acts there was to conduct brief training sessions in preventive psychiatry for doctors, nurses, mental health workers, clergymen, policemen, businessmen, schoolteachers, bartenders, barbers, and beauticians. The latter three groups, seemingly improbable community mental health workers, were included because "they are among the few who automatically hear the troubles of the community in their work and have a chance to help people talk about their problems." Today it is almost impossible to walk through the streets of Xenia, or to sit in one of its two remaining bars, and not hear some resident pouring out his troubles to a sympathetic listener.

Dr. Falls compares the town's reactions to those of a grief-stricken person mourning the death of a loved one. "We are going through the same stages," he says, "learning to live with the reality of the loss, talking about it, remembering."

But for the families of 32 Xenians, the mourning has a far more poignant quality. They are not only grieving the loss of the town, of their homes, and, in some cases, their jobs: They also are suffering the greater loss of loved ones. One of them, whose ordeal recently came to light in the *Dayton Daily News*, is Mrs. Waltraude DeJarnette, who lived through Allied bombing raids against her native Germany during World War II. "It was the same smell, the same look on people's faces," she recalls.

On the afternoon of April 3, Waltraude DeJarnette was at work in the institutional studies department of the University of Dayton. She heard the tornado watch on the radio and called home. Xenia was in the midst of a thunderstorm. She talked to three of her children, Sabina, 12, Mike, 16, and Christina, 13. She told them to stay at home and not to worry. If the storm did develop into a tornado, she told them to take the precautions that they had discussed previously. A fourth child, Alexander, was at a friend's house. She called him and told him to stay there.

Christina later explained what happened. Mike saw the huge tornado cloud approaching and hurried his two sisters into a hallway of the house. Because Christina was the most frightened, Mike and Sabina spread themselves protectively on top of her. "I saw the wall ahead of us and mother's bedroom door collapse," Chris says. "The last thing I remember was that the dog jumped over my head."

Mrs. DeJarnette drove through the storm between Dayton and Xenia. When she arrived, her house was gone. Only the concrete slab remained. Alexander was safe but was not found until two days later. The friend's house had also been destroyed. That night she found Christina in Miami Valley Hospital in serious condition. The next day she found Mike's body in a Dayton morgue. A day later she found Sabina, also in Miami Valley Hospital, gravely injured. Sabina, who never regained consciousness, died in the hospital on Sunday.

A few days later at her office Mrs. DeJarnette found this prescient poem that Sabina had written two years before for a school class:

> As I was in the meadow
> I saw the clouds dance
> and heard the wind hum
> and the trees sing.
> But suddenly the clouds grew wild
> and the wind got mad
> and the thunder growled
> and the trees got sad.
> The clouds went away
> and the wind went to sleep
> and the trees sang their happy melody
> as I was in the meadow.

Article 32 Crisis Intervention After a Natural Disaster

FREDA BIRNBAUM

JENNIFER COPLON

IRA SCHARFF

More than a year has passed since tropical storm Agnes laid waste to parts of eastern United States. The writers of this article were among those who went to Wilkes-Barre, Pennsylvania, to participate in the delivery of casework services to the Jewish community in the aftermath of the flood that resulted from the storm. Because there has been an absence of literature on social work crisis intervention at times of mass disaster, the writers believe that their experiences may be informative as to the kind of need that arises when rivers overflow or earthquakes occur.

DESCRIPTION OF THE DISASTER

On the morning of June 23, 1972, the Civil Defense Squad of Wyoming Valley was sent to the Wilkes-Barre area to announce through bullhorns and by other means that the Susquehanna River was rising to a dangerous level in the vicinity. People were told to vacate their homes within twenty minutes and to take nothing with them. There was almost no panic, because the Susquehanna River rises periodically, and people leave their homes for a few hours until the danger is past. Few thought back to 1936 when there had been a flood of somewhat major proportions. People walked and drove in orderly fashion across bridges connecting Wilkes-Barre to Kingston, heading for higher ground. The supply

FREDA BIRNBAUM, A.C.S.W., is director, family life education and mental health consultation, Jewish Family Service, New York, New York.

JENNIFER COPLON, A.C.S.W., was a caseworker, Jewish Family Service, at the time this article was written.

IRA SCHARFF, M.S.W., is a caseworker, Jewish Family Service.

Source: Social Casework, Vol. 54, No. 9 (November 1973), 545–51. Reprinted by permission of the publisher, Family Service Association of America.

of regular sandbags had been exhausted, and makeshift plastic bags and pillow cases filled with sand were being used in an attempt to reinforce the dikes and hold back the rising waters.

As the rain continued to fall and the river continued to rise, the sandbag operation was halted in order not to jeopardize the lives of those who were making last-ditch attempts to hold back the river. When the dikes gave way, the breaks occurred on both sides of the river and water rushed through at a force estimated at ninety miles an hour, carrying automobiles along like toys and moving houses off their foundations. The river crested at over forty feet and spread over an area almost two miles wide. The muddy waters rushed into basements, rose into ground floors, and often reached the second stories of houses.

Days passed before the water receded and streets were declared safe enough for people to return to see what remained of their homes. Inches of foul-smelling, dangerously slippery, contaminated mud coated every surface where the waters had been, indoors and out. Doors and window frames were warped and stuck, and many people had to break into their homes to face the devastation within. Little that had soaked for days in the muddy water could be salvaged. Wood was swelled and rotted, metal was rusted or corroded, and fabrics were shrunken and stained. Along with utilitarian items, precious paintings and lifetime mementos were ruined or lost, and family heirlooms and the collections of family photographs were washed away or soaked beyond recognition. For weeks the great heaps of debris piled in front of each home consisted of everything from sofas to teddy bears. Where the river had been, homes and shops had to be gutted; the water-clogged plaster walls had to be stripped down to the basic foundation planking. Few houses in the area were habitable for months.

PERSONAL IMPRESSIONS

Four weeks after the flood, one of the writers found a town still under military patrol, a population in shock, and a community uncertain of its ability to rebuild itself. People were living doubled up in hotels and motels, in evacuation centers, or crowded into the houses and apartments of friends, relatives, and even strangers. The mobile homes and campers promised by the Department of Housing and Urban Development (HUD) had not yet arrived, nor had the promised state unemployment insurance checks. For thousands of people, the only clothing they owned was that on their backs. Many jobs had been "flooded out," but those able-bodied men who could work on the debris and mud removal obtained low-salaried positions in the massive clean-up operation. Some children had been sent to stay with relatives in distant communities, and a few were fortunate enough to have been sent to camp.

Wherever people gathered, the flood was the sole subject of conversation. People would tell their stories over and over, as though still dazed by their experience. No one was immune from the impact of the disaster. For some people, memories reverted to the displacement during the Nazi holocaust and even to

Noah's flood and the biblical exodus from Egypt. Elderly people were particularly vulnerable. They often had had neither the strength nor the mobility to evacuate their homes quickly and thus had been forced to leave by boat after the waters had risen. Aged couples who had lived frugally all their lives and saved painstakingly to purchase a home found all that they had struggled for destroyed in the flood; they despaired at having neither the time nor the strength to start again.

Widows with young children were particularly frightened by unemployment and the lack of housing. Owners of small businesses felt helpless with their plants wiped out and inventories unsalvageable. College students mourned the loss of valued books and records; many faced the prospect of leaving school because family resources were depleted and their help was needed at home. Children were upset by the loss of toys and frequently asked to sleep with their parents for fear of recurrent nightmares. The fearful knowledge of what rising water could do and the uncertainty as to whether the Susquehanna River might rampage again caused people to wonder if the hurricane season in the late summer or the melting snows the next spring would bring renewed flooding.

As people emerged from shock and initial immobilization, they began to piece together interim plans with the aid of emergency provisions of governmental and voluntary agencies. By mid-August, a small proportion of people were in campers or mobile homes and others were awaiting delivery of the temporary housing facilities. Under the stress of homelessness, some people had felt compelled to sign leases on apartments in nearby communities and subsequently struggled with the dislocation from familiar schools, houses of worship, and other neighborhood facilities. People began supplementing their clothing and equipping new living quarters as they acquired them. A few shops had opened. This phase can be characterized by the omnipresent slogans on shop windows, on doors of residences, and on automobile bumper stickers—"The Valley with a Heart Coming Back Better than Ever" and "Rebuild We Will."

More than three months after the flood and with the chill of Fall compounding the hardships, many mobile homes still were without gas, electricity, water, and sewage lines. A significant number of housing units had not even been delivered. Applications for Small Business Administration (SBA) loans for restoring houses and businesses were extremely slow in being processed, and those that were granted were often for less money than requested; in some cases the loans were denied. An added factor that made difficult the planning for rebuilding was the uncertainty about which areas would be designated for demolition. As a result, hope and optimism waned. People were working many additional hours in reopening businesses, families were feeling cramped and trapped in their mobile homes with minimal space for movement or storage, and children experienced isolation from friends and boredom from the lack of recreational facilities. Depressive reactions became more manifest.

Although different reactions to the flood were observed at different periods following the flood, these reactive phases should not be considered universal.

People respond to crises in a variety of personal ways that are only partly determined by the type and extent of the crisis itself. In other words, although the flood created certain universal feelings of loss and depression, the ability to readjust was largely determined by one's previous functioning—physically, emotionally, and socioeconomically. While increased solidarity and a positive response to fight were noted in some families, the spectrum of responses included total family breakdown, and this response seemed primarily correlated with the family's level of adjustment prior to the flood.

STRUCTURE OF SOCIAL WORK INTERVENTION

The Jewish community was among the hardest hit in Wilkes-Barre. Jewish residents' homes were clustered along both sides of the river, as were such Jewish institutions as the Jewish Community Center, the synagogues, and the United Hebrew Institute. Of the approximately fifteen hundred Jewish families in the Wyoming Valley, only twenty families reportedly were not affected and only fifty families were able to reoccupy their homes when the water receded. Consequently, the National Council of Jewish Federation and Welfare Funds decided to supplement the services provided by the federal government and the American Red Cross. Experts were sent in to offer additional social services to the Jewish population, and sectarian programs serving other groups in the community were formed concomitantly.

What evolved in the series of meetings between local leaders, outside experts, and representatives of the Jewish community at large was a tripartite structure— a committee on restoration of synagogues and institutions, a committee on restoration of businesses, and a committee on individual needs. Each committee formulated policy in keeping with its goal and devised a way to implement it. The focus of this article is the work of this third committee.

PROFESSIONAL EXPERIENCE

In order to put into operation the plans of the committee on individual needs, Jewish family agencies within a reasonable radius of Wilkes-Barre were asked to donate the services of professional social workers for approximately two-week periods so there would be sufficient staff to interview the flood victims and consider their requests for financial aid and other help. A schedule was devised for the amount of financial assistance that could be given for specific items, such as beds, kitchen table and chairs, linens, utensils, and other household essentials. The operation, which came to be called the Action Center, was run by the Flood Emergency Committee of the Wyoming Valley Jewish Center. The committee established a policy to meet the flood victims with empathy and a readiness to give along designated but flexible guidelines.

Provision for financial assistance and other practical aid, such as information on available resources, was the backbone of the operation of the Action

Center's committee on individual needs, but in instance after instance people needed more than the concrete help that they requested. The caseworker could provide a differentiated service based on his ability to recognize the applicant's underlying feelings and respond in a way that might best enable the flood victim to take the next step in coping with the effects of the disaster.

There was, for instance, a widow who came in for a third time because she and her oldest child had moved into their newly delivered trailer home, although it was not connected to utilities. She was in a mild panic at the prospect of her two younger children returning from camp in two days and having to live without plumbing, refrigeration, electricity, or gas. This woman was a healthy, able person who had coped well with repeated family crises during the aftermath of the flood. At this time a talk with the caseworker diminished her sense of isolation, and the caseworker's attempt to intercede with HUD on her behalf cut down her sense of impotence and her rage at the impotence. Even though the caseworker did not succeed in getting the trailer connected to utilities and water immediately, the applicant, having been able to share her feelings and obtain a response to them, was less panicked and felt her situation was somewhat more manageable.

A man came in attacking the Action Center because he anticipated it would not give the amount of financial help he wanted. The caseworker got beneath his chip-on-the-shoulder attitude to a recognition of his upset feelings when he was in a dependent position, and he acknowledged how overwhelmed he felt. Practical help was given; he received one-half of the sum he had requested (he had asked for twice the maximum amount). This financial assistance plus an opportunity to be heard enabled him to formulate plans for obtaining the rest of the money.

An adult daughter came in requesting continued financial assistance for food and homemaker service for an invalid, aged mother who had been moved to the home of a son and a rejecting daughter-in-law in a nearby town. This applicant's need for casework help focused on her involved attachment to her mother; this fact was of even greater importance than the continuation of the financial support for the interim plan of having the mother housed with the working daughter-in-law. The worker responded to both the tangible need for money for a homemaker and to the need to deal with the daughter's guilt about the forced "abandonment" of her mother.

In another troubled relationship between a woman and her daughter-in-law, it was the mother-in-law who applied for assistance. Her feeling of desertion and loss when her husband died and when her only child married was exacerbated by the reality losses of the flood. More than ever, this dependent woman with a deeply entrenched pattern of martyrdom viewed her daughter-in-law as having taken away or alienated the only thing she had, her son. The caseworker's response to this applicant, in addition to financial assistance which had some affect on alleviating her sense of deprivation, was to encourage the woman to find a part-time job; it was hoped that work would relieve her from excessive rumination about the wrongs her daughter-in-law had done her and give her a chance for positive feedback from fellow employees.

One family that came into the Action Center for help talked so incoherently that it was difficult to discern the reality situation. They were asking for housing in Wilkes-Barre to enable the son to start his freshman year in college there the following week. They had been staying in a room in New York City and commuting to Wilkes-Barre to deal with the authorities on their applications for SBA loans and HUD housing. In this situation, the social worker made several telephone calls to uncover the snarl of red tape: The request for a trailer had been suspended because the family had applied for a trailer home both on a group site and on an individual site, expecting to take whichever came through first. The processing of both had been stopped until one request was withdrawn, but the family had not understood this fact. The caseworker talked with the family and helped them decide which application should be retained. Shortly after the trailer was delivered, the family came in to request financial assistance with equipping the mobile home with a few of the missing essentials. The previous incoherent speech and projection that made the family members appear limited had disappeared. Instead, there was a working family unit that felt able to take the steps necessary to establish itself in its new home and to find its way in reestablishing the father's employment.

Much of the time the caseworker saw the damaging effects of the displacement and uprootedness on people, particularly the elderly, who had had to locate in nearby communities. One seventy-five-year-old man in such a position requested help in rebuilding his small business as a salesman in household supplies. His warehouse with its small inventory had been flooded as well as his home, and, although none of his creditors was pressing him, he wanted to retain the dignity of the spotless reputation he had always had as a salesman and businessman. Aware of this applicant's profound sense of aloneness, the caseworker was able to learn that the adult son of this man was one of the few who had moved totally away from the area soon after the flood. Once the caseworker had explored this applicant's sense of being trapped and alone in a strange town in his late years, he discovered what was of real importance to the applicant—returning to his familiar synagogue for the High Holy Days and being greeted warmly and caringly by people who knew him. He did not really have much desire to start over in business and withdrew the request which had been a cover-up for his loss.

In another application overtly based on the need for financial assistance to reestablish a business, two partners came in together. In talking with them, the caseworker was able to discern that one partner had lost his house in the flood and that the house of the other partner was intact. The two partners were unable to agree on the form the business was to take if it were reestablished. The partner who had suffered the tremendous personal loss was experiencing an overwhelming sense of hopelessness and panic that made him seem at odds with the whole world, including his partner. The caseworker, responding to what the distressed partner was feeling, helped clarify that this man's primary concern was to attend to his family's pressing problems before he could turn his attention to the business. In the course of the interchange between the social worker and the two partners, the more fortunate partner became sensitive to his partner's pain and told him gently to take care of his family before worrying about the business.

Occasionally, a person came in asking for something for someone else in his family, such as a wife or a brother. As the caseworker explored the situation, it became evident that the projection of need was covering other feelings or needs. In one instance, the request from one man was for a portable heater for his mobile home to assure his family of adequate heat during his absence on business trips. The caseworker was able to help this man become aware of his feelings of guilt about leaving his family, which felt to him like abandonment at this difficult time. The combination of obtaining the heater and discussing his situation with the caseworker helped this applicant to alleviate the anxiety under which he was operating.

OUTREACH EFFORTS UTILIZED

At times the workers at the Action Center, hearing of someone who had not applied for assistance, would reach out. In one instance, the caseworker asked an older woman to have her daughter call because the daughter might be entitled to some assistance even though she was working for a local governmental agency and consequently drawing a salary. When the young woman came in, she was deeply moved at having someone concerned about the emotional drain on her at work. The sense of having someone there for her, even briefly, and of being "given to" left her better able to go back to the endless round of evaluating requests for aid from those the disaster had left without resources.

In the final phase of the Action Center operation, outreach efforts included visits to homes and stores. The purpose of the outreach was multiple—to act as the friendly visitor supporting and reinforcing the gains the residents had made and to actually uncover and respond to a desperate need when a prospective applicant did not come to the Action Center.

In October when there were fewer cases to process, time permitted another outreach approach. Two mothers' discussion groups were formed. It was thought that group discussion could meet emotional needs on many levels because of the commonality of experiences and concerns and because of the loss of usual communicative contacts, for example, husbands' unavailability resulting from long hours at work while restoring businesses, dislocation of neighbors and relatives, and preoccupation with so many flood-related meetings and programs. With the aid of the recreational program coordinator at the Jewish Community Center, who also had an investment in group activities and was well known by local residents, the groups were planned to meet both during the day and in the evening, in order to reach as many mothers as possible. Although the attendance at both groups was small, there was much dynamic interchange, including a sharing of each individual's personal experiences and a real attempt to work on issues that were raised: adjustment problems resulting from cramped quarters and isolation from friends and family, emotional reactions of children identified by the mothers as related to the children's preflood functioning and adjustment, concerns about children not adequately expressing feelings of sadness and loss and questions of whether this reaction was normal, fears of future ramifications regarding emotional health, increasing appreciation of different personality styles and ways of coping, and an awareness of the emotional overload that taxes energy so greatly that there is little left to direct toward spouses and children.

The response to these groups was so positive that two of the women in attendance assumed responsibility for seeking out more potential group members; they organized two follow-up meetings in places closer to their homes, one in a woman's mobile home and the other in the Jewish Community Center of an adjacent community where many Wilkes-Barre residents were temporarily living. These groups were short-lived, but they served the therapeutic purpose of bringing people together to support and empathize with one another and to begin to help each other with problem-solving.

Repeatedly, the caseworker saw substantiated the fact that those who were the least sturdy in terms of physical health or emotional well-being prior to the flood were those who were the most vulnerable and in the weakest position to cope with the ramifications of emergency living. The aged, with their assorted infirmities, who had lost a lifetime's belongings and the homes they had saved for years to buy, were the least able to be without eyeglasses, dentures, orthopedic shoes, or lounging chairs that gave comfort to aching limbs or backs. The less the capacity for adaptability and the less the capability of coping, the greater the need was for some intervention on the part of the caseworker to shore up the limited ability to function in the face of the multifaceted stresses brought on by the disaster.

PROBLEMS IN THE DELIVERY OF SERVICES

While many applicants were relieved at seeing an unfamiliar, impartial person to whom they could give vent to their feelings of loss, anger, guilt, and grief, others objected to what they deemed outside interference. The latter group often expressed the feeling that the workers could never understand their tragedy.

However, the antagonistic reaction of some flood victims was not the only obstacle that prevented the Action Center staff from providing the community with a more comprehensive casework service. There was also the pressure of time and continuity, for each social worker worked a limited time in Wilkes-Barre. When the individual assistance program first began, scores of applicants came for financial aid. Consequently, it was impossible to spend much time with each person to explore his or her total needs. When the number of applicants diminished and there was more time to spend with each person, the discontinuity of casework service was still significant. It was disconcerting for returning applicants to see a new interviewer every two or three weeks. After flood victims had secured food, clothing, and shelter, many wanted to talk about their personal and family problems but, with the rotation of temporary workers, a sense of ongoing trust in a working relationship was not feasible. Consequently, some people expressed resentment about revealing painful feelings without the assurance of being able to talk to the same worker the next time they came for help. In addition, the Action Center program was defined by the community as a primarily financial aid service, and some people assumed that the workers would not be interested in more emotion-laden difficulties. There was little orientation or education toward informing the Jewish community of the variety of services potentially available at the Action Center.

PROJECTIONS FOR THE FUTURE

A natural disaster is sudden and can not be totally prepared for. Nevertheless, there has been a general but developing format for aiding stricken communities. Traditionally, such voluntary agencies as the American Red Cross and the Salvation Army have moved in quickly to provide the most immediate necessities of food, clothing, and temporary shelter. Recently, the government has assumed increased responsibility to assist in recovery efforts, and such agencies as HUD, SBA, and the General Services Administration provided a major portion of the rescue and rebuilding operations in Wilkes-Barre, with the newly formed Office of Emergency Preparedness expediting the use of federal aid. As has been shown, private philanthropy may also step in with casework services based on immediate concrete need and individual assistance related to emergent overall need. Yet social work has not begun to tap its rich experience in crisis intervention. Historically, social work has been more involved than any other profession in outreach to individuals in trouble. It would, therefore, be appropriate and natural that social work enlarge its concept of crisis intervention to include intervention in mass disasters. This intervention could take the form of delivery of concrete services, individual and group counseling, consultations with local leadership, and community organization and social action.

The writers contend that there was in Wilkes-Barre an underutilization of social work expertise except for delivery of concrete services and some response to individual emotional needs. This phenomenon was not peculiar to Wilkes-Barre but exists broadly. Local leaders tend not to be familiar with the role social work can play in the recovery efforts following a natural catastrophe. Therefore, social workers must educate local community leaders about the potentiality and value of social work services in a crisis situation. Social workers can also be available for consultation on an ongoing basis to help local leaders anticipate and understand the underlying dynamics and reactions of community residents to a disaster. Additionally, in the capacity of consultant, the social worker can help the local leader, overwhelmed by the double load of personal loss and community responsibility, to effectively resume his much-needed leadership role.

The social worker helps those in positions of leadership to resume control at a time when people are feeling helpless.

In relation to direct service to residents, the group modality, thus far little utilized in mass disasters, is a natural and expedient means of reaching people. A well-prepared, alert, professional social worker could begin his work at the on-set of a mass crisis as people arrive in evacuation centers. His function there would be to inform people about community resources as they develop, as well as to respond therapeutically to immediate emotional problems. He might also encourage social action by beginning to organize people to be advocates for themselves, a step that encourages self-mobilization in order to get beyond apathy and depression. As people leave evacuation centers, the social worker could enlarge his outreach services to meet people in usual gathering places such as community centers or places of worship. Children, too, can be seen in groups, with or without parents, where they would have an opportunity to deal with some of their fears and to learn ways of communicating with parents at a time of stress.

The writers' experience demonstrated that professional casework intervention was a valuable and necessary service in the aftermath of a particular disaster. Nonetheless, the vast potential for social work intervention remained unrealized in Wilkes-Barre. Although there are limitations to preparations that can be made for mass disaster, the writers believe that there is a professional responsibility not yet adequately undertaken to apply theoretical and practical knowledge of crisis to how the challenge created by a natural disaster could be met so that social to large-scale emergencies. In this article, the writers have made suggestions as workers could move quickly into an afflicted area and become a visible and effective force through the entire period of rescue, relief, and restoration.

Unusual Stresses
of Modern Life

XII   Modern society, with increasingly sophisticated technology, and with rapid and unsettling changes in social and moral values, presents some individuals with unusual stresses. The situations discussed in this section are examples of some of the challenges faced by individuals in this age of complex technology and momentous change. The adaptive skills necessary to withstand these unusual stresses derive from coping techniques developed in response to earlier challenges.

In the first article Sylvia Jacobson describes her experience, and that of the other passengers and the crew of a plane skyjacked and held captive in the Desert of Jordan for almost a week. Anxiety over their fate, concern for their waiting families, and the physical discomfort caused by a shortage of water and toilet facilities, the oppressive heat, and the sustained inactivity all contributed to an atmosphere of tension and irritability. Subgroups began to form immediately —families, seatmates, college students, mothers of infants—with overlapping memberships and conflicting interests on certain issues. The crew offered some leadership, intervening for example when some non-American passengers began to agree with, and repeat, the anti-American accusations made by the Arab captors. The unexplained removal of six men one night added to the general anxiety, but the arrival of the Red Cross brought some hope. With the removal to hotels of all non-Jewish women and children and the men of certain nationalities, the remaining group became smaller, more homogeneous, and more cohesive. When bands of Arab soldiers came through the plane gawking at the captives, the passengers affected disdain and became still more united in their common feelings of humiliation, revulsion, and anxiety.

The captives found the most effective means of coping with their anxiety and sense of helplessness to be self-organization directed at regaining some control over their lives. Even relatively simple tasks like keeping the children occupied and taking turns regulating the use of water and toilet facilities helped relieve the stress and improved group morale. A study of six thousand people trapped in an airport for two days by a snowstorm showed that the absence of any task around which to organize self-help activity was the principal basis of people's feelings of dependency, resignation, impotence, and frustration.[3] With no visible leadership and no obvious project the stranded airport passengers became passive, self-centered, and withdrawn. Jacobson maintains that increased group cohesiveness improved general understanding, promoted the readiness of people to act together constructively, and reduced the level of anxiety.

Many new and exotic environments have been developed through recent advances in scientific technology. Spaceships, nuclear submarines, Antarctic research stations, and other remote scientific outposts have in common the characteristics of confinement, isolation, monotony, and limited stimuli. In the second article Jim Earls examines life aboard a nuclear submarine. There are six major stresses to which the sailors must adjust: extremely limited privacy and personal territory, isolation from the outside world (radio broadcasting is forbidden when submerged), inability to help their families if needed, a monotonous environment, few rewards for good job performance, and the hazards of having nuclear weapons aboard. Earls, who was a medical officer on a submarine, noted a cycle of responses. After an initial period of good mood as people get established in the submergence routine, psychosomatic complaints, sleep disturbance, anxiety, and irritability begin to increase. By the middle of the submergence period the majority of the crew are experiencing psychosomatic illnesses like headaches, changes in appetite, and sleep disturbance. Depression and withdrawal are common. Three-quarters of the way home personal interactions begin to increase, and depression and physical complaints subside. In the last few days most submariners become possessed by a hypomanic "channel fever."

Earls outlines several strategies used in coping with these stresses. Denial of feelings is used to control anger, as is the intensely sarcastic humor found in the half-way phase. The withdrawal characteristics of that period may be seen as an attempt to increase one's personal space by keeping others more distant. Sublimation (for example, the abundant use of sexual humor) and isolation of affect (seen in the inability of submariners to discuss nuclear weapons on any but the most technical level) are common coping tools. Studies of research stations in Antarctica have reported similar, cyclical symptoms—headaches, sleep disturbance, depression, impaired memory and concentration, and irritability— and similar coping strategies.[2,4]

As an example of the serious problems of modern urban living, such as poverty, racial and ethnic prejudices, and the rise in violent crimes, the third article offers a discussion of women who have been victims of the crime of rape. Sandra Sutherland and Donald Scherl discuss the typical patterns of response, the adjustment these women must make, and the usefulness of professional intervention. Immediate reactions include shock, disbelief, and dismay. This "acute phase," in which the woman is often agitated and incoherent, can last from a few days to a few weeks. Burgess and Holmstrom[1] found two different emotional styles in the acute period: "expressed," when anger and tenseness are openly displayed (for example, by crying), and "controlled," when feelings seem subdued and the outward appearance is calm. The second phase is one of "outward adjustment," when the woman returns to her normal routine and temporarily denies or suppresses her feelings about the rape. Finally a time of integration and resolution begins, signified by an inner depression and a need to talk. At this point the woman must deal with two major issues. She must accept a new view of herself in light of her experience and her role, if any, in the rape, and she must resolve her feelings about her assailant.

Intervention by a health professional, mental health worker, or other counselor can be very helpful to the rape victim. In the acute phase it can take the form of aid in solving immediate practical issues (such as whether to press charges, how to tell family or close friends, how to get pregnancy and venereal disease tests), of information on what to expect emotionally (descriptions of phases two and three), and of help in expressing feelings like fear, humiliation, and anger. Anxiety usually decreases when the woman can discuss her experience with her family. Many rape victims also deal with their fear by changing their residences or their telephone numbers.[1] It is often of value for the counselor to make some effort to help friends or relatives understand the rape and their own attitudes to it and to the victim. While the need for outside help will usually be denied in the pseudo-adjustment phase, the counselor should remain available for consultation in the third phase. Even with an experience as brutal and senseless as rape, successful coping in the integration and resolution phase can be a positive and strengthening process.

REFERENCES

1. BURGESS, A. W., AND HOLMSTROM, L. L. Rape trauma syndrome. *American Journal of Psychiatry*, 1974, *131*, 981–86.
2. GUNDERSON, E. K. E. Emotional symptoms in extremely isolated groups. *Archives of General Psychiatry*, 1963, *9*, 362–68.
3. HAMMERSCHLAG, C. A., AND ASTRACHAN, B. M. The Kennedy Airport snow-in: An inquiry into intergroup phenomena. *Psychiatry*, 1971, *34*, 301–8.
4. POPKIN, M. K.; STILLNER, V.; OSBORN, L. W.; PIERCE, C. M.; AND SHURLEY, J. T. Novel behaviors in an extreme environment. *American Journal of Psychiatry*, 1974, *131*, 651–54.

Article 33 Individual and Group Responses to Confinement in a Skyjacked Plane

SYLVIA R. JACOBSON

Virtually no time elapsed between warning and threat, as skyjackers rushed forward, holding some 149 passengers and ten crew at gunpoint, while an American plane slowly circled in midair over Brussels, shortly after noon, Sunday, September 6, 1970.[1] Passengers' reactions varied. Shock; fear; excitement; extreme depression; non-comprehension; the thought that someone had gone berserk; that this was a Labor Day hoax; stunned disbelief that this could really be happening, were among the responses of the instant—seeming to support, for the most part, Withey's[8] observation that new stimuli tend to be interpreted within the context of the known and familiar, and as non-threatening until such interpretation is no longer sustainable. Stewards and first-class passengers, faced by the male skyjacker, were gun-herded backwards, hands up, mouths open, faces ashy-green, into the already full tourist section. Shortly, a woman's voice over the intercom announced the seizure of the plane in the name of the Popular Front for the Liberation of Palestine. The plane began to retrace its path southeast.

During the ensuing five hour flight to Jordan, the male skyjacker stood facing the passengers, gun leveled toward them. Within the rigid confines of the plane, sub-groupings began subtly to form. Families clung more closely to their children. Seatmates and those adjacent, hitherto strangers for the most part, dropped normal psychological barriers; in the sudden sense of unity in danger noted by Jacobson,[5] they whispered together in sudden communion, exchanging conjectures, anxieties, fears. The fourteen college students scanned seat-pocket maps, speculating among themselves, trying to foretell destination, evidencing a sense of adventure. Almost at once, they formed and remained a subgroup.[7] One elderly passenger turned to her seatmate and spoke quickly, regretfully of a

SYLVIA R. JACOBSON, M.A., is Associate Professor, School of Social Welfare, Florida State University, Tallahassee.

Source: *American Journal of Orthopsychiatry*, Vol. 43 (1973), No. 3 (April), 459–69. Copyright © 1973 the American Orthopsychiatric Association, Inc. Reproduced by permission.

quarrel with her only kin: "No time now to make amends!" A voice hissed harshly that this was as it had been in the Nazi Holocaust: "People were led helplessly to their deaths, like sheep." Some prayed; some wept silently; some read; some sat seemingly frozen within themselves; some dozed. I found my own life running in review in my mind, with regret at the many opportunities life had offered me and from which I had turned aside.

Around what would ordinarily have been the time of landing at Kennedy Airport, self-orientation and self-concern shifted to expressions of alarm for expectantly waiting families. How shocked they would be on learning that this plane would not land! Seatmates and passengers within whispering reach of one another allied themselves, exchanging names and addresses, making promises that survivors would somehow notify the next of kin of those who might not make it.

With darkness, foreboding swept the plane. From the cockpit, the woman's voice announced that the plane was approaching the Desert of Jordan, where "friendly people are waiting to welcome you." Silent, anxious, those at windows peered down, minds struggling to make sense out of dimly perceived scenes of moving trucks, knots of running figures, red flares. The Captain's voice asked for four volunteers to move forward to balance the plane for landing. The male skyjacker slowly backed, gun still pointed, to make way for two silent men, a woman, and a girl who rose and moved in line to the forward section. After agonizing circlings, a perfect landing was made on the desert floor. Ladders were quickly raised to the forehatch. By lantern and flashlight, armed, uniformed women and men climbed aboard, a woman *fedayeen* taking and retaining command.

"Now you know why you are here," she announced in English, "but none will leave the Desert until the just demands of my government have been met. It is up to you to pressure the Red Crescent to act, through the International Red Cross and then through your families and your governments to expedite your release."

For the first time we realized that we were *all* held hostage. We had become a community of our own, in an unfamiliar environment under threat and stress.

Landing cards for the Popular Front for the Liberation of Palestine were distributed for completion. One woman reached down swiftly in the dimness to write her passport number in the lining of her shoe. When I, recalling the passivity that had led so many to concentration camps, asked why my passport was wanted, a sibilant, "Sshh—don't make trouble," was murmured from all sides. With the demand for passports, an apprehension that was never to leave flooded the cabin. It became the first divisive wedge, setting apart those who held dual passports of two different nations, and families in which parents and children held passports of different nations. The stewards urged compliance. One woman, separated from one of her children, became hysterical, demanding that her child be seated beside her. About this time, the sound of another plane was heard but it was not seen. All slowly settled in for a cold, restless night, four in many seats built for three, without lights except for lanterns at the fore and rear hatches, while armed men paced the aisle.

Monday's four o'clock dawn revealed the Swissair plane, its nose crumpled, directly behind the American, Arabian tents pitched beside the planes, and rings of armed men, of armed half-trucks, and of tanks in concentric circles surrounding the planes. The desert stretched beyond in all directions. Light and warmth were wanted, but stewards urged keeping the blinds down against the coming day's heat. Some complied. Some, perceived as isolates, pulled up their blinds. A uniformed Arabian physician came aboard. He listened to complaints. He was asked to arrange for access to baggage in the hold for those with personal medicines. This evoked a fierce, whispered quarrel. Access to held baggage was opposed by passengers whose luggage contained second passports. Sharply hissed quarreling about the rights of individuals to meet their own needs versus the implicit threat to others ended for a while when the former complied, ceasing their requests. Sub-groups existed in uneasy, distrustful proximity to one another. Complaints and fears were voiced. The physician offered tranquilizers *ad lib*. One elderly woman whispered that she would save as many as she could for a quick death, as everyone knew that these days old women were raped and killed before younger ones.

Jordanian Army officers came aboard. They told of the concerted air piracies of the day before. They regretted the obvious discomforts of the captives and their own powerlessness to help unless there were a direct attack upon the planes. They promised to send in a box meal. The factual statements and the promise of food brought a surface calm and unity that was broken when the food came. Food introduced a second divisive element that was to continue. Jewish parents who urged their children to eat, to stay nourished, were sharply criticized by parents who refrained from eating non-Kosher food, setting examples for their children. Existing sub-group boundaries of age and of dual passport-holders were invaded. Despite these splitting factors, the adolescents managed to be and to remain a definite, almost closed sub-group, immersed in companionship and self-organization for available satisfactions. They continued to evidence a certain sense of adventure. By noon, however, the mothers of babies had spontaneously become another sub-group, concerned with the practical problems of caring for their infants and themselves. Formula preparation, bottle heating, the improvisation of cribs against the night cold engaged them. The need for diapers and for sanitary napkins was acute. The doctor brought cotton batting. Scissors and powder were lent or contributed. Cutting and slicing to size occupied time. While most passengers continued to speculate on their fate, to read, to pray, to doze, half-heartedly to play cards, the mothers, with busy hands, remained purposefully and enviably engaged and united. Along with the stewardesses, women passengers moved to look after the six children who were traveling alone. It seemed as though this, as other life crises of birth through death, called upon the work of women and their hands.

Lack of running water and water shortage, the already malodorous and over-used, non-functioning lavatories, heat and airlessness, and sustained inactivity began undermining adult morale. Some conjectured aloud, Job-like, on what they must have done to deserve this fate. Some showed irritation at the running play of small children. Episodes of bickering occurred. Aisle-seated passengers resented being brushed against. Window-seated passengers stared silently outward, then suddenly complained that they had not been informed of what was going on.

At dusk, in order, first Indian and then other nationals, then mothers and their children, then all women were ordered to prepare to leave for hotels in Amman. Each was required to sign name, place, and date of birth, nationality, religion, and places of embarkation and debarkation. Each then climbed down the ladder to board waiting minibuses, some crying softly at the unanticipated separation from husbands. All were soon ordered out of the buses and into a rude circle in the dark. By torchlight, names were called and religions asked. Those who answered "Jewish," along with two non-Jewish women who pleaded not to be separated from their husbands, were gun-pointed back into the plane, while the others returned to the buses that left for Amman. Apprehensive and dejected, families split, passengers tried to arrange themselves as best they could aboard the cold, filthy, smelly, waterless plane, using coats and cut-down cabin curtains for blankets. During the night, six men were quietly removed by the captors.

Tuesday's dawn found the passengers temporarily united in the shared shock and sorrow of the bereft women, and the enhancement of the secret fears of the men whose wives and children had already been removed to Amman. The re-minder of the Holocaust was heard again: "This was the way men were silently taken at night to die in Buchenwald." [3] Realization that the Arabian physician slept in the tent outside, available on call, brought a definite, universal sense of comfort. A plastic sack, containing *pitta*, hard boiled eggs, cheese, and fruit was brought to each. With tea made on an emergency gas ring, this became the standard meal.

The arrival of the International Red Cross, their offers to try to send messages to the next of kin of the captives, to deal with male and female passenger representatives around meeting the greatest needs, and to ask consent to negotiate with the leaders of the PFLP and the several governments concerned for the release of all, raised spirits generally. Messages were quickly written. A sense that things were really starting to happen for the better re-established group cohesion. Two stipulations, however, quickly thinned the cohesion. We were to agree to ask no questions about the negotiation proceedings and to ask no questions about the fate of the men who had already been taken away. We soberly agreed.

By afternoon, armed men of various North African armies began to pass through the plane in a steady stream, hour after hour, staring curiously at its interior and at its occupants. This aroused feelings of revulsion, degradation, de-humanization, and vague threat. Distinctly, we felt impotent, confined, con-stricted, and like animals in a zoo. Even the adolescents resented the feeling of being "on display." [6] Divisiveness was subdued. Despite differences of nationality, religion, race, sorrows, needs, values, resentments, and fears, there was unity in a showing of proud indifference, aloofness, and disdain to the inquisitive stares. Some of the soldiers tried to communicate friendly interest by smiles and bows. One glanced at a child making a "cat's cradle" out of string and gestured to show that his children did the same. A complement of a Desert Guard, in contrast to the khaki, desert-camouflaged uniforms of others, wore long, brown *caftans*, red and white *kefiyehs*, scarlet leather ammunition belts, and bandoliers. Each wore a silver dagger thrust through the breast of his *caftan*. Their dress, their boldness of bearing and expression appeared to me to be as they must have been, un-changed, two thousand years ago, and inspired a cold chill of fright. This was doubtless the intent.

Suddenly, all passengers were ordered off the plane, ostensibly for air, were handed down the vertical ladder to the desert floor and allowed to circle slowly within a ring of armed men. The stewards, navigator, and engineer had all warned passengers against talking politics among themselves, and they circled silently, stiffly, listlessly in the heat. Signs painted on the planes under draped flags of the PFLP could plainly be seen: "Down with Imperialism; Down with Zionism; Down with Israel." The Swissair and its people could be seen but no contact was allowed. On return to the plane, it was apparent that cabin baggage had been searched. The female *fedayeen* leader announced that passports had been found and demanded others thought to be still hidden. The stewards again urged compliance. Hushed arguments about the rights of the few versus the haz-ards to the many broke out again. The female soldier stated coldly that the British, American, and Swiss governments had been given a 72 hour ultimatum to meet her people's demands. These, as before and subsequently, were self-contradictory and unclear. Veiled threats were made. The captain of the plane announced that food had been found in the search, that health was endangered by hoarding food in the desert heat, and that greater care must be taken with the lavatories, which, even with the steady, exhausting efforts of the crew, could not be kept func-tioning.

A general rise in anxiety showed itself in querulous remarks: "If it were not for the Jews, none of us would be in this spot," and in a new sub-group wave of anger directed against passengers who had appeared to pass the time by talking in friendly fashion, exchanging details of different life styles, with the physician and with the female guards. Such "fraternization with the enemy must cease," was muttered. Hill and Hansen[4] have commented, in this connection, on the uses of rationalization to settle conflicts in threatening situations, and on the loss and inadequacy of response in systems where there is lack of energy of input. The boundaries of sub-groups continued to shift and to overlap as succeeding and different crises pressed new sub-groups into being.

That evening, some passengers began to talk spontaneously of specific planning towards organization and more self-direction of their confined life. Under the leadership of the few men and women who had responded to the request of the International Red Cross for passenger representation, various plans were considered for a man and a woman, in turn, to supervise the use of the lavatories by children, to allot and distribute water and food, and to collect refuse. None of these plans was acted on for another two nights and a day. A few sporadic and abortive efforts at leadership were met with contemptuous remarks such as: "Who gave *you* any special rights?" Meantime, the plastic airline cups saved from the last meal or salvaged from the rubbish became valuable, guarded possessions. A generator was brought from the Amman Airport, but after ten minutes of electric power for the functioning of lights and facilities and for cool, circulating but lavatory-odor laden air, the generator had to be allowed to cool down for fifty minutes. It was considered not powerful enough for this plane, and taken away.

Wednesday marked a change in captors' attitudes. They were abrupt and talked rapidly among themselves. The doctor was quick and curt. The morning food sack was late, and two packets each were distributed, suspicion-arousing in itself. The irritating parade continued. Two Danish neutrals passed quickly through the aisle, affirming the fact of the other air piracies and stating that newspaper headlines were telling the world what had happened. Only a few at a time could hear what was said, and the plane rapidly became a rumor factory of misinformation and accordingly mixed emotions. As group interaction increased, individuals tended more strongly to establish affiliates among those expressing, at the moment, understandings and values similar to their own.

A severe desert storm arose. Even with all hatches closed, heavy brown dust filtered through. Lavatory odors permeated the stifling heat. There was evidence of increasing diarrhoea and vomiting. In the oppressive, thickened air it seemed that we did not even have the privilege of perspiring individually. Suddenly, word was passed that still another plane had been captured. In intervals between blown brown masses of dust, the captured BOAC plane appeared, disappeared, circled, and slowly dropped to land at an angle across and ahead of the American airship's prow. The *fedayeens* exulted, embracing, pounding one another's backs, and, leaving only a skeleton crew on guard, raced to wave their guns in salute to the two skyjackers proudly standing in the open BOAC hatch. Ambivalent feelings were spontaneously expressed by those aboard the American plane. These included relief in the notion that in numbers lay safety; guilt at this relief; empathy with the anxieties of the newcomers: dismay at the loss of secret hopes that the American plane might somehow take off and escape. As to whether the landing of the BOAC boded ultimate good or ill for all, the college group was most articulate. A propaganda speech was made by a PFLP spokesman as propaganda materials were distributed to the passengers for reading. The guards alternately threatened to blow up the plane if their ultimatum was not met, and taunted their captives with the low regard in which their governments must hold them to let them sit so long in such distress. To complaints of the stifling dust, airlessness, and heat, there were reminders that they had to live with these constantly and that Americans, especially, criticized and wasted the food that they were saving for the passengers out of their own rations.

One college girl, Leah, Jewish, born in the Sudan, an American citizen and speaking Arabic, had been the subject of special taunts for living on "Capitalistic fat" while her fellow Arabs "lived like rats in the desert, dying without complaint, struggling for a better future" for their children. They sought her alliance with them, with promises of personal safety. Leah signaled to me at one point. Under promise of secrecy, she told me that she heard talk that the plane had already been wired for demolition—only the time had not yet been set. Selected hostages were taken, one by one, to tents outside the plane for hours of interrogation about Israeli addresses, photos, hotel receipts, or letters written in Hebrew that had been found in the searches. Singing by small children was brusquely interrupted by guards. Mothers got some relief when adolescents began spontaneously to help in caring for babies and children. One mother had virtually given up all care of her infant to a devoted teen-ager, almost as though apprehension for herself was too great to allow her to remain responsive to her child's needs. Night came at once with five o'clock sunset. It seemed to me to be beyond my capacity to endure another night. Wakeful and restless, beset by desert asthma, my coughing disturbed what peace others might have had.

By Thursday's dawn, the guards were more curt. They taunted us anew with the "lies" of the "deserting" Red Cross and our worthlessness in their eyes and in the eyes of our governments. Three children were feverish. The doctor was abrupt. He announced that he could bring no more medicines from Amman due to the civil war raging there.* Increasing anti-American feelings were expressed by the guards. A new split emerged as some non-American passengers, aggressively irritable, muttered assent, identifying with the *fedayeen* in expressing both anti-American and anti-Jewish feelings. One man announced that although he was a naturalized American citizen, earning a good living in America, he despised all its values and customs and was only waiting for retirement to return to his native country. The crew, together with a black American soldier passenger, effectively cooled such remarks. Knowledge of the impending demolition of the plane had spread throughout. The only reactions seemed numbness, apathy, depressed isolation on the part of some, increased tendencies among affiliates for relatively idle talk, bolstering of self defenses in fitful scapegoating, and the on-going bickering around eating.

With news that water would be chlorinated and rations still further limited,

---

* To complaints about our distress, we were told by our guards that matters were far worse on the other planes. Later, I learned that the contrary was true. We, on the American plane, had the worst of the three situations. Aboard another plane, greater total cohesiveness was sustained, and a passenger was heard to say: "Our fear was controlled; our anxieties uncontrolled." [6]

passengers began to organize themselves. College students set up activities for small children, "school" for the others. A water detail was formed and drinking allowances stipulated. Water rations for lavatory flushing and plans for lavatory supervision were made. Arrangements for the communal use of one basin of water for hand washing and another for rinsing were set. A garbage collection detail was organized. Adolescents assumed direct responsibility in distributions. This led to whispered expressions from the old: "The young will care only for themselves; they will take and have no concern for us." Food hoarding began again. Some older persons began a despairing weeping, acknowledging a longing to die by the graves of their spouses. A committee was organized to confront the parents of one child who had been especially annoying. On parental insistence that, "He is only a child," the father was told that, if necessary, the child would be seat-confined.

Passengers were again ordered out. Legs swollen from prolonged sitting, all stumbled down the ladder, into the shade of the wings. Girls set up jump-rope games; children played ball; adults had setting up exercises. At the request of some of the soldiers, a life raft was pulled out and inflated. It immediately exploded. Another was soon drawn out, inflated, shaded, made into a baby's play pen. Soft drinks were distributed from cases in the open hold. A man had a heart attack but refused to leave his wife for hospitalization in Amman. On return to the plane, it was apparent that possessions had again been searched. All were warned that a search of garbage and lavatory wastes had disclosed torn-up passports. Renewed demands for concealed passports, along with threats, were made. The open air, the fresh drink, the activities, the fact that the day was passing, lifted spirits. Conversation, except for the isolates, became general. More passenger movement up and down the aisle was allowed. Gallows' humor jokes were exchanged. Veterans of Buchenwald scorned the complaints of others. Compared with what they had lived through, this was a "Hilton Hotel." A young girl acknowledged that it was her birthday. Some canned pears were found. A stewardess broke out two bottles of wine. A gift was managed. We sang. Finally, we made ready for the night, which was broken by the armed guards' pacing and by my endless coughing. During the night ten more men were taken away.

Friday morning, with the exception of a few withdrawn elderly and consistent isolates, all were aghast at the pallid, ravaged faces of wives and mothers of the missing men. Their grief was controlled with dignity. Tears that appeared were quickly checked by the comforting hand of another. A second dust storm forced the closing of the hatches. The doctor stated that he could no longer carry responsibility for our health. The female *fedayeen* in command suddenly announced that all cabin possessions, including handbags, were to be taken away at once. We were sure now of the ultimate destruction of the plane, but not when. A plan was made to try to send a message to President Nixon, with urges that it be dignified and nondemanding. The crew rejected first efforts, feeling that the wording was not strong enough. Compromise was reached. There was a general sensing of the rapidly spreading deterioration of health and morale. The atmosphere grew more physically and emotionally oppressive. The remaining men were paler; those of military age showed and complained of gastric disturbances; vomitings increased. Babies were irritable; half-hearted attempts at card-playing and checkers were abandoned or increased; more sudden sobbing occurred and stopped; more intimate expressions of loss of hope and longing for death by older ones were heard. Grief over guilt towards others in the past emerged with a seeming lack of coherence. An older person clutched wordlessly at the doctor as he passed her seat. A stewardess sat with and comforted those whom she could. The Red Crescent official came to announce that he had no knowledge of what was going on; that he had been told that the soldiers could supply no more food; that he had requested emergency food supplies from his headquarters. Our guards, silent and withdrawn most of the day, suddenly appealed to us to help them in whatever might happen: "We have helped you, been kind and generous to you; now you help us!"

At dusk, the stewardesses passed their cart with a cold spread of cheese, salad and fruit brought by the Red Crescent. Permission was given for a candle to be lit by an older woman in token of the Sabbath. Shortly after, the remaining men, passengers and crew alike, were taken. When the last jeepload of men had pulled away, stewardesses, who said they had no information about what was to happen next, urged the deeply frightened, silent women and girls to put the children to sleep and to bed down themselves. Leah, another woman, and I decided to stand watch in turn, in the area where the children slept. We whispered together softly, companionably, curiously alone and curiously united; occasionally we were joined for a smoke by another wakeful woman. At one point, the rear-hatch guard became ill and was led off the plane by the forehatch guard. This alarmed me. Was this a ploy to get the *fedayeen* off the plane before it exploded? When, some half hour later, he was returned, we were newly disquieted. Having been ill, was he thought worthless enough to be blown up with us? In an access of paranoia, we asked one another, where was Leah? Was she still on board, or had she slipped away to safety? Whom could we trust? What was to happen? In the darkness, I counted the aisle seats by hand to Leah's seat and found her resting quietly. My news brought obvious reassurance, and immediate guilt over our suspicions, to the wide-eyed woman watchers. Before dawn, another guard came aboard, seemingly having been drinking, demanding Leah. He pushed us back as he found her. She pretended to be half asleep. They exchanged brief remarks in Arabic. As he left, he bade an amorous farewell, in broken English, to the silently staring women.

Saturday's dawn found the women and children alone, without food, water, or usable facilities. Under leadership of a woman whose husband had been taken, they organized themselves into details to achieve the impossible. At mid-morning, the first buses began to arrive to take them to Amman. One devout woman hesitated. This was the Sabbath. Riding was forbidden. With one accord, the last group of women turned on her, saying that, if necessary, they would take her by force. If they could not, they would take with them the child she was caring for.

All was confusion at the final debarkation into buses driven and guarded by armed men. The buses set off in a wide fan of separated courses across the desert to partial reunion at the Amman Hotel. There, those who could, gave names and descriptions of the women and girls still missing.

In subsequent rescue stages and on the long flight westward to America, recriminations were more open and bitter between the women who, with their children, had refrained from eating, and those who had not. Side-taking was sought. Bitter feelings among some were solidified. Contempt was expressed for those who mourned lost or stolen luggage. The women whose men were "lost" maintained their sad, quiet dignity, deeply concerned over the separation from family members still held captive somewhere, and anxious about the hardships anticipated in the many implications for their futures of this sustained and not yet ended stress.[2] The sense of release and homecoming expectations were tempered by the realization of girls and women also still held captive. Affectional ties among some were made more binding, especially among those who felt deeply connected that final night aboard the plane. Again it was apparent that the individual's functioning in the group, under and through crises, continued to reveal past and habitual coping patterns, their timely uses and their flaws.

IMPLICATIONS FOR EDUCATION

1. Emergencies, threat, crises, and stress, with concomitant feelings of anger, fear, and anxiety, expressed directly and through the defenses of projection and denial, are on-going aspects of our current society. Educational measures to offset the personally and socially undesirable effects of these responses may be undertaken.

2. Extensive literature, from 1920 on, deals with strategies and tactics involved in anticipating, understanding, and managing human behavior under stress. Wider and more general utilization of selected appropriate aspects of this knowledge should be at the disposition of parents, teachers, youth leaders, and persons responsible for the safety of large numbers of people. As Wilson[9] has noted:

> Civilian populations . . . may be prepared for disaster situations by training and education, so individuals may be prepared to maintain their mental health during extreme situations if they can be taught to cope with varieties of normal stress. (p. 135)

3. The "normal" life crises to which children and youth become exposed in our society may well be utilized for educationally constructive crisis purposes.

This may involve opportunities for talk between children and accepting adults whom children and youth like and admire, in which the children can be helped to identify, to express, and to *objectify* their own feelings, and their own perception of their needs, and behavior. This also involves giving to children and youth honest, factual, simple explanations, at their levels of comprehension and assimilation, in order that they may integrate their own cognitive and perceptual activities.

4. Camping, Scouting, Hosteling, and other outdoor equivalents tend to cultivate in children and youth the setting up of problems in living and survival, for "fun" and "adventure." The experiences of patterns of mastery of such problems, individually and in groups, can be further deliberately objectified to have the effect of desensitization to later crises and stresses and to promote composure and an enhanced ability to master threat and its effects.

5. When danger threatens, people naturally seek the company, understanding, and aid of others, and join with others in protective efforts. It is important, therefore, to be aware of those factors that tend towards group divisiveness—with deceleration, diffusion, and diminishment of efforts to reach group goals—and of those that tend towards enhancement of group cohesiveness—with consequent reduction of anxiety, comprehension of the relevance of the problems to be surmounted in conforming to group needs, development of feelings of warmth, especially *trust*, in affiliation with group members, and readiness to participate freely on constructive group protective measures. This, too, can be consciously and deliberately prepared for in discussions of group experiences.

6. Appropriate forms of non-insight oriented encounter groups for children, youth, and adults may be used to help people become objectively acquainted with their own feelings and behavioral responses to threat, especially as these are directed both towards others and towards group-oriented tasks and goals. This technique can further be educationally preparatory for more personally and socially constructive responses to crises and threats.

REFERENCES

1. BAKER, G. AND CHAPMAN, D., eds. 1962. Man and Society in Disaster. Basic Books, New York.
2. CHAPMAN, D. 1962. Dimensions of models in disaster behavior. *In* Man and Society in Disaster, G. Baker and D. Chapman, eds. Basic Books, New York.
3. FRIEDSAM, H. 1962. Older persons in disaster. *In* Man and Society in Disaster. G. Baker and D. Chapman, eds. Basic Books, New York.
4. HILL, R., AND HANSEN, D. A. 1962. Families in disaster. *In* Man and Society in Disaster, G. Baker and D. Chapman, eds. Basic Books, New York.
5. JACOBSON, G. 1965. Crisis theory and treatment strategy: some sociocultural and dynamic considerations. J. Nerv. Ment. Dis. 141(2):209–218.
6. SALENGER, H. 1972. Vice-Principal, Upper Cheektowaga High School, Cheektowaga, New York. Personal communication.

7. SALENGER, H. 1972. A study of groups in an extreme stress situation. Unpublished paper.
8. WITHEY, S. 1962. Reaction to uncertain threat. *In* Man and Society in Disaster, G. Baker and D. Chapman, eds. Basic Books, New York.
9. WILSON, R. 1962. Disaster and mental health. *In* Man and Society in Disaster, G. Baker and D. Chapman, eds. Basic Books, New York.

Article 34 Human Adjustment to an Exotic Environment

The Nuclear Submarine

JIM H. EARLS

Man has, in the current century, been making tremendous technological advances, and these advances have permitted him to explore new or "exotic" environments. Examples of these explorations are the year-around colonization of the Antarctic, the adventures into outer space, and the developing exploration of and attempts to inhabit the inner space, the oceanic subsurface. The inherent stresses man encounters in attempting habitation of the polar ice caps or outer space are items of common knowledge; the stresses encountered in man's effort to explore, inhabit, and use the "inner" space are not, however, well documented. Efforts made toward understanding and using the submarine space have been primarily military ones. This restrictiveness may be one reason for the scarcity of published material dealing with the adjustment patterns of the men who make these efforts.

My intention here is to provide observational data on one aspect of the submarine environment: the adjustment of men to prolonged submergence aboard a nuclear-propelled polaris-missile-firing submarine. These observations were made while I was serving as the medical officer aboard two polaris submarines. Discussions with fellow submarine medical officers led me to believe that adjustment patterns reported herein are not isolated occurrences but are perhaps common to many polaris submarine crews. It is recognized, however, that human adjustment is a complex function and is affected by many variables. It is not my intention to claim that the adjustment pattern described in this paper applies to all submarine crews.

The submarine is not a device which came into being in this century, as might commonly be thought. Leonardo da Vinci designed a submarine but never tried to build it. The first operational submarine was apparently designed and built in 1624. It was a wood and leather vessel which was rowed at a depth of

JIM H. EARLS, M.D., is with Associates in Psychiatry in Oklahoma City.

Source: *Archives of General Psychiatry*, Vol. 20 (January 1969), 117–23. Copyright 1969, American Medical Association.

from 12 to 15 ft in the Thames River. One hundred fifty-two years later the American colonial navy used the *Turtle*, a one-man submarine, against a British vessel in New York harbor. The Confederate navy built a series of submarines; the Confederate submarine, *Hundley*, was the first submarine to destroy an enemy ship. On Feb. 17, 1864, the *Hundley* sank the USS *Housatonic*, but also sank itself as well. The use of submarines in World Wars I and II is well known.

Prior to 1954, the submergence capability of a submarine was determined primarily by mechanical factors and the submarine's atmospheric dependence and not by human factors. The launching of the USS *Nautilus*, SSN571, in 1954 changed the determining factors. Nuclear power truly freed, for the first time, the submarine from its atmospheric dependence. The submarine became capable of prolonged submergence periods, and man's capacities for adjustment and endurance became the new limiting factors. The US Navy, in apparent recognition of man as the limiting factor, has elected to man each polaris submarine with two complete crews. These crews alternate between being on the submarine for about 90 days and receiving refresher training at a US Navy base. This method of manning the polaris submarine has apparently been adopted to obtain the maximum submerged patrol time on a continuing basis.

THE SUBMARINER

The submariner considers himself to be essentially different from his Navy-wide peers. He is, first of all, a volunteer for hazardous duty. He is young, healthy, and considers himself to be intellectually brighter and more educated than a non-submarine-going peer. Some of the sailor's verbalized motivations for submarine service frequently given are (1) extra pay, (2) good food, and (3) opportunities to learn interesting skills.[5] He has had to pass a number of screening tests which examine his intellectual level, emotional stability, and physical status. The evolution of these tests has been described elsewhere by Weybrew.[4] After the intensive screening, the submarine candidate receives additional training in his specialty area and a formal school period in the US Navy Submarine School. One screening device which has not been reported elsewhere is a crucial one and is used after the sailor's graduation from school and after he reports to his first submarine assignment. This device is the "Qualification in Submarines" procedure. The new submariner is required to demonstrate his theoretical and practical knowledge of the submarine to a group of senior shipmates. Should he be considered deficient in either area he may be dropped from the submarine program. Although the qualification phase does not explicitly examine the candidate's ability to make a satisfactory interpersonal adjustment, such ability is of great importance to his shipmates and is "unconsciously" evaluated and acted on by them. The crew is able to quickly isolate and reject any new member who is unable to make a satisfactory interpersonal adjustment. Finally, the submariner is characterized by a much higher level of career motivation and reenlistment rate than is found in the navy at large.

THE SUBMARINE ENVIRONMENT AND ITS STRESSES

The group of about 140 men making a polaris submarine patrol live in an encapsulated world. Moreover, they live in that environment for about 60 days at a time and during that period are relatively isolated from the external world. Since the submarine operates more or less independently of the surface, some loss of circadian and geographical orientation occurs. The only practical circadian clues routinely available to the submariner are the type of food being served at meal time and the length of time until the evening movie. Whereas the sailor may well know the date he may well *not* know the day. The environment consists of continual and relatively nonvarying auditory, olfactory, and thermal sensations which soon assume a "white background" characteristic. These inputs become important only when they suddenly alter and the submariner begins his reflexive quest for "what's wrong."

The submarine's environment is a densely populated one. Personal territory is highly limited and is, in effect, defined as one's bunk, the only space that the submariner is not forced to share with others. It has been estimated that the individual submariner has 5 cubic yards of private space,[4] the space occupied by a 6-ft-tall man standing on an area of 57 sq in. The crew's mess hall and the officer's wardroom are used as social areas and eating areas but also as lecture halls, movie theaters, and even as surgical suites. All other spaces are primarily work spaces.

The degree of the submariner's isolation is shown by the fact that he is unable to communicate with the external world. To keep the secrecy of its position inviolate, the submarine is enjoined from initiating radio transmissions. Normally the submarine will communicate with the outside world only when it experiences a major emergency. The submarine is, however, in continual receipt of radio messages. Some of the material received is general news which may be passed on to the crew. In addition to the general news material, the individual submariner is allowed to receive a 15-word message from his family on each of three times during the patrol. Although the contents of these messages is bland, the failure to receive a "family gram" at the expected time is a stressful situation for the submariner.

The submariner is well aware of the fantastic destructive power present in the nuclear missiles on board. He is also aware that these missiles are to be fired only in retaliation to a nuclear attack on the United States. The implications of these weapons and the significance of an actual launching situation are relatively clear to all personnel on board. It is a rare occasion when the missiles and their nuclear armament are discussed in any way other than a highly technical fashion. This very avoidance reflects the stressfulness of their presences.

Finally, there are two other areas of stress which are not necessarily an inherent part of the submariner's environment but which are frequently seen to be operative. The first of these areas is involved with the submariner's professional life and the rewards for his expended efforts. The polaris submariner is repeatedly told that he is working to prevent war, primarily a nuclear war; the submariner has some difficulty integrating this goal with his knowledge of the ongoing world conditions. In addition, the submariner is unlikely to perceive much inherent reward in his job. Most of the personnel on board are involved in either the constant monitoring of machinery operation or the preventive maintenance of machinery. The crewman may receive little praise at all if his machinery functions well but may receive instant criticism when a piece of the machinery fails. The second area of stress is centered in the external world and is crystallized in the submariner's concern about the general welfare of the family and the conduct of his wife. Whereas the submariner may continue to feel a major obligation for the welfare of his family unit, he also recognizes that while on patrol he is impotent in assisting the family. He may become overly concerned about his family's welfare—as if he must create the image of a family problem which is as ominous as his impotence to deal with the problem. Some of these fears, including the conduct of his wife, may be well founded or they may be unrealistic and may approach a delusional intensity.

GENERAL ADJUSTMENT PATTERN

The polaris submariner's time on the submarine is clearly divided into two unequal parts. The first part is about a 28-day period and the second part is about 60 days in duration. Each period has its characteristic stresses.

Upkeep

First is the "upkeep" period for the polaris submarine, which lasts about four weeks. The submarine will have just returned from a patrol, and the upkeep period begins with the new crew coming aboard. During this period consumable stores and spare parts must be replenished, any external hull work accomplished, the repair or replacement of malfunctioning machinery effected, and any possible land-based social activity enjoyed. This four-week period also embraces much time spent in the crew's refamiliarization with the submarine and the testing of the submarine and the crew by operational trials at sea. It is a period characterized by marked physical and emotional demands on the submariner to accomplish the work, to control his anger generated over frustration in attempting to get the submarine ready for sea, and to enjoy the traditional sailor's pleasures ashore. The initial portion of the "upkeep" period appears to be characterized by a slight elevation of the group mood as compared with that of the period immediately preceding the upkeep. However, this mood soon begins to turn downward, and before the termination of the four-week period the crew's prevailing mood is one of mild depression characterized primarily by irritability, restlessness, and a mild depressive affect.

Submergence

The polaris deterrent patrol starts when the submarine leaves its tender and cruises on the surface to its diving area. The initial submerged period appears to be one of relief for the sailor. He experiences relief from the physical and emotional demands of upkeep. He is setting up the patrol routine which he will follow for the next eight weeks. His routine day is becoming established and is probably not yet interrupted by the drills, lectures, and other "all hands" evolutions which will later disturb his off-duty hours or his sleep. There is a general rebounding of the mood from the mild depression which existed at the end of upkeep to a mild or moderate elation during the first several days of submergence. The sailor explains this change by the aphorism, "Sailors belong on ships and ships belong at sea." The group mood soon, however, changes again, and a depressive trend becomes apparent.

The first definable period of adjustment to submergence may be called the "one-quarter-way syndrome" and is usually well established by the second week of submergence. A marked increase in sick call visits occurs. The submariner's complaints are usually subjective ones. Headaches, nuchal and occipital in location, constipation, and "chest and head colds" comprise the bulk of the complaints. Many statements are made concerning feelings of anxious expectancy. Crew members begin to develop disturbances of their sleep patterns. These deviations encompass both poles, insomnia and hypersomnia. The crew displays some tendency to adolescent gang behavior. The submarine crew represents a well-organized and formal group structure, with the crew comprising a missile-men gang, sonarmen gang, navigation gang, etc. Each of these naturally occurring subgroups is formed of peers having a specific technical skill. There develops between these gangs a feeling of "friendly competition"; the competition may center about attempts to kidnap the totem of another gang—the missile gang's plastic bird, for example. While observing this behavior one is reminded of gang behavior of early pubescent boys. This is also a period when sexual humor is very evident—usually in conjunction with the recounting of personal exploits ashore. This adjustment pattern continues to develop and then fades into the next phase.

The next definable period is at the end of four weeks—the "half-way syndrome." This period represents the low point in the group's mood. A feeling of depression, of varying intensity, appears to be experienced by all the crew during this phase. Many complaints are verbalized and many of these have a familiar depressive quality. Changes in appetite, bowel function, complaints of headache and muscle ache, difficulty in concentration, and sleep disturbance occur in a major portion of the crew. Also, various crew members verbalize another common depressive position, pessimism: because the past has been unpleasant and unrewarding the future will be the same. The submariner looks back on four weeks of routine, of boredom, and of increasing depression, and he looks ahead to four weeks more of the same feeling. The individual man shows much introspection and intrapersonal withdrawal. He becomes primarily concerned with himself and the welfare of his family. His social contact with his fellow crew members is primarily maintained by involvement in structured situations—card games, watch standing, and meals. Another striking change that occurs during this phase of adjustment is the appearance of intensely sarcastic humor in place of the previous sexual humor. This change in humor style appears to serve two functions. One is the discharge of hostile and aggressive affect which is personally and culturally unacceptable and which might otherwise be physically acted out. The second is the keeping of the other shipmates at a comfortable distance, resulting in the temporary expansion of an individual's personal territory. Occasionally, during this period, a few individuals begin to complain about the loss of normal circadian clues and may eventually report brief derealization or depersonalization-like episodes.

The transition from the half-way syndrome to the next describable point, "three-quarter-way syndrome," is not as smooth as the other transitions. The half-way syndrome is terminated and the transition to the three-quarter-way syndrome is marked by a sudden, but short-lived, elevation of the group's mood. There then sets in a progressive remission of the depressive mood. By the end of about six weeks of submergence, there is once again affectual lability; interpersonal approach no longer requires a structured program and the bulk of the depressive somatic equivalents have been relinquished. The type of humor has once again changed, with sexual material again becoming predominant. Along with the sexual humor there is an increasing tendency to physical contact. This, rather than homosexual acting out, seems to serve the function of preparation for anticipated heterosexual activity.

The transition from the three-quarter-way syndrome to the final-week syndrome is rather rapid. The majority of the crew continues to show a lifting from the depressive mood. However, for the first time during the submerged period, a definite split in the affective tone of the crew can be seen. While the majority of the crew shows a continuation in the remission of depression, some crew members show a sudden reversal and exacerbation of the depressive syndrome. These last-week depressives may be classified into two subgroups. One group readily verbalizes a fear of returning home to resume the masculine role, both sexual and social, in the family. They seem to view their on-board position as much less threatening than their position at home. This group, as one might guess, shows a predominance of the passive-dependent character. The other subgroup is composed of highly motivated, career-oriented, and compulsive individuals who started the patrol with definite goals to be achieved. While his goals may not have been excessive in view of a 60-day submergence period, the individual's effectiveness in working to achieve the goals was unfavorably affected by the general depressive experience. The sailor fails to achieve his self-imposed goal by the end of the patrol. These individuals, unlike the first subgroup, do not desire to remain at sea indefinitely but only long enough to "finish the project." In addition their degree of untoward response to the approaching termination of the patrol is not as marked as that of the former group.

The final-week syndrome is terminated in the last several days before surfacing by the development of what the sailor would call "channel fever." A feeling develops which is somewhat comparable to a hypomanic state in that the sailor has a general sense of well being, a feeling of being capable of an excessive amount of work, and a feeling of diminished need for sleep. This state persists until the submarine "ties up" alongside its tender, and then the hypomanic state tends to rapidly revert into a mild depressive one, with the principle somatic equivalent being a feeling of fatigue. This final depressive position is normally resolved only after some time is spent away from the submarine.

COMMENT

The polaris submariner lives in a monotonous and crowded environment. His personal safety and the accomplishment of the submarine's mission depend on the successful blending of diverse personalities into a smoothly functioning team. The major operative stresses in the environment which interfere with the team formation are the lack of objective reward for expended efforts, the inability to communicate with persons in the outside world, the lack of sufficient personal territory, the nonvariability of physical environmental stimuli, the concern for the conduct and welfare of the family, and, finally, the presence of nuclear weapons on board.

The mental mechanisms used to deal with the operative stresses are essentially those that Grinker considers to be a normal part of the psychic structure of young adult males.[1] They are denial, as frequently seen in the control of anger or the dealing with the interpersonal conflicts; sublimation, as readily seen in the employment of humor as a means of dealing with sexual impulses or aggressive feelings; and isolation of affect, as may be seen in the apparent inability to discuss the submarine's nuclear armament in any other than a technical and bland manner. The defense mechanisms, however, do not appear to be sufficiently strong to assist all individuals in maintaining normal adjustment status. Instead, individual defense mechanisms may become more pronounced in the submariner's adjustment to the submergence. Should the individual, for example, tend to externalize his intrapersonal problems he may be expected to develop some degree of paranoid ideation during the submerged period. The degree of decompensation is, of course, determined by a complex of variables, but the range of the more pathological decompensations is from a transient situational reaction to a gross psychotic disorder.

In effect the sailor on a polaris submarine patrol is the same individual he is when ashore, i.e., he has the same ego balance of strength and of weakness. The chronic stresses of the patrol test the equilibrium of the submariner's ego balance. The outcome of the testing is most likely the same whether the test occurs while the sailor is on patrol or ashore.

As described above, the common mode of adjustment to the exotic environment of the polaris submarine appears to be a depressive one. The crux of the various forces leading to this depressive position would appear to be the anger experienced by the various members of the crew. The anger is an outgrowth of the frustrations experienced by the submariner in dealing with his environment. However, there appears to be no personally or culturally acceptable means of discharging this anger. The paternalistic organization of the military system is one which does not permit the direct expression of anger and aggression toward the military system. In addition, there is the personal fear that the overt expression of anger may lead to a socially isolated position within an already isolated community. The individual has little opportunity to handle his hostile affect by sublimation, except through humor. The submariner is then forced to deal with his anger by denial, suppression, or turning against himself. The hostile affect becomes internalized, but it ultimately manifests itself as a depressive phenomenon.

The terminal group depressive response is of interest in that its genesis is most likely different from the one I have used to explain the patrol depression. The terminal depression appears to be explained most readily by the psychiatric concept of separation anxiety. The termination of patrol represents, to the submariner, an object loss—both real and symbolic. At the "real" level the termination of patrol means the dissolution of the crew group. A marked decrease in social contact will occur among the various crew members during the three months ashore. Some crew members will either be transferred or be leaving the navy at the end of each patrol, clearly showing the interpersonal loss. The symbolic loss is, perhaps, more significant than the real loss. This intrapersonal loss involves the surrender of a dependent position. The sailor has been living in an emotionally stressful environment but an environment which also regularly, reliably, and abundantly met his physical needs—with the singular exception of sex. The sailor has not had to employ even the normal terrestrial maneuvers needed to obtain food, shelter, clothing, etc. By necessity, all has been provided for him.

It is not uncommon, during the submerged patrol, to hear the submariner make joking references to "returning to the womb," and he is capable of recognizing the omnipresent justification for his remark. Thus the termination of his time on the submarine requires that the submariner surrender this passive mode of existence. The surrender of the passive, dependent role is not necessarily an easy thing, for the submariner or for others, and is accompanied by the depression which is frequently seen when people are forced to renounce their infantile strivings.

Observations which tend to support the above formulation have been reported by A. Blackburn in a personal communication to me. He has "wintered out" in a small Antarctic station occupied by equal numbers of military personnel and civilians. The military personnel showed an adaptation strikingly similar to the one described in this communication. The civilians, on the other hand, appeared relatively free from the depressive phenomena. The civilians freely expressed their anger. The military personnel were apparently either unable or unwilling to express their anger openly. Gunderson[2] and Mullin[3] also report very similar adjustment patterns in their writings on adjustment to Antarctic living. Gunderson's study was unable to show any significance between the group's size and its adjustment. His study, which was apparently based on a written questionnaire, also showed that symptoms of dysadjustment were most prevalent at the midwinter point, as were the submariner's at the midpatrol point. The subjective complaints given by the men were sleep problems, irritability, loneliness, headaches, and depression—essentially the complaints of the submariner. Gunderson does not, however, speculate on the genesis of this adaptational syndrome. Mullin has stated rather clearly what he and I consider to be one of the main problems in human adjustment to an isolated environment. He stated, "We were impressed by the relative absence of overtly expressed hostility" and "Group and individual tensions and irritations are ever present, but the most important lesson a wintering-over man learns is that he cannot afford to alienate the group; that in this tight little society he is dependent in large measure upon the goodwill of the next man and of the group as a whole for his vital feelings of security, worth and acceptance." Mullin also suggested, as this communication does, one motivation to volunteer for duty in an isolated and uni-sexual environment: "On the other hand, for a few men it was obvious that separation from home, wife, children, and family responsibility meant for them the subtraction of an element of stress in *their* personal adjustment."

It would seem that with the prospect of space travel and inhabitation of some portion of space so closely at hand, at least as represented by the manned orbital laboratory, that studies to determine the adjustment of groups to isolated environments are imperative. The polaris submarine represents an environment in which many important studies could be carried out. The situation is one which would lend itself to careful and long-term study and one which should be productive of significant results.

The polaris submarine offers a unique opportunity to observe and study the adjustment pattern of men as they cyclically move between the "normal" terrestrial life adjustment and the "exotic" submarine life adjustment. It should be a fertile ground for investigations in group dynamics and adjustment.

SUMMARY

The polaris submariner is a highly screened individual placed into a chronically stressful and frustrating environment. When the individual begins to develop feelings of anger in response to the frustrations, he is faced by a cultural structure which does not readily permit the expression of anger. He is then forced to turn the anger inward and then experiences a depressive phenomenon in reaction to operative stresses. The course of this depressive phenomenon is believed to be a ubiquitous phenomenon among the polaris submarine crews. A similar adjustment pattern has been reported from other isolated environments. It is believed that the polaris submarine represents an ideal laboratory in which to study the dynamics of group adjustment to unusual environments.

REFERENCES

1. GRINKER, R. R., SR., AND GRINKER, R. R., JR.: "Mentally Healthy" Young Males (Homoclites): A Study, *Arch. Gen. Psychiat.* 6:405–453, 1962.

2. GUNDERSON, E. K. E.: Emotional Symptoms in Extremely Isolated Groups, *Arch. Gen. Psychiat.* 9:362–368, 1963.
3. MULLIN, G. S.: Some Psychological Aspects of Isolated Antarctic Living, *Amer. J. Psychiat.* 117:323–325, 1960.
4. WEYBREW, B. B.: "Psychological Problems of Prolonged Marine Submergence," in Bruno, N. M.; Chambers, R. M.; and Hendler, E. (eds.): *Unusual Environments and Human Behavior: Physiological Problems of Man in Space*, Glencoe, Ill.: Free Press of Glencoe, 1963.
5. YOUNISS, R. R.: An Investigation of Motivation for Submarine Duty and Its Relation to Submarine School Success, *U.S. Naval Med. Res. Lab. Rep.* 15:(7, whole No. 278) (Nov.) 1956.

Article 34 Patterns of Response Among Victims of Rape

SANDRA SUTHERLAND

DONALD J. SCHERL

Large numbers of young people today have chosen to live and work with poor people. As a result, they are exposed to the risks and difficulties inherent in living in low-income neighborhoods. We have had the opportunity to study a number of young women who have suffered as a result of one such risk, that of sexual assault.

The President's Commission on Law Enforcement and Administration of Justice reports the incidence of forcible rape is increasing nationally. The frequency of reported rape per 100,000 population was approximately 3 in 1933, 9.2 in 1960, and 11.6 in 1965. A detailed interview study undertaken by the National Opinion Research Center of the University of Chicago (NORC) found forcible rape actually occurred at more than $3\frac{1}{2}$ times the reported rate or 42.5 per 100,000. "These figures suggest that, on the average, the likelihood of a serious personal attack on any American in a given year is about 1 in 550. . . . The actual risk for slum dwellers is considerably more; for most Americans it is considerably less." [11]

The victim's adjustment following sexual assault has received little attention in the literature. Specific references to the young woman most frequently discuss the possibility of her conscious or unconscious participation in the incident.[1,5,7] Other reports discuss child victims of sex offenders,[3,13] the personality and motivation of the person who commits the crime,[15] and the necessary medical and legal procedures to be followed in obtaining the information which will be required if the matter is taken to court.[2,3,6,8,10,14]

Over the period of a year we have studied a group of 13 victims of rape.

SANDRA SUTHERLAND FOX, PH.D., A.C.S.W., is co-director of the Metropolitan Mental Health Skills Center, Washington, D.C.

DONALD J. SCHERL, M.D., is Assistant Professor of Psychiatry and Director of Community Health Services, Massachusetts Mental Health Center, Harvard Medical School.

Source: American Journal of Orthopsychiatry, Vol. 40 (1970), No. 3 (April), 503–11. Copyright © 1970 the American Orthopsychiatric Association, Inc. Reproduced by permission.

These women ranged in age from 18 to 24 and each had a background consistent with accomplishment, independence, and apparent psychological health. The study was undertaken in a context similar to that of a crisis intervention team at a community mental health facility. The nature of the setting permitted us to see most of the young women within 48 hours of the assault and to follow the acute course of the victim. It was our purpose to identify a specific predictable sequence of responses to the rape, viewed as a psychologically traumatic event, and to design a pattern of short-term mental health interventions placing particular reliance on techniques of anticipatory guidance and crisis intervention. Rape as defined here refers to forced sexual penetration of a woman by a man accomplished under actual or implied threat of severe bodily harm.[9]

Each of the victims was a young white girl who had moved into a low-income (not necessarily black) community to implement her conviction about "doing something real" in contemporary society. While it is possible to speculate with respect to the relevance of this to the occurrence of the rape, to do so seems unprofitable. These cases were drawn from a total population at risk of several thousand women. The incidence of rape among this group at risk was no higher than the incidence among all American women of the same age.* It seems clear that the victims responded to the assaults in a pattern that may hold for other women under other circumstances. For our group, however, the content (though not the pattern) of the response was no doubt a product in part of the same internal factors which led the women to live in low-income areas in the first place.

THREE PHASES OF THE VICTIM'S REACTION TO RAPE

Phase One: Acute Reaction

In the moments, hours, and days immediately following the rape, the victim's acute reaction may take a variety of forms including shock, disbelief, and dismay. She often appears at the police station or the hospital in an agitated, incoherent, and highly volatile state. Frequently she is unable to talk about what has happened to her or to describe the man who has assaulted her. Sometimes the victim will initially appear stable only to break down at the first unexpected reminder of the incident. The following case illustrates such a situation:

Case 1-A. The day following that on which she had been raped, Louisa was taken by two police detectives to establish the exact location of her assault. As they were returning to town, a call on the police radio reported the automobile described by Louisa as that of the abductors had been located. They immediately drove to where the auto had been found, unaware of the presence there of two of the assailants. As Louisa got out of the police car to make positive identification of the auto, she saw the men. She began sob-

* An interview study by NORC records the following rates for victims of forcible rape: Age 10 to 19: 91 per 100,000 population. Age 20 to 29: 238 per 100,000 population.[11]

bing loudly and uncontrollably and had to be assisted in returning to the police car.

The immediacy with which the assault is reported and the persons who are first notified are of diagnostic interest. Those women who feel there has been no invitation, seduction, or willing compliance on their part generally make an immediate telephone call to the police or go to the nearest emergency medical facility. This was true of eight of the 13 young women studied. The remaining five women experienced an inner sense of guilty involvement confirmed by data emerging in subsequent interviews. This was reflected in the manner they chose to reveal the assault: One told her landlady the following day but refused to go to the hospital because she feared the doctors would be required to report the incident to the police. One, later found to have been two months pregnant at the time of the rape, told her boss the following day but refused to notify police. One told her roommate six days after the rape, saying she was afraid there would be publicity and stating she was sure she was pregnant since it was her impression all intercourse resulted in pregnancy. One did not reveal her rape by three men until two months later when she was hospitalized for a psychotic episode. And finally, one reported her rape two months after its occurrence when pregnancy resulting from the incident was confirmed. When and to whom the young woman reports the rape thus appears to provide some early clues about her conscious or unconscious perception of the role she has played in what has happened to her.

In this early phase of response to rape, shock and dismay are often succeeded by gross anxiety. This frequently occurs at the time the victim must first deal with the consequences of the assault. Notification of parents is one of the early issues to arise during phase one. A very common statement during the initial interview with the women studied was: "My parents must not know." Immediately following this came one of several other statements of explanation such as: "They told me this would happen" or "Now they will make me come home" or "It will kill them." Implicit in these responses seems to be the girl's fear that her own poor judgment precipitated the crisis. However, she is often unable to consider this in any depth until a later time. Seven of the young women studied chose parental illness as the reason why they felt unable to notify their families. In most instances, a marked decrease in the patient's anxiety and other symptoms occurred after she had been able to discuss the incident with her family.

In addition to the concerns the victims raised about telling their parents, each of the women also focused on a number of other issues during this period of acute anxiety: Should she press legal charges? Will she be able to identify the alleged rapist and how will she feel about seeing him? Will all of her friends and neighbors find out what happend? What will be the nature of the publicity about the incident? Will she become pregnant? Should she tell her boyfriend or fiance what has happened? What will her clergyman think of her?

The example which follows illustrates a typical Phase One reaction:

Case 1-B. Noreen, a 22-year-old woman of middle-class background, was living and working in the ghetto area of a large city. During the night a man broke into her house and raped her at gunpoint while her roommate stood helplessly by. Noreen immediately notified the police and was taken to a local hospital for medical attention.

The following day she told her employer what had happened. She expressed no feelings about the rape even when questioned. At her employer's suggestion she came to the mental health clinic three days later. She was initially calm and controlled, much as she apparently had been since the assault. However, as she described what had happened, she began to cry, saying she wanted to tell her parents but it would "kill" her mother who had cancer. After a lengthy discussion of possible alternatives, Noreen decided to notify her married sister who lived approximately 40 miles away. By previous arrangement, the girls were to go home for a holiday celebration in about ten days and Noreen thought she could then talk with her parents.

During the time before she went home, Noreen's awareness of her own anxiety increased. She had refused to consider alternate living arrangements during this period, but one night she and her roommate called the police at 3 a.m. saying they were unable to sleep and were afraid the assault might be repeated. The following day both girls moved to a friend's apartment in another part of town.

At Noreen's request, she and her sister were seen a few days later to discuss the possible reactions the family might have when they were told about the rape, and the ways in which the girls could deal with these responses. Noreen then revealed her recent fear and anger and her previous loss of appetite and sleeplessness.

The Phase One reaction normally resolves within a period of a few days to a few weeks. There is the expected decline of nonspecific anxiety as the victim turns increasingly away from fantasy and, with support, toward handling the realistic consequences and problems created by the sexual assault. The woman is given as many appointments as she wishes with one or several members of the professional team (psychiatrist and social worker) and the collateral resources of the clinic (lawyer, gynecologist, clergyman, etc.) and is helped to talk about what has happened to her. It is sometimes tempting to support any defenses the young woman has already adopted rather than to explore or discuss something that may embarrass or upset her. Often a retreat from anxiety on the part of the mental health worker only confirms the girl's suspicion that what has happened is so terrible that no one wants to hear about it. If she has talked to the police and/or medical personnel, their concern has usually centered on the physical facts. Rarely has anyone been able to let her share her feeling and fears.

As issues and problems arise, the clinic staff considers alternative courses of action and their possible outcome with the young woman. If she wishes, for example, a member of the staff will help her notify her family, will accompany her to the police station, or will arrange for legal or medical consultation. In addition, the victim's attention is directed toward the future and she is encouraged to consider how she will feel in a few weeks or months. The reactions experienced by others in similar situations are discussed with her. Possible feelings of anger, depression, and fear are reviewed, along with concerns that may arise in association with dating, engagement, or marriage. The victim is told these reactions are generally to be expected although they may or may not occur in any specific situation. She is encouraged to seek further professional help when it seems useful to her with the knowledge that to do so need not imply any underlying neurosis or psychosis.

Phase Two: Outward Adjustment

The second phase in the victim's emotional response to rape is a logical outgrowth of the resolution of the preceding phase. After the immediate anxiety-arousing issues have been temporarily settled, the patient generally returns to her usual work, school, or home pursuits. This seeming adjustment is reassuring to those who have been involved with her, for it looks as if she has dealt successfully with the experience and has properly integrated it. The woman announces all is well and says she needs no further help.

It is our impression that this period of pseudo-adjustment does not represent a final resolution of the traumatic event and the feelings it has aroused. Instead, it seems to contain a heavy measure of denial or suppression. The personal impact of what has happened is ignored in the interest of protecting self and others.

During this phase the victim must deal with her feelings about the assailant. Anger or resentment are often subdued in the interest of a return to ordinary daily life. The victim may rationalize these feelings by attributing the act to blind chance ("it could have happened to anyone"), to "sickness" on the part of the assailant, or to an extension of the social struggle of black against white or of poor against rich. In similar fashion and for the same reasons the victim's doubts about her role in the assault are also set aside.

The following example illustrates the reaction of the victim during Phase Two:

> *Case 2.* Denise, a 20-year-old woman, was working as an "indigenous" neighborhood aide for the local poverty program in the city where she had moved following her high school graduation. One afternoon as she was walking through a deserted park she met a young man who had attended several meetings at the neighborhood center. He introduced her to a friend who was with him, and the three chatted briefly before the first man left. As Denise began to walk away, the remaining man demanded she have intercourse and threatened to kill her if she refused. Denise submitted.
>
> As soon as she was released, Denise reported the incident to her neighborhood center director, who notified the police and arranged for medical care. That evening she talked at length with the director and then called her parents who were concerned but supportive. She refused professional mental health assistance.
>
> Denise insisted after a two-day "vacation" that she wanted to continue her work at the neighborhood center and refused an offer to be transferred to a

similar job in another part of the city. She commented that changing jobs would be an admission of failure and also said it would "let the guy know he really got to me." The neighborhood center director was concerned that perhaps Denise was being "too brave." At his suggestion, he and Denise discussed this question with the clinic social worker who served as a consultant for the neighborhood center. An evaluation at the clinic was recommended. Denise agreed, provided evaluation, and not treatment, was the intended purpose.

Denise was outwardly calm, composed, and reasonable during her clinic interview. She felt a transfer to another part of the city was unnecessary because she would be more careful in the future and thus an assault of this type would not occur again. She decided against pressing legal charges because she felt to do so might limit her ability to work effectively with the people in the neighborhood. Denise said she was not a virgin at the time of the rape, so she was not upset by the incident itself. She denied any sleeping or eating difficulties and said she was annoyed that no one seemed to think she "had the strength to take something like this."

The worker accepted Denise's statements but anticipated with her the likely development of further questions and worries. Shortly after this Denise failed to keep a followup appointment, left her job at the neighborhood center, and returned to her parents' home in another state.

The victim has little if any interest in gaining insight through treatment during Phase Two. The woman often strongly asserts she must get back to school or work as if nothing more traumatic than an ankle sprain had occurred. Whatever the explanations and rationalizations offered by the victim, they have inherent within them components of the fears from which they spring. Thus, for Denise, an admission of her fear and concern would have seemed to be a sign of her own weakness.

Phase Three: Integration and Resolution

Phase Three begins when the victim develops an inner sense of depression and of the need to talk. It is during this period that the resolution of the feelings aroused by the rape usually occurs. Concerns which have been dealt with superficially or denied successfully reappear for more comprehensive review. The depression of Phase Three is psychologically normal and occurs for most young women who have been raped. While careful evaluation is always indicated, the depressive feeling should not be interpreted immediately as a sign of illness.

There are two major themes which emerge for resolution in this phase. First, the victim must integrate a new view of herself. She must accept the event and come to a realistic appraisal of her degree of complicity in it. Statements such as "I should have known better than to talk to him or to open the door" "I should have had the lock on the window fixed" or "I should never have been out alone" emerge at this time. Second, the victim must resolve her feelings about the assailant and her relationship to him. Her earlier attitude of "understanding the man's problems" gives way to anger toward him for having "used her" and anger toward her self for in some way having permitted or tolerated this "use."

Phase Three may begin with a specific incident or discovery or with a more general deterioration and breakdown of the successful defenses of Phase Two. Diagnosis of pregnancy, the need to go to a police line-up for identification of the assailant, a marriage proposal, a glimpse of someone who resembles the rapist —these or many other situations may introduce Phase Three. Frequently, however, it is not possible to identify a specific precipitant. Instead, the victim finds herself thinking increasingly about what has happened to her and functioning progressively less well.

The following example illustrates the case of a young woman who handled Phases One and Two by herself and sought professional help only when she encountered the difficulties of Phase Three:

> *Case* 3. Alice, age 23, was employed as an adult education teacher at an inner-city school. One night her principal's 20-year-old son called to say his mother had ready the books she was donating to the night school and wanted Alice to pick them up and have coffee. Alice had met the son once and accepted his offer to pick her up and drive her to the home. Soon after getting into the car, Alice realized they were not going in the right direction. When she questioned the young man, he stopped the car and raped her. Afterward Alice jumped out of the car and ran home, telling no one what had happened. She went to work the next day and learned the principal's son was moving out of town at the end of the week to take a new job.
>
> Except for occasional nightmares and crying spells, Alice later reported she had rapidly been able to get herself under control. Eight weeks later, after two menstrual periods, she discovered she was pregnant.
>
> Alice called the mental health clinic three days later, asking to talk with someone about a therapeutic abortion. Following an initial interview with a psychiatrist, she was seen on a daily basis for one week by the social worker and had two additional appointments with the psychiatrist. During this time she realized she had chosen to handle the incident according to her usual pattern of protecting others at great expense to herself. For example, Alice said she could not tell her parents what had happened because they would tell her fiance and "they would all be hurt." Alice was adamant that she could not bear the child, both because it was interracial and because of her abiding faith in the sanctity of conception only within circumstances of love and marriage. She was referred to an obstetrician and two weeks later a therapeutic abortion was performed.
>
> While there was no suicidal intent, Alice's week of clinic interviews was marked by feelings of depression with symptoms of frequent crying and loss of appetite. As she considered her recent experience, she revealed how overwhelming it had been to her and how scared she was to think about it for fear she would "fall apart." The depression resolved as Alice began to talk about what had happened and to make some decisions about her future. She felt she had not handled things too well in the past but said perhaps this was "the beginning of the new Alice."
>
> Alice returned to the clinic following the abortion to say she was going

home. She planned to discuss the situation with her family when she arrived and asked if her parents could call the clinic to talk further. Alice also asked for a referral to a mental health agency in her hometown for further discussion of problems less immediately connected with the rape.

For Alice, the resolution of her depressive reaction involved recognition of a lifelong pattern of punitive self-sacrifice. In addition, the process of recognizing, considering, and then terminating the pregnancy consolidated ego capacities and a sense of esteem and self-reliance badly fractured by the traumatic assault.

CONCLUSIONS

For the young adult victim of rape we have studied, a clear pattern of responses to the assault emerged. As with other traumatic events, an understanding of the victim's reactions permits the design of supportive mental health interventions specific to the pattern of responses evoked by the event.[12]

In the initial phase of the response, including the time immediately following the assault, the victim will exhibit signs of acute distress. The immediacy of the report, the person notified, and the decision about informing the family raise practical issues that must be considered with the patient. She should receive a thorough medical examination including relevant laboratory studies during Phase One.[3,7,8,14] Legal and pastoral counseling should be made available. During this phase psychological attention for the young woman is best placed on helping her deal with the realistic concerns consequent to the rape. It is particularly useful to describe for her the likely cycle of future responses.

Phase Two, often mistakenly thought to represent a successful resolution of the reaction to the rape, includes denial of the impact of the assault and is characterized by pseudo-adjustment and return to usual activity. Denial, suppression, and rationalization replace shock and dismay. This replacement of nonspecific anxiety by adaptive responses is reassuring to the victim. During Phase Two the mental health worker remains available to the victim on an as-needed basis. Insight-oriented psychotherapy is not indicated unless specifically and knowledgeably requested by the young woman.

The third phase, frequently unrecognized or misdiagnosed, includes depression and the need to talk. There are often obsessive memories of the rape at this time. The victim also has concerns about the influence the assault will have on her future life. The affect associated with the original event now becomes available for integration and resolution. The woman often develops a need to consult with a professional and this is to be encouraged.

There is a tendency on the part of the victim of rape to focus a spectrum of life problems around the single specific event, and the mental health worker who is initially consulted about the sexual assault often finds himself engaged at the end in a broader process. Forewarned during Phase One of the likely recurrence of feelings in Phase Three, the patient is able to approach these feelings adaptively rather than with panic and flight. The intensity of the feelings is kept at a tolerable level and the victim need not feel, therefore, that she is losing control.

Work with the victim of rape can be organized with relative confidence that one can predict for oneself and the patient the general course of future emotional events related to the assault. While predictable, these responses are colored and determined by the life experience of the victim as it is focused through the immediate crisis.

The reactions to rape we have observed among these 13 victims and defined here cannot, without further study on a less narrow population at risk, be considered generic responses. While we suspect the pattern we have delineated will hold for other populations of victims, the specific content of the illustrative case materials reflects the age and life circumstances, including all those internal factors motivating youth to work in low-income areas, of those we have studied. We have found foreknowledge of the sequence of emotional events in the victim to be useful in the design of a rational plan of supportive mental health services.

REFERENCES

1. Barnes, J. 1967. Rape and other sexual offences. Brit. Med. J. 2:293–295.
2. Bornstien, F. 1963. Investigation of rape: medico-legal problems. Med. Trial Technique Quar. 9:61–71.
3. Capraro, V. 1967. Sexual assault of female children. Ann. N. Y. Acad. Sci. 142:817–819.
4. Devereux, G. 1957. The awarding of a penis as a compensation for rape: a demonstration of the clinical relevance of the psychoanalytic study of cultural data. Int. J. Psychoanal. 38:398–401.
5. Factor, M. 1954. A woman's psychological reaction to attempted rape. Psychoanal. Quar. 23:243–244.
6. Graves, Jr., L. and Francisco, J. 1962. A clinical and laboratory evaluation of rape. J. Tenn. Med. Assn. 55:389–394.
7. Halleck, S. 1962. The physician's role in management of victims of sex offenders. JAMA 180:273–278.
8. Hayman, C. et al. 1967. A public health program for sexually assaulted females. Pub. Health Rpts. 82:497–504.
9. Matsumoto, F. 1967. Unpub.
10. Medical testimony in a criminal rape case, showing the direct and cross-examination of the bacteriologist and psychiatrists, and including the court's instructions to the jury. 1963. Med. Trial Technique Quar. 9(Mar.):73–124 and 9(June):89–106.
11. President's Commission on Law Enforcement and Administration of Justice. 1967. The Challenge of Crime in a Free Society. Govt. Printing Off., Washington. Data abstracted by Edith J. Jungblut.
12. Scherl, D. and Mack, J. 1966. A study of adolescent matricide. J. Amer. Acad. Child Psychiat. 5:569–593.
13. Sexual Assaults on Children. 1963. Brit. Med. J. 2:1146–1147.
14. Sutherland, D. 1959. Medical evidence of rape. Can. Med. Assn. J. 81:407–408.
15. Tappan, P. 1955. Some myths about the sex offender. Federal Probation 19:7.

Author Index

Subject Index